Pests of
Ornamental
Trees, Shrubs and
Flowers

A Colour Atlas of
Pests of Ornamental Trees, Shrubs and Flowers

David V Alford
BSc PhD FRES
Head of Entomology
Agricultural Development & Advisory Service
Ministry of Agriculture, Fisheries & Food
Cambridge, England

MANSON
PUBLISHING

Copyright © D. V. Alford, 1991 & 1995
This edition published by Manson Publishing Ltd, 1995
A CIP catalogue record for this book is available from the British Library
First published by Wolfe Publishing Ltd, 1991
Printed and bound in Spain by Grafos SA, Barcelona
ISBN 1-874545-34-0

This edition is published and distributed in North and South America by Halsted Press, an imprint of John Wiley & Sons, Inc., 605 Third Avenue, New York, NY 10158-0012, USA, 1995.
Library Of Congress Cataloging-in-Publication Data available upon request.
ISBN 0-470-23494-6

For a full list of Manson titles please write to:
Manson Publishing Ltd, 73 Corringham Road, London NW11 7DL, England

Errata.
Pages 14 and 382. References to *Halictidae* should read *Megachilidae*.
Pages 268 and 269. References to *Phycita roborella* should read *Acrobasis consociella*.

CONTENTS

To Inge, Ingaret, Kerstin
and Michael

PREFACE

Ornamental trees, shrubs and flowers are important components of modern life, lending great attraction to our- domestic, leisure and working environments. The market for ornamentals is, therefore, very wide and includes a demand for alpines, bedding plants, cacti, cut-flowers, house plants and pot plants, as well as herbaceous plants, ornamental grasses, shrubs and trees for gardens, parks and other amenity areas.

The care and cultivation of ornamental plants will often lead to the discovery of pests or pest damage which, unless correctly diagnosed and controlled, may prove disastrous. This book provides a means of recognizing the various pests associated with ornamental plants in the British Isles and much of continental Europe from the Alps northwards. Biological details, and information on the importance of such pests and how to control them, are also given. Many of the pests included here also occur in other parts of the world, including North America, Japan and Australasia, several as accidental introductions.

In recent years, regulations governing the use of pesticides for the control of plant pests have become far more stringent. In the United Kingdom, for example, it is now illegal to use any pesticide except as officially approved under the Control of Pesticides Regulations 1986, such legislation forming part of the UK Food and Environmental Protection Act (FEPA). Under FEPA, the use, sale, storage and supply of pesticides are also required to conform to published Codes of Practice. Similar regulations exist in other countries but, even within the European Economic Community, pesticides approved in some member states may not be registered for use in others. Specific recommendations for the chemical control of pests also vary from country to country, and are frequently updated as new materials are developed and recommendations for existing products are revised. For these reasons, chemical control recommendations cited in the present work are given in general terms only, and names of actual pesticides have been omitted. Readers who require details of pesticides available for use against any particular pest should refer to current editions of annually revised publications, such as *The UK Pesticide Guide* (published jointly by CAB International and the British Crop Protection Council) or the British Agrochemical Association's *Directory of Garden Pesticides*. Relevant up-to-date literature produced by pesticide manufacturers or state advisory services should also be consulted.

In addition to domestic markets, international trade in ornamental plants is also a major industry, and there is an ever-increasing demand for novel and high-quality products to import and export. As a direct result of such trade, alien pests are constantly being intercepted by importing countries. The likelihood of non-indigenous pests gaining a foothold in a new country is lessened by stringent plant health (hygiene) regulations and inspections, but the risk cannot be entirely eliminated or total success claimed; indeed, over the years, many non-indigenous pests have become established in Europe, especially on glasshouse plants. Chrysanthemum cuttings despatched to Europe from subtropical or tropical countries pose a particular threat, many consignments having to be rejected or destroyed because of the presence of alien pests, especially leaf miners and thrips.

It would be impractical to include details of every pest likely to occur on ornamental plants, but information is provided for those that most commonly cause damage and those that, although of little or no economic importance, are often noticed upon such plants (perhaps because of their large size, or colourful or unusual appearance). Limitations of space, however, have precluded consideration of parasites and predators. Details are given of various non-indigenous pests, including those most often discovered on imported ornamentals, and some of those which, from time to time, have become temporarily if not permanently established in northern Europe. Various forestry pests that cause damage to ornamental trees or shrubs are also mentioned. However, many forestry pests do not pose a threat to ornamentals; this includes a vast assemblage of wood-boring insects (including beetles of the families Bupestidae and Cerambycidae), primarily of significance to the timber industry.

Scientific names of insects follow the Kloet & Hincks (1945) *Check List of British Insects*, including later-published additions and amendments; these works should be consulted for full details of synonyms. Common names are based largely on Seymour's (1989) list of *Invertebrates of Economic Importance in Britain*.

In compiling this account, I am grateful to many friends and colleagues for their help in various ways. Particular thanks are due to R. W. Brown, Dr J. H. Buxton, D. J. Carter, J. V. Cross, G. R. Ellis, B. J. Emmett, C. Furk, Miss M. Gratwick,

A. J. Halstead, N. J. Hurford, A. W. Jackson, M. J. Lole, D. MacFarlane, H. Riedel, Dr G. Rimpel, P. R. Seymour, S. J. Tones, R. A. Umpelby, F. Wellnitz and Dr K. B. Wildey, many of whom provided essential live samples of pests or plants for me to photograph. Finally, I acknowledge the help and encouragement of my wife and family, all of whom have been in some way inconvenienced (if not disadvantaged) by my literary activities; without their forbearance, this book would never have been written.

David V. Alford
Bristol

INTRODUCTION

Ornamental plants are attacked by a wide variety of pests, most of which are arthropods (phylum Arthropoda). Arthropods are a major group of invertebrate animals characterized by their hard exoskeleton or body shell, their segmented bodies and their jointed limbs; of these, insects and, to a lesser extent, mites are of greatest significance.

INSECTS

Insects differ from other arthropods in possessing just three pairs of legs, usually one or two pairs of wings (all winged invertebrates are insects), and by having the body divided into three distinct regions: the head, the thorax and the abdomen (Fig. **1**).

The outer skin or integument of an insect is known as the cuticle. It forms a non-cellular, waterproof layer over the body and is composed of chitin and protein, the precise chemical composition and thickness determining its hardness and rigidity. The cuticle has three layers (epicuticle, exocuticle and endocuticle) and is secreted by an inner lining of cells which forms the hypodermis or basement membrane. When first produced the cuticle is elastic and flexible, but soon after deposition it usually undergoes a period of hardening or sclerotization and becomes darkened by the addition of a chemical called melanin. The adult cuticle is not replaceable, except in certain primitive insects. However, at intervals during the growth of the immature stages, the 'old' hardened cuticle becomes too tight and is replaced by a new, initially expandable one secreted from below. Certain insecticides have been developed that are capable of disrupting chitin deposition; although ineffective against adults, they kill insects undergoing ecdysis (i.e. those moulting from one growth stage to the next).

The insect cuticle is often thrown into ridges and depressions, is frequently sculptured or distinctly coloured, and may bear a variety of spines and hairs. In larvae, body hairs often arise from hardened, spot-like pinacula (often called tubercules) or larger, wart-like verrucae. In some groups, features of the adult cuticle (such as colour, sculpturing and texture) are of considerable value for distinguishing between species.

The basic body segment of an insect is divided into four sectors (a dorsal tergum, a ventral sternum and two lateral pleurons) which often form horny, chitinized plates called sclerites. These may give the body an armour-like appearance and are either fused rigidly together or joined by soft, flexible, chitinized membranes to allow for body movement.

Segmental appendages, such as legs, are developed as outgrowths from the pleurons.

The head is composed of six fused body segments and carries a pair of sensory antennae, eyes and mouthparts. The form of an insect antenna varies considerably, the number of antennal segments ranging from one to more than a hundred. The basal segment is called the scape and is often distinctly longer than the rest; the second segment is the pedicel and from this arises the many-segmented flagellum. In a geniculated (elbowed) antenna, the pedicel acts as the articulating joint between the elongated scape and the flagellum; such antennae are characteristic of certain weevils and wasps. Many insects possess two compound eyes, each composed of several thousand facets, and three simple eyes called ocelli, the latter usually forming a triangle at the top of the head. Compound eyes are large and well developed in various predatory insects, where good vision is important. The compound eye provides a mosaic, rather than a clear picture, but is well able to detect movement.

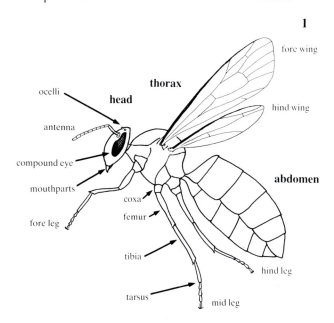

1 Main body features of an insect.

The ocelli are optically simple and lack a focusing mechanism. They are concerned mainly with registering light intensity, enabling the insect to distinguish between light and shade. Unlike nymphs, insect larvae lack compound eyes but they often possess several ocelli, arranged in clusters on each side of the head. Insect mouthparts are derived from several modified, paired segmental appendages; they range from simple biting jaws (mandibles) to complex structures for piercing, sucking or lapping. Among phytophagous insects, biting mouthparts are found in adult and immature grasshoppers, locusts, earwigs, beetles, etc. but may be restricted to the larval stages, as in butterflies, moths and sawflies. Some insects (e.g. various dipterous larvae) have rasping mouthparts which are used to tear plant tissue, the food material then being ingested in a semiliquid state. Stylet-like, suctorial mouthparts are characteristic of aphids, mirids, psyllids and other bugs; such insects may introduce toxic saliva into plants and cause distortion or galling of the tissue. Certain insects (notably aphids) can also transmit plant viruses to host plants.

The thorax has three segments — the prothorax, the mesothorax and the metathorax — whose relative sizes vary from one insect group to the next. In crickets, cockroaches and beetles, for example, the prothorax is the largest section and is covered on its upper surface by an expanded dorsal sclerite called the pronotum; in flies, the mesothorax is greatly enlarged and the prothoracic and metathoracic segments are much reduced. Each thoracic segment bears a pair of jointed legs. Their form varies considerably but all legs have the same basic structure (see Fig. 1). Wings, when present, arise from the mesothorax and metathorax as a pair of fore wings and a pair of hind wings respectively. In many insects, the base of each fore wing is covered by a scale-like lobe, known as the tegula. Basically, each wing is an expanded membrane-like structure supported by a series of hardened veins, but considerable modification has taken place in the various insect groups. In cockroaches, earwigs and beetles, for instance, the fore wings have become hardened and thickened protective flaps, called elytra, and only the hind wings are used for flying; in true flies, the fore wings retain their propulsive function but the hind wings have become greatly reduced in size and are modified into balancing organs known as halteres. Details of wing structure and venation are of importance in the classification and identification of insects, and the names of many insect orders are based upon them.

The abdomen is normally formed from 10 or 11 segments but fusion and apparent reduction of the most anterior or posterior components is common. Although present in many larvae, abdominal appendages are wanting on most segments of adults, their ambulatory function, as found in various other arthropods, having been lost. However, appendages on the eighth and ninth segments remain to form the genitalia, including the male claspers and female ovipositor. In many groups, a pair of cerci are formed from appendages on the last body segment. These are particularly long and noticeable in primitive insects, and in crickets, stoneflies, mayflies and earwigs, but are absent in the most advanced groups. Abdominal sclerites are limited to a series of dorsal tergites and a ventral set of sternites; these give the abdomen an obviously segmented appearance.

The body cavity of an insect extends into the appendages and is filled with a more-or-less colourless, blood-like fluid called haemolymph. This bathes all the internal organs and tissues, and is circulated by muscular action of the body and by a primitive, tube-like heart which extends mid-dorsally from the head to near the tip of the abdomen.

The brain is the main co-ordinating centre of the body. It fills much of the head and is intimately linked with the antennae, the compound eyes and the mouthparts. The brain gives rise to a central nerve cord which extends back mid-ventrally through the various thoracic and abdominal segments. The nerve cord is swollen at intervals into a series of ganglia, from which arise numerous lateral nerves. These ganglia control many nervous functions (such as movement of the body appendages) independent of the brain.

The gut or alimentary tract is a long, much modified tube stretching from the mouth to the anus. It is subdivided into three sections: a fore gut, with a long oesophagus and a bulbous crop; a mid-gut, where digestion of food and absorption of nutriment occurs; and a hind gut, concerned with water absorption, excretion and the storage of waste matter prior to its disposal. A large number of blind-ending, much convoluted Malpighian tubules arise from the junction between the midgut and the hind gut. These tubules collect waste products from the body and pass them into the gut.

The respiratory system comprises a complex series of branching tubes (tracheae) and microscopic tubules (tracheoles) which ramify throughout the body in contact with the internal organs and tissues. This tracheal system opens to the outside through segmentally arranged valve-like breathing holes or spiracles, present along either side of the body. Air is forced through the spiracles by contraction and relaxation of the abdominal body muscles. Spiracles also occur in nymphs and larvae (they are often very obvious in butterfly and moth caterpillars) but they are often much reduced in number. In some groups (e.g. various dipterous larvae) the tracheae open via a pair of anterior spiracles, commonly located on the first body segment (prothorax), and a pair of posterior spiracles, usually located on the anal segment; these spiracles are often borne on raised processes. Morphological details of the spiracular openings and processes are often used to distinguish between species (e.g. in agromyzid leaf miners).

Female insects possess a pair of ovaries, subdivided into several egg-forming filaments called ovarioles. The ovaries enter a median oviduct and this opens to the outside on the ninth abdominal segment. Many insects have a protrusible egg-laying tube, called an ovipositor. The male reproductive system includes a pair of testes and associated ducts which lead to a seminal vesicle in which sperm is stored prior to mating. The male genitalia may include chitinized structures, such as the claspers which help to grasp the female during copulation. Examination of the male or female genitalia is often essential for distinguishing between closely related species.

Sexual reproduction is commonplace in insects but in certain groups fertilized eggs produce only female offspring and males are reared only from unfertilized ones. In other cases, male production may be wanting or extremely rare and parthenogenesis (reproduction without a sexual phase) is the rule.

Although some insects are viviparous (giving birth to active young), most lay eggs. A few, such as aphids, reproduce viviparously by parthenogenesis in spring and summer but may produce eggs in the autumn (after a sexual phase). Insect eggs have a waterproof shell and many are capable of surviving severe winter conditions in exposed situations on tree bark or shoots.

Insects normally grow only during the period of pre-adult development, as nymphs or as larvae, their outer cuticular skin being moulted and replaced between each successive growth stage or instar. The most primitive insects (subclass Apterygota) have wingless adults, their eggs hatching into nymphs that are essentially similar to adults but smaller and sexually immature. The more advanced, winged or secondarily wingless, insects (subclass Pterygota) develop in one of two ways. In the Exopterygota, there is a succession of nymphal stages in which wings (when present) usually develop externally as buds that become fully formed and functional once the adult stage is reached. In such insects, nymphs and adults are frequently of similar appearance (apart from the wing buds or wings), and often share the same feeding habits. This type of development is termed incomplete metamorphosis. The most advanced insects (Endopterygota) show complete metamorphosis, development including several larval instars of quite different structure and habit from the adults. Here, wings develop internally and the transformation from larval to adult form occurs during a quiescent, non-feeding pupal stage. Insect larvae are of various kinds. Some, commonly called caterpillars, have three pairs of jointed thoracic legs (true legs) and a number of fleshy, false legs (prolegs) on the abdomen. Many sawfly, butterfly and moth larvae are of this type. The prolegs of butterfly and moth larvae are usually provided with small chitinous hooks known as crochets. Some larvae, including many beetle grubs, possess well-developed thoracic limbs but lack abdominal prolegs. Other insect larvae are totally legless (apodous); wasp and various fly larvae are examples.

Classification of insects
Class **INSECTA**
Subclass APTERYGOTA
    Order Thysanura   Bristle-tails, silverfish
    Order Diplura   Diplurans
    Order Protura   Proturans
    Order Collembola   Springtails
Subclass PTERYGOTA
 Division *EXOPTERYGOTA*
    Order Ephemeroptera Mayflies
    Order Odonata   Dragonflies
    Order Plecoptera   Stoneflies
    Order Grylloblattodea Grylloblattodeans
    Order Orthoptera   Crickets, grasshoppers
    Order Phasmida   Stick-insects, leaf-insects
    Order Dermaptera   Earwigs
    Order Embioptera   Web-spinners
    Order Dictyoptera   Cockroaches, mantids
    Order Isoptera   Termites
    Order Zoraptera   Zorapterans
    Order Psocoptera   Psocids or booklice
    Order Mallophaga   Biting lice
    Order Anoplura   Sucking lice
    Order Hemiptera   True bugs
    Order Thysanoptera Thrips
 Division *ENDOPTERYGOTA*
    Order Neuroptera   Alder flies, lacewings, etc.
    Order Coleoptera   Beetles
    Order Strepsiptera   Stylopids
    Order Mecoptera   Scorpion flies
    Order Siphonaptera   Fleas
    Order Diptera   True flies
    Order Lepidoptera   Butterflies, moths
    Order Trichoptera   Caddis flies
    Order Hymenoptera   Ants, bees, wasps, sawflies, etc.

Pests of ornamental plants are found in many different insect orders, the main groups being characterized as follows:

Collembola: small, wingless insects, often with a forked springing organ on the ventral side of the fourth abdominal segment; biting mouthparts; antennae with usually, four segments; metamorphosis slight: *family Sminthuridae* (p. 19); *family Onychiuridae* (p. 19).

Orthoptera: medium-sized to large, stout-bodied insects with a large head, large pronotum and usually two pairs of wings, the thickened fore wings called tegmina; either the fore wings or hind wings may be reduced or absent; femur of hind leg often modified for jumping; tarsi usually three- or four-segmented; chewing mouthparts; cerci usually short and unsegmented; metamorphosis incomplete: *family Gryllotalpidae* (p. 20); *family Gryllidae* (p. 20).

Dermaptera: elongate, omnivorous insects with biting mouthparts; fore wings modified into very short, leathery elytra; hind wings semicircular and membranous, with radial venation; anal cerci modified into pincers; metamorphosis incomplete: *family Forficulidae* (p. 21).

Dictyoptera: small to large, stout-bodied but rather flattened insects with a large pronotum and two pairs of wings, the thickened fore wings called tegmina; hind wings folded longitudinally like a fan; chewing mouthparts; antennae very long and thread-like; legs robust and spinose, and modified for running; the tarsi usually three- or four-segmented; cerci many-segmented; metamorphosis incomplete: *family Blattidae* (p. 22).

Hemiptera: minute to large insects, usually with two pairs of wings and piercing, suctorial, needle-like mouthparts; fore wings frequently partly or entirely hardened; metamorphosis usually gradual and incomplete.

Suborder Heteroptera: usually with two pairs of wings, the fore wings with a leathery basal area and a membranous tip; hind wings membranous; wings held flat over the abdomen when in repose; the beak-like mouthparts arise from the front of the head and are flexibly attached; prothorax large; some species are plant-feeders but many are predacious: *family Tingidae* (p.000); *family Miridae* (p.000).

Suborder Homoptera: a large group of mainly small insects with one or two pairs of membranous wings; wings held over the back in a sloping, roof-like posture when in repose (not in the Phylloxeridae); the beak-like mouthparts arise from the posterior part of the head and are rigidly attached; prothorax usually small; all members are plant-feeders.

Superfamily Cercopoidea — tegulae absent; hind legs modified for jumping, with long tibiae bearing one or two long spines: *family Cercopidae* (p. 26). ***Superfamily Cicadelloidea*** — tegulae absent; hind legs modified for jumping, with long tibiae bearing longitudinal rows of short spines: *family Cicadellidae* (p. 28). ***Superfamily Psylloidea*** — antennae usually ten-segmented; tarsi two-segmented and with a pair of claws: *family Psyllidae* (p. 34); *family Triozidae* (p. 37); *family Carsidaridae* (p. 39); *family Spondyliaspidae* (p. 39). ***Superfamily Aleyrodoidea*** — antennae seven-segmented; wings opaque and whitish: *family Aleyrodidae* (p. 40). ***Superfamily Aphidoidea*** — females winged or wingless; wings, when present, usually large and transparent, with few veins; abdomen usually with a pair of siphunculi on the abdomen; tarsi two-segmented and with a pair of claws: *family Lachnidae* (p. 43); *family Chaitophoridae* (p. 49); *family Callaphididae* (p. 51); *family Aphididae* (p. 57); *family Hormaphididae*(p. 80); *family Mindaridae* (p. 80); *family Pemphigidae* (p. 80); *family Thelaxidae* (p. 86); *family Adelgidae* (p. 87); *family Phylloxeridae* (p. 92). ***Superfamily Coccoidea*** — females always wingless; males usually with a single pair of wings and vestigial mouthparts, developing through a pupal stage; tarsi, if present, one-segmented and with a single claw: *family Diaspidae* (p. 92); *family Coccidae* (p. 97); *family Eriococcidae* (p. 102); *family Pseudococcidae* (p. 103); *family Margarodidae* (p.105).

Thysanoptera: small or minute, slender-bodied insects with short antennae and asymmetrical, rasping and sucking mouthparts; a protrusible bladder at the tip of each tarsus; wings, when present, very narrow with hair-like fringes and greatly reduced venation. The nymphs are similar in appearance to adults but are wingless; metamorphosis includes two or three inactive pupa-like stages: *family Thripidae* (p. 106); *family Phaeothripidae* (p. 111).

Coleoptera: minute to large insects with biting mouthparts; fore wings modified into horny or leathery elytra which usually meet in a straight line along the back; hind wings membranous and folded beneath the elytra when in repose, but often reduced or absent; prothorax normally large and mobile; metamorphosis complete. The largest insect order, with more than a quarter of a million species worldwide.

Superfamily Scarabaeoidea — a large group of often very large, brightly coloured insects, some of which possess enlarged horns on the head and thorax: *family Scarabaeidae* (p. 112). ***Superfamily Elateroidea*** — elongate beetles with a hard exoskeleton, the head sunk into the prothorax, toothed or comb-like antennae; hind angles of the prothorax sharply pointed and, often, extended: *family Elateridae* (p. 114). **Superfamily Cucujoidea** — beetles usually with five visible abdominal segments and, often, clubbed antennae: *family Nitidulidae* (p. 115); *family Byturidae* (p. 116). **Superfamily Chrysomeloidea** — mostly plant feeders, adults with four-segmented tarsi (the fourth segment very small); larvae usually with well-developed thoracic legs: *family Chrysomelidae* (p. 116). **Superfamily Curculionoidea** — a very large superfamily, including weevils and bark beetles; weevils have an often very elongated snout (rostrum), which bears the mouthparts and antennae; in most weevils, the antennae are elbowed (geniculate) with an elongated basal segment (scape) but in the Attelabidae all antennal segments are of a similar length; larvae usually apodous: *family Attelabidae* (p. 130); *family Curculionidae* (p. 133); *family Scolytidae* (p. 149).

Diptera: minute to large insects with a single pair of membranous wings; hind wings reduced to small, club-shaped halteres; mouthparts suctorial and sometimes adapted for piercing; larvae apodous, and usually with a reduced, retracted head; metamorphosis complete.

Suborder Nematocera: antennae of adults with a scape, pedicel and flagellum, the latter comprising numerous, similar-looking segments, each bearing a whorl of hairs. The larvae usually (not in the Cecidomyiidae) with a well-defined head and horizontally opposed mandibles: *family Tipulidae* (p. 151); *family Bibionidae* (p. 153); *family Chironomidae* (p. 154); *family Sciaridae* (p. 154); *family Cecidomyiidae* (p. 155).

Suborder Cyclorrapha: antennae of adults with a scape, pedicel and flagellum, the latter usually forming an enlarged, compound segment tipped by a short, bristle-like arista. The larvae are maggot-like, often tapering anteriorly; they possess distinctive, rasping 'mouth-hooks', but the head is small and inconspicuous; pupation occurs within the last larval skin, which then forms a protective barrel-like puparium from which the adult eventually escapes by forcing off a circular cap (the operculum): *family Syrphidae* (p. 170); *family Tephritidae* (p. 173); *family Psilidae* (p. 174); *family Ephydridae* (p. 174); *family Drosophilidae* (p. 175); *family Agromyzidae* (p. 175); *family Anthomyiidae* (p. 187).

Lepidoptera: minute to large insects with two pairs of membranous wings; cross-veins few in number; body, wings and appendages scale-covered; adult mouthparts suctorial but those of larvae adapted for biting; the larvae are mainly caterpillar-like and plant-feeders; metamorphosis complete.

Superfamily Eriocranioidea — adults with a short proboscis; females with a piercing ovipositor; pupae with functional mandibles: *family Eriocraniidae* (p. 189). **Superfamily Hepialoidea** — adults with non-functional, vestigial mouthparts and short antennae: *family Hepialidae* (p. 191). **Superfamily Nepticuloidea** — adults with wing venation reduced; ovipositor soft: *family Nepticulidae* (p. 192); *family Tischeriidae* (p. 195). **Superfamily Incurvarioidea** — small day-flying moths; antennae of males often very long: *family Incurvariidae* (p. 196). **Superfamily Cossoidea** — heavily-bodied moths with primitive wing venation: *family Cossidae* (p. 197). **Superfamily Tineoidea** — primitive moths with narrow or very narrow wings: *family Lyonetiidae* (p. 198); *family Hieroxestidae* (p. 202); *family Gracillariidae* (p. 203); *family Phyllocnistidae* (p. 213). **Superfamily Yponomeutoidea** — an indistinct and rather diverse group: *family Sesiidae* (p. 214); *family Choreutidae* (p. 215); *family Yponomeutidae* (p. 215). **Superfamily Gelechioidea** — a large group of moderately small moths: *family Coleophoridae* (p. 226); *family Oecophoridae* (p. 227); *family Gelechiidae* (p. 230); *family Blastobasidae* (p. 232). **Superfamily Tortricoidea** — a major group of moderately small moths with mainly rectangular fore wings; the larvae are mainly leaf-folding or leaf-rolling: *family Tortricidae* (p. 233). **Superfamily Pyraloidea** — a very large group of mainly slender-bodied, long-legged moths, often with narrow, elongate fore wings: *family Pyralidae* (p. 266). **Superfamily Papilionoidea** — adults with clubbed, but unhooked antennae: *family Pieridae* (p. 269). **Superfamily Bombycoidea** — often large to very large moths, with non-functional mouthparts; male antennae strongly bipectinated: *family Lasiocampidae* (p. 270). **Superfamily Geometroidea** — mainly slender-bodied moths with broad wings; larvae with a reduced number of functional abdominal prolegs: *family Thyatiridae* (p. 273); *family Geometridae* (p. 274). **Superfamily Sphingoidea** - large-bodied, strong-flying moths, often with a large proboscis; larvae usually possess a characteristic dorsal horn on the eighth abdominal segment: *family Sphingidae* (p. 292). **Superfamily Notodontoidea** — a small group of moths, sometimes included within the Noctuoidea: *family Notodontidae* (p. 294); *family Dilobidae* (p. 298). **Superfamily Noctuoidea** — the largest group of lepidopterous insects, with a wide variety of forms: *family Lymantriidae* (p. 299); *family Arctiidae* (p. 303); *family Noctuidae* (p. 308).

Trichoptera: small, medium to large insects, with two pairs of wings which are held in a roof-like position when in repose; wings with few cross-veins and coated with small, inconspicuous hairs. The larvae have biting mouthparts and are omnivorous; they live submerged in water, most species occupying characteristic cases constructed from silk and pieces of vegetation or grains of sand. Development includes egg, larval and pupal stages; metamorphosis complete: *family Limnephilidae* (p. 331).

Hymenoptera: minute to large insects with, usually, two pairs of membranous wings, the hind wings smaller and interlocked with the fore wings by small hooks; mouthparts adapted for biting but often also for lapping and sucking; females always possess an ovipositor, modified for sawing, piercing or stinging; metamorphosis complete.

Suborder Symphyta: includes the sawflies, insects with a well-developed ovipositor, and the abdomen and thorax joined without a constriction or 'waist'; the larvae are mainly caterpillar-like and plant-feeders.

Superfamily Megalodontoidea — a small group of primitive sawflies, with a flattened abdomen: *family Pamphiliidae* (p. 332). **Superfamily Tenthredinoidea** — the major group of sawflies, adults with a saw-like ovipositor: *family Argidae* (p. 336); *family Cimbicidae* (p. 336); *family Diprionidae* (p. 336); *family Tenthredinidae* (p. 338).

Suborder Apocrita: the main group of hymenopterous insects, the first abdominal segment being fused to the thorax and separated from the rest of the abdomen (known as the gaster) by a wasp-like 'waist'. The suborder is composed of two groups: the *Parasitica* (most of which are parasites and have the ovipositor adapted for piercing their hosts) and the *Aculeata* (in which the ovipositor is modified into a sting).

Superfamily Cynipoidea — minute or very small, mainly black insects, including gall wasps: *family Cynipidae* (p. 372). **Superfamily Scolioidea** — a large, primitive group of Aculeates, including the ants: *family Formicidae* (p. 380). **Superfamily Vespoidea** — wasps, the pronotum extending back to the tegulae; the larvae are fed on meat: *family Vespidae* (p. 381). **Superfamily Apoidea** — generally hairy insects (solitary or social bees), with broad hind tarsi and the pronotum not extending back to the tegulae; the larvae are fed on nectar and pollen: *family Andrenidae* (p. 381); *family Halictidae* (p. 382).

MITES

Mites and ticks (the subclass Acari) form part of the Arachnida, a major class of arthropods. Unlike insects, they have no antennae, wings or compound eyes, are usually eight-legged and possess chelicerate mouthparts adapted for biting or piercing. The body is composed of a gnathosoma, which bears a pair of sensory pedipalps and paired chelicerae, and a sac-like idiosoma with no obvious segmentation (Fig. **2**); they are thus readily distinguished from other arachnids that have the body divided into a distinct cephalothorax and a usually (not spiders) clearly segmented opisthosond. The respiratory system in the Acari often includes a pair of breathing pores, also known as stigmata; their position on the body, or their absence, forms a basic character for naming the various acarine orders.

Unlike members of other arachnid groups, the Acari includes many phytophagous species, mainly in the order Prostigmata. The chelicerae of most prostigmatid mites are needle-like and are used to penetrate plant cells; also, the idiosoma is subdivided by a subjugal furrow into the propodosoma and the hysterosoma, each region bearing two pairs of legs (see Fig. **2**). The body and limbs of a mite possess various setae, the arrangement and characteristics of which are of considerable value in the classification and identification of species.

Development from egg to adult usually includes a six-legged larva and two or three eight-legged nymphal stages: proto-, deuto- and tritonymphs. Larvae and nymphs are generally similar in appea-

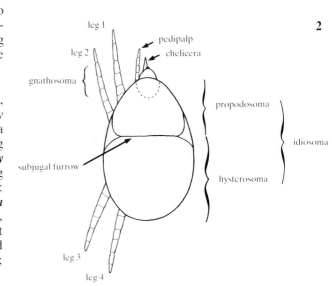

2 Main body features of a prostigmatid mite.

rance and habit to the adults but are smaller and sexually immature.

Many phytophagous mites are free-living but others (notably various members of the Eriophyoidea) inhabit distinctive plant galls formed in response to toxic saliva injected into the host during feeding. A few species are important vectors of plant virus diseases.

Classification of Mites

Subclass **ACARI**

Superorder OPILIOACAROIFORMES
Order **Notostigmata**	weakly sclerotized, brightly coloured, harvestman-like mites with four pairs of abdominal stigmata

Superorder PARASITIFORMES
Order **Tetrastigmata**	a small group of large, heavily sclerotized mites
Order **Mesostigmata**	a large, diverse group of mites, including many predacious species
Order **Metastigmata**	ticks

Superorder ACARIFORMES
Order **Prostigmata**	an extremely varied group of minute to large, usually lightly sclerotized mites, including many phytophagous pests
Order **Astigmata**	soft-bodied, lightly sclerotized mites without stigmata
Order **Cryptostigmata**	dark, strongly sclerotized mites, known as oribatids or 'beetle mites'

Brief details of the main groups containing pests of ornamental plants are given below:

Prostigmata: mites with the stigmata placed between the chelicerae, and often with one or two pairs of sensory trichobothria on the propodosoma.

Superfamily Eriophyoidea: minute, sausage-shaped or pear-shaped mites with two pairs of legs, each leg terminating in a branched feather-claw; body with a distinct prodorsal shield; abdomen (hysterosoma) more-or-less annulated with a dorsal series of tergites and a ventral series of sternites (Fig. **3**).

Family Phytoptidae (p. 383): prodorsal shield bearing three or four setae; feather-claw simple — genus *Phytoptus*.*

Family Eriophyidae (p. 384): similar to the Phytoptidae but prodorsal shield bearing two or no setae; feather-claw either simple or divided. *Subfamily Cecidophyinae* — elongate mites without prodorsal shield setae: genus *Cecidophyopsis*. *Subfamily Eriophyinae* — elongate, worm-like mites with a pair of setae on the prodorsal shield; abdomen (hysterosoma) subdivided into numerous tergites and sternites, typically subequal anteriorly: genera *Acalitus, Aceria, Artacris, Eriophyes.** *Subfamily Phyllocoptinae* — elongate, cigar-shaped to pear-shaped mites with a pair of setae on the prodorsal shield; abdomen (hysterosoma) subdivided by relatively few, broad tergites and several, narrow sternites: genera *Acaricalus, Aculus, Epitrimerus, Phyllocoptes, Tegonotus, Vasates*.

Superfamily Tarsonemoidea: mites with short, needle-like mouthparts: *family Tarsonemidae* (p. 404): hind legs of female without claws; hind legs of male modified into claspers.

Superfamily Tetranychoidea: spider-like mites with long, needle-like chelicerae: *family Tetranychidae* (p. 408); *family Tenuipalpidae* (p. 404).

Superfamily Eupodoidea: includes tarsonemid-like species with claws on each pair of tarsi: *family Pyemotidae* (p. 415).

Astigmata: soft-bodied, semitransparent mites; chelicerae forceps-like.

Superfamily Acaroidea: features as for order: *family Acaridae* (p. 415).

3

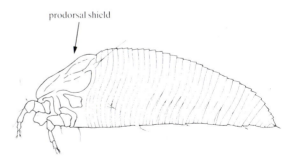
prodorsal shield

3 Main body features of an eriophyid mite.

* Application of the generic names *Phytoptus* and *Eriophyes* follows that in general use before 1971, and not that proposed by Newkirk, R. A. & Keifer, H. H. (1971) *Eriophyid studies* C5, California Department of Agriculture, Sacramento.

WOODLICE

Woodlice *(phylum Arthropoda: class Crustacea)* are terrestrial, 14-legged crustaceans, forming a distinct order, the Isopoda. They feed on decaying vegetation, but also sometimes attack the roots, stems and leaves of healthy plants; animal matter is also included in the diet, individuals commonly feeding on dried-blood fertilizer: *family Armadillidiidae* (p. 417); *family Oniscidae* (p. 418); *family Porcellionidae* (p. 418).

MILLEPEDES

Millepedes *(phylum Arthropoda: class Diploda)* usually have elongate bodies composed of a variable number of double abdominal segments, most of which bear two pairs of legs. The head is armed with biting mouthparts, and bears simple eyes and a pair of short, clubbed antennae. Millepedes are secretive, light-shy creatures, usually inhabiting moist, sheltered situations. Although some species are scavengers, and a few can be phytophagous or predacious, most feed on decaying vegetation. Millepedes can damage germinating seeds and seedlings but are of little or no importance on older plants.

SYMPHYLIDS

Symphylids *(phylum Arthropoda: class Symphyla)* are small, soft-bodied creatures with three pairs of mouthparts, 12 pairs of legs and a posterior pair of cerci. Most species inhabit the soil and feed on decaying vegetation; a few attack the underground parts of plants.

NEMATODES

Nematodes *(phylum Nematoda)* are unsegmented worm-like invertebrates which lack circulatory and respiratory systems, and are devoid of cilia (both external and internal); they also possess a stiff but flexible external cuticle which, unlike that of insects, lacks chitin. Nematodes are often abundant in soils, feeding on various micro-organisms, but many species are parasitic. Those attacking vertebrates (including man) are commonly known as 'roundworms', while those associated with plants are often known as 'eelworms'. Plant-parasitic nematodes are microscopic, commonly no more than 0.1–0.5 mm long; they are unique in possessing a distinctive spear-shaped structure in the oesophagus, with which they pierce the walls of plant cells. The detailed form of this spear is often useful for distinguishing between genera. To be active, a nematode is dependent upon the presence of moisture; individuals usually travel through the soil or over plant tissue in films of water, progressing with serpentine movements of the body.

SLUGS AND SNAILS

Slugs and snails *(phylum Mollusca: class Gastropoda)* are soft-bodied, non-segmented invertebrates with the body composed of three regions: head, foot and visceral mass. The latter is covered by a layer of epithelial cells, called the mantle; this secretes a shell of calcium carbonate and encloses a mantle cavity. The mouth usually contains a rasping tongue or radula, armed with thousands of minute chitinous teeth. Most molluscs, such as clams, cuttlefish, octopuses, oysters, sea-slugs and squids, are marine animals; several species live in freshwater habitats but only certain slugs and snails in the order Pulmonata are able to survive on land.

Terrestrial slugs and snails are hermaphroditic creatures with a slimy, asymmetrical body. Their mantle cavity is vascularized and functions as a lung, with an aperture on one side of the body. The visceral mass is often contained within a hard, helical shell. The head is well developed and bears a long and a short pair of retractile tentacles, each longer tentacle having a simple eye at its tip. The foot is muscular, broad and flattened, and functions as the propulsive organ, the animal gliding along on a trail of slime. Slugs and snails are most active in warm, humid conditions. They feed on plant material of various kinds, and some species are regarded as important pests.

EARTHWORMS

Earthworms *(phylum Annelida: class Oligochaeta)* are well-known hermaphroditic creatures with long, thin and distinctly segmented bodies. They burrow in the soil and feed mainly on decaying vegetative matter, thereby contributing to soil fertility, aeration and drainage. Although primarily beneficial, a few species can be a nuisance in lawns and sports turf.

BIRDS AND MAMMALS

Birds *(class Aves)* and mammals *(class Mammalia)* are of only minor significance as pests of ornamental plants but a few species can cause damage.

Birds are of particular significance as pests of flower buds or open blossom, and mammals as pests of young plants, new shoots, bulbs, corms and seeds.

PEST DAMAGE

The kind of damage inflicted upon ornamental plants by pests is extremely varied, being related to feeding habit and to feeding methods (e.g. whether the pest's mouthparts are adapted for biting, piercing, rasping or sucking). Some pests attack the roots or other underground parts, but most affect the leaves, stems, shoots, buds or flowers. Damage symptoms vary from minor, often imperceptible blemishes, colour changes or loss of vigour, to complete death of plants. Leaves, for example, may become blistered, chewed, discoloured, disfigured, distorted, dwarfed, galled, malformed, mined, punctured, ragged, skeletonized, speckled, thickened, webbed, wilted or withered. Symptoms are sometimes in themselves sufficiently characteristic to enable (at least with experience) the causal organism to be identified; leaf mines formed by certain insect larvae are good examples. However, in many cases the cause of plant damage cannot be determined with confidence unless the pest itself is actually located and identified.

Some pests are indiscriminate feeders, attacking a wide range of plants, but others are more specific and often able to feed on only a restricted group of hosts (perhaps those from a single family or genus, or even a single species); plant susceptibility to pests can also vary at the subspecific level and sometimes differs from cultivar to cultivar.

Ornamentals with identical or closely related wild equivalents can suffer considerable damage from pests, but others may be largely, if not entirely, immune. Exotic plants introduced from abroad, for example, are often (but by no means always) unsuitable hosts for European pests. However, such exotics can suffer damage from pests introduced with them from their country of origin; the presence on *Eucalyptus* of eucalyptus sucker (p. 39) is an example.

In some instances (e.g. aphids, leaf beetles, mites and slugs), all active stages (adults and juveniles) of a pest can cause similar damage. In others, the type of damage caused by adults and juveniles may be different; chafer adults, for example, attack the leaves of plants but the grubs are root-feeders. In many cases (e.g. dipterous and lepidopterous pests), damage is caused only by the larvae; in a few instances (e.g. leaf-cutting bees), only the adult is damaging to plants.

Several pests (e.g. certain aphids, midges, wasps, mites and nematodes) produce characteristic galls on host plants; such gall-formers inject a toxin into the plant cells, thereby stimulating unusual development of the plant tissue. Some galls are brightly coloured and very spectacular in appearance. In other cases (e.g. certain aphids, capsids, tarsonemid mites and nematodes), injection of toxins causes distortion of the plant tissue, affected shoots, leaves or flowers becoming malformed and often discoloured. Feeding by other pests (e.g. thrips, spider mites, rust mites) affects the general appearance of leaves, the tissue often developing a distinctive speckling or an overall bronzing, the latter commonly leading to premature leaf-fall and loss of vigour.

Many pests of ornamental plants have little or no direct effect on growth but their depredations can be disfiguring; such damage is often of little or no consequence on established plants but, on young hosts (particularly in commercial nurseries), may have a significant effect on plant quality. The mere presence of certain pests (e.g. wax-secreting scale insects or woolly aphids) can be unacceptable, even at low levels of infestation.

CONTROL OF PESTS

Good husbandry will reduce the likelihood of pest problems developing on ornamentals but, in some instances, specific control measures may be necessary to protect plants from attack or to keep pests and their damage within acceptable bounds. Pest attacks can be lessened by using traps or physical barriers (e.g. greased tree-bands for pests such as winter moth, and netting for birds or mammals) but such methods are not always practical and are certainly not available for combating the majority of pests.

Attention to hygiene is important for lessening the impact of pests, especially in glasshouses; plant debris should be cleared as soon as cropping is completed, and buildings, pots and other equipment disinfected before new plants are introduced. Efficient weed control, both within glasshouses and out-of-doors, will reduce the range of places where pests can find shelter and will also limit the number of possible alternative host plants upon which certain pests can survive or increase. Regular cultivation will help to control weeds and will also keep soil pests in check, either destroying them directly or exposing them to desiccation or to the attention of birds and other predators.

Wherever practical, plants should be examined regularly for signs of pests, so that appropriate action can be taken at the earliest possible stage. Newly acquired plants, including the roots and adhering soil or compost, should always be inspected to prevent the accidental introduction of pests into clean sites; this is of particular importance for combating insidious pests such as nematodes.

On a small scale, some pests can be controlled by hand, any egg clumps, larvae or other stages found on plants being squashed or picked off and destroyed; in some cases affected parts of plants, such as shoots containing galls or webbed by caterpillars, can be cut off and destroyed. Prunings and other plant debris, whether thought to be harbouring pests or not, should never be left lying around but should be gathered-up immediately and burnt.

Various pesticides are recommended for use against pests of ornamental plants. Some are broad-spectrum materials (capable of killing various kinds of pest) but others are more selective and some very specific; modes of action also vary. Choice of product will depend on many different factors. Systemic materials (which are absorbed through the leaves or taken up by the roots, and then translocated through the plant in the sap) are particularly effective against sap-feeding pests, especially aphids and leaf miners; contact materials have a variety of uses, and stomach poisons are useful for killing pests such as caterpillars and weevils. In some situations, fogs and fumigants may be useful but, in others, granules, pellets or sprays will be more appropriate. Some pests (e.g. spider mites, and certain glasshouse aphids and thrips) have developed resistance to pesticides, and this has limited the effectiveness of many products. Whichever pesticide is selected, the directions on the manufacturer's label should be followed, and care taken to ensure that treatments are applied both effectively and efficiently.

Some ornamental plants, such as *Begonia, Calceolaria* and *Hydrangea*, can be intolerant of pesticides, differences in susceptibilities sometimes varying from cultivar to cultivar. In some cases, growth is checked, perhaps imperceptibly; in others, tissue becomes discoloured or distorted and, in extreme cases, may be killed. Where information regarding the safety of a pesticide to any particular plant species or cultivar is lacking, or if doubt exists, a few plants should be treated first and these later checked for signs of phytotoxicity before larger-scale treatment is undertaken. Young, tender plants are especially susceptible to chemicals; also, certain sprays otherwise considered safe can have an adverse effect on open blooms, causing a range of undesirable symptoms such as specking or discoloration of the tissue; spray damage of this type is well-known on glasshouse-grown chrysanthemums. As a general rule, spraying of open flowers should be avoided, not only because of the risk of phytotoxicity but also to safeguard pollinating insects

which might be foraging upon them. Further, spraying with pesticides should not be undertaken during bright sunlight, the risk of damage from excessive temperatures being especially serious in glasshouses and when plants are under stress. Problems of phytotoxicity are more likely to occur on protected plants than on those grown out-of-doors.

Most pesticides recommended for use on ornamental plants are available only to commercial growers. However, some products are specially formulated and recommended for use in private gardens. These pesticides will protect plants against the majority of important pests but non-chemical methods are often preferable and frequently just as effective.

The application of pesticides in amenity areas poses particular problems and is often impractical. Not only are there increased risks of killing non-target species, but potential hazards to the public and wildlife in general must also be considered. A few insecticides are specifically recommended for control of pests in amenity areas; such treatments are, of course, 'safe' when used as recommended by the pesticide manufacturers, but on environmental grounds their use should be kept to the absolute minimum.

In some situations, both in amenity areas and elsewhere, it is possible to use a biological control agent rather than a chemical pesticide, and this has obvious attractions; the application of *Bacillus thuringiensis* to kill caterpillars and the use of predatory mites to combat spider mites are examples. Biological control of pests is especially appropriate under glasshouse conditions.

Pests are subject to attack from a wide range of natural enemies, including a vast array of parasitic wasps, predatory beetles, predatory mites, and so on. Pests also succumb to other controlling agents, such as bacterial, fungal and viral diseases. Indeed, in many situations, pest populations will remain below economically important levels unless the balance of this natural control is overturned. Although some pesticides are intrinsically safe to beneficial insects and mites, many can have significant adverse effects upon them. It is prudent, therefore, to restrict the use of chemicals and to ensure that, when treatment with a pesticide is required, the one chosen from the list of those available will have the least deleterious effect on non-target organisms.

An ability to correctly identify pests or the symptoms of damage is an essential starting point for good pest management. A knowledge of the habits and biology of the various pests, and of the risks they pose, is also necessary if correct decisions concerning their possible control are to be made.

INSECTS

Order **COLLEMBOLA** (springtails)

1. Family **SMINTHURIDAE**

Globular-bodied springtails with thorax and abdomen fused; antennae long; ocelli usually present.

Bourletiella hortensis (Fitch)
syn. *signatus* (Nicolet)
Garden springtail
An often common pest, especially in wet, acid soils, causing damage to seedling plants including various ornamentals; conifer seedlings in forest nurseries, especially beach pine *(Pinus contorta),* can be seriously affected. Widely distributed in Europe; also present in North America.

DESCRIPTION **Adult:** 1.5mm long; black to dark green, often spotted with white; head large, with long antennae and prominent, black, yellowish-bordered eyes; abdomen globular with a small ventral tube-like sucker and a forked springing organ.

LIFE HISTORY Eggs are laid in the soil, usually in small groups, each female depositing up to 100 in about three weeks. The eggs swell rapidly after laying and hatch shortly afterwards. Under favourable conditions maturity is reached in two to three months but development can take much longer, individuals continuing to moult even after the adult stage is attained. Breeding is continuous throughout the year but reaches a peak in the spring, the insects being most numerous from late April to the end of June.

DAMAGE **General:** the hypocotyl and cotyledons of seedlings are pitted, and holes are formed in the young leaves, but damage rarely occurs after July. **Conifer seedlings:** damage to the hypocotyl and cotyledons results in stunted seedlings with a brush-like mass of swollen, distorted needles; these seedlings develop into useless multi-stemmed plants.

CONTROL To reduce the risk of attack, avoid overwatering and excessive soil acidity. If required, apply a soil insecticide and work this into the top layers of soil before planting; alternatively, incorporate an insecticide into the seed compost. Soil sterilization is also effective. If attacks develop in seedbeds, apply an insecticide at regular intervals throughout the germination period until all seedlings have emerged.

2. Family **ONYCHIURIDAE**

Springtails without ocelli, but with complex sensory organs on the antennae and with mandibulate mouthparts.

Onychiurus spp.
White blind springtails
Various species of *Onychiurus* (e.g. *O. nemoratus* Gisin and *O. stachianus* Bagnall) cause damage to seedlings, pitting the cotyledons, hypocotyl and roots, and chewing off the root hairs and rootlets; attacked seedlings may collapse and die, often keeling over at about soil level. On older plants, leaves in contact with the soil may also be holed and skeletonized. Damage occurs on various outdoor and glasshouse plants, including ornamentals. The springtails are abundant in wet soil with a high organic content, and often gain entry into pots and seedboxes if these are placed directly onto infested ground. The pest can also be introduced into containers in unsterilized compost. Individuals (up to 3mm long) are white and stout-bodied, with a large head, short antennae and legs, and six abdominal segments; the springing organ is reduced or absent. They breed continuously in favourable conditions, development from egg to adult taking several months, the insects continuing to undergo various moults even after the adult stage is reached.

Order **ORTHOPTERA** (crickets and grasshoppers)

1. Family **GRYLLOTALPIDAE** (mole crickets)

Crickets with the fore legs greatly enlarged and modified for burrowing.

Gryllotalpa gryllotalpa (Linnaeus)
syn. *vulgaris* Latreille
Mole cricket

A large, soil-burrowing insect (Fig. 4), sometimes causing damage to glasshouse and outdoor plants, including ornamentals, by biting or gnawing the roots and basal parts of the stems. Damage normally occurs at or just below the soil surface but, unlike that caused by cutworms (e.g. *Agrotis segetum*, p. 309), tends to be indiscriminate. Adults are 35–50 mm long and greyish-brown to yellowish-brown, coated in fine, velvet-like hairs; the prothorax is elongate, and the front tibiae much enlarged and distinctly toothed. This insect occurs in North Africa and Western Asia but, nowadays, is rarely found in Europe.

CONTROL Application of a soil insecticide can be effective; poison baits are also recommended.

4 Mole cricket, *Gryllotalpa gryllotalpa*.

5 House cricket, *Acheta domesticus*.

2. Family **GRYLLIDAE** (true crickets)

Body relatively broad and somewhat flattened, the fore wings being held more-or-less horizontally; antennae longer than body; ovipositor and anal cerci long.

Acheta domesticus (Linnaeus)
House cricket

Although of only minor importance, this widely distributed species occurs occasionally in heated glasshouses. The insects hide by day in dark crevices, emerging at night to feed. They may then cause damage to the stems, flowers or foliage of plants; house crickets will also attack the aerial roots of orchids and other ornamentals. The adult males frequently strigilate, producing their characteristic 'song' by rubbing their fore wings together. Individuals are 15–20 mm long, yellowish-brown to greyish-brown, and clothed with fine hairs (Fig. 5).

CONTROL Avoid the build-up of debris which might act as shelter for these insects, particularly around heating pipes. Infestations can be eliminated by careful hygiene and application of contact insecticides.

Order **DERMAPTERA** (earwigs)

1. Family **FORFICULIDAE**

Forficula auricularia Linnaeus
Common earwig

A useful predator of aphids and various other pests, but also a frequent pest of flowers such as carnation (*Dianthus caryophyllus*), *Chrysanthemum, Cineraria, Clematis, Dahlia, Delphinium* and pansy (*Viola*); buds and leaves are also attacked. Cosmopolitan. Widely distributed in Europe.

DESCRIPTION **Adult female:** 12–14 mm long; chestnut-brown; hind wings, when folded away, projecting beyond elytra; pincers slightly curved. **Adult male:** 13–17 mm long; similar to female but pincers distinctly curved (Fig. **6**). **Egg:** 1.3×0.8 mm; pale yellow. **Nymph:** whitish to greyish-brown.

LIFE HISTORY Adults of both sexes overwinter in sheltered situations in the soil, mating in the early winter. Eggs are laid in December or January, each female depositing a batch of up to 100 in an earthen cell, and guarding over them until after they hatch in February or March. Earwigs are omnivorous insects, the nymphs feeding throughout the spring to reach maturity by the early summer; there are four nymphal stages. Overwintered adult females may deposit a second batch of eggs in May or June. Nymphs from these eggs develop from late June or early July to September. Earwigs are nocturnal, hiding by day within damaged flowers, in crumpled leaves, under loose bark and so on. Although occurring mainly out-of-doors, attacks are sometimes reported in glasshouses.

6 Male common earwig, *Forficula auricularia*.

DAMAGE Damaged flower petals become ragged, spoiling their appearance; attacks on leaves are usually unimportant, but chewed buds may die, resulting in blind shoots; most damage occurs from June to September.

CONTROL On a small scale, earwigs can be trapped by placing inverted pots or plastic beakers stuffed with straw on the tops of canes, or by placing rolls of corrugated cardboard or sacking alongside vulnerable plants; such traps should be checked regularly and any earwigs present killed. If necessary, the ground and lower parts of plants can be sprayed with a contact insecticide. Keeping beds free of weeds and general garden rubbish will reduce the likelihood of attacks developing.

Order **DICTYOPTERA** (cockroaches and mantises)

1. Family **BLATTIDAE** (cockroaches)

Distinguished from mantises (family Mantidae) by the unmodified front legs and broad pronotum which covers the head.

Blatta orientalis Linnaeus
Common cockroach

This generally common, well-known bakery and warehouse insect often occurs in glasshouses, destroying seeds and seedlings, and also causing damage to the aerial parts of older plants. Individuals hide by day but at night become active and then move rapidly over the floors and beds of infested houses. Eggs are deposited in groups in purse-like oothecae. These egg cases eventually split open to release young nymphs, the incubation period of the eggs lasting for one or more months according to temperature. The nymphs feed for nine months, or more, before becoming adults. Under suitable conditions, breeding is continuous. Adults are 20–30mm long, rather flattened, shiny blackish-brown, with long, many-segmented antennae, long legs and a pair of anal cerci; the wings are poorly developed.

7 German cockroach, *Blattella germanica.*

CONTROL Good hygiene within glasshouses will reduce the likelihood of infestations developing. Due to the development of resistance, cockroaches can be difficult to kill with pesticides, but certain contact insecticides and poison baits can be effective. On a limited scale, the insects can be trapped in small, baited containers (pit-fall traps) sunk into the soil.

Blattella germanica (Linnaeus)
German cockroach

Infestations of this relatively small (*c.* 12–14mm long), yellowish-brown cockroach (Fig. 7) are sometimes established in heated glasshouses and hot-houses. In common with other species, they feed at night and sometimes cause damage to ornamental plants.

Periplaneta americana (Linnaeus)
American cockroach

A relatively large (38–42mm long), reddish-brown cockroach; often present in heated glasshouses, where it may cause damage to ornamentals and various other plants. Unlike *Blatta orientalis*, the wings are fully developed and reach beyond the tip of the abdomen.

Periplaneta australasiae (Fabricius)
Australian cockroach

Minor infestations of this cockroach have also become established in heated glasshouses in Europe. Individuals are smaller than the previous species (30–36mm long) and, as adults and nymphs, more extensively marked with yellow.

Pycnoscelus surinamensis (Linnaeus)
Surinam cockroach

A typically parthenogenetic cockroach, unusual in retaining its ootheca within a brood sac so that the eggs hatch whilst still within the mother's body. Probably of Oriental origin but now cosmopolitan; in Europe infestations occur widely in heated glasshouses, where cultivated plants, including ornamentals, can be damaged. Adults (21–23mm long) are dark brown with paler wings and a pale band along the front of the pronotum; the wings are fully developed.

Order **HEMIPTERA** (true bugs)

1. Family **TINGIDAE** (lace bugs)

Flattened bugs, the pronotum and wings ornamented with a netted, lace-like pattern; pronotum usually covering the scutellum.

Stephanitis rhododendri Horváth
Rhododendron bug

A locally important pest of *Rhododendron*, especially hybrids of *Rhododendron arboreum*, *R. campanulatum*, *R. campylocarpum*, *R. catawbiense* and *R. caucasicum*; probably of North American origin. Infestations occur in various parts of mainland Europe; also found in southern England and Wales.

DESCRIPTION **Adult:** 4 mm long; body brownish to blackish; wings with a creamish, net-like venation. **Nymph:** yellowish, ornamented with numerous brown spines.

LIFE HISTORY Adult bugs first appear in June, and may survive on host plants until the early winter. Although fully winged, they are relatively sedentary and do not fly. Once established, therefore, infestations tend to persist and build up on individual plants. In the autumn, females deposit eggs along the mid-rib of young leaves. These eggs overwinter *in situ*, and eventually hatch in the following spring. Nymphs then feed on the underside of the leaves, typically occurring in groups of up to 50 individuals (Fig. **8**). There is a single generation annually.

DAMAGE This pest causes a yellow mottling of the foliage (Fig. **9**), the underside of damaged leaves also developing a rusty-brown appearance. Heavy attacks lead to extensive discoloration of plants, and may cause leaves to wilt. Attacks are especially severe on plants growing in sunny, dry situations.

CONTROL Affected branches can be pruned out in the early spring and burnt. If required, apply an insecticide in the early summer, ensuring that the spray reaches the underside of the leaves; two or three sprays at about three week intervals are recommended.

8 Nymphs of rhododendron bug on underside of leaf of *Rhododendron*.

9 Rhododendron bug damage to upper surface of leaf of *Rhododendron*.

Corythucha ciliata (Say)
Sycamore lace bug

An important North American pest of plane (*Platanus*); also now present in mainland Europe, causing extensive damage to both nursery and mature trees. Since 1964 it has spread throughout Italy and has also become established in neighbouring countries, including Austria and France; also now present in southern Germany.

DESCRIPTION **Adult:** 3 mm long; body blackish; wings transparent, with a network of white veins, the fore wings with a dark central patch; antennae brownish-white, hairy and with a slightly clubbed apex (Fig. **10**). **Nymph:** mainly black.

LIFE HISTORY Adults overwinter under the bark of trees, usually congregating on the north-west side of the trunks. They emerge in the spring and commence feeding on the new foliage. Eggs are laid on the underside of the leaves, the nymphs then feeding to reach maturity in late June or July. A second generation occurs during the summer, new adults appearing from late August onwards. The young adults then enter hibernation. In favourable, more southerly districts, three generations are reported annually.

DAMAGE Infested foliage becomes discoloured (Fig. **11**), heavily infested leaves dropping prematurely and reducing plant vigour.

CONTROL Apply an insecticide to nursery plants in mid-June and again in late July, aiming to kill the nymphs and young adults. Treatment of mature trees is impractical.

2. Family **MIRIDAE** (capsid or mirid bugs)

Very active, soft-bodied bugs with elongate elytra and long, probing, needle-like mouthparts.

Lygocoris pabulinus (Linnaeus)
Common green capsid

An often abundant pest of trees, shrubs and herbaceous plants, including ornamentals such as *Caryopsis*, *Chrysanthemum*, *Clematis*, *Dahlia*, *Forsythia*, *Fuchsia*, *Geranium*, *Hydrangea*, morning glory (*Ipomoea*), *Pelargonium*, rose (*Rosa*), *Salvia* and sunflower (*Helianthus*). Present throughout Europe.

DESCRIPTION **Adult:** 5.0–6.5 mm long; bright green with a dusky-yellow pubescence; pronotum lightly punctured and with moderate callosities; antennae comparatively long (Fig. **12**). **Egg:** 1.3 mm long; banana-shaped, cream, smooth and shiny. **Nymph:** pale green to bright green; tips of antennae orange-red.

LIFE HISTORY The winter is passed as eggs inserted in the bark of first- or second-year shoots of woody hosts such as crab apple (*Malus*), currant (*Ribes*),

10 Sycamore lace bugs, *Corythucha ciliata*.

11 Sycamore lace bug damage to leaf of *Platanus*.

12 Common green capsid, *Lycocorus pabulinus*.

hawthorn (*Crataegus*) and lime (*Tilia*). They hatch from April onwards, the young and very active nymphs feeding for a few weeks on the new foliage. The nymphs then migrate to herbaceous hosts to complete their development, adults appearing in June and July. These adults lay their eggs in the stems of herbaceous hosts, including various ornamentals and potato, and weeds such as bindweed (*Convolvulus*), dandelion (*Taraxacum*), deadnettle (*Lamium*), dock (*Rumex*) and groundsel (*Senecio vulgaris*). Nymphs of the second generation feed on these summer hosts to reach the adult stage by the autumn. There is then a return migration to woody hosts, where winter eggs are laid. Although most abundant on outdoor hosts, the pest is sometimes introduced into glasshouses during the summer on infested chrysanthemum plants.

DAMAGE Foliage at the tips of the new shoots becomes speckled with reddish or brownish-red, or tattered and distorted (Fig. **13**), and often peppered with small holes, symptoms appearing on herbaceous hosts from May onwards; in some cases (e.g. clematis and dahlia) flowers are distorted, and in others (e.g. fuchsia, on which damage is often severe) flower buds are aborted.

CONTROL Apply an insecticide as soon as damage or active bugs are seen. Sites should also be kept free of weeds, which can act as alternative hosts.

13 Common green capsid damage to leaves of ornamental *Prunus*.

14 Tarnished plant bug, *Lygus rugulipennis.*

Lygus rugulipennis Poppius
Tarnished plant bug

A common polyphagous pest of annual and herbaceous plants, including African daisy (*Arctosis*), *Chrysanthemum, Dahlia*, Michaelmas daisy (*Aster*), nasturtium (*Tropaeolum*), poppy (*Papaver*) and *Zinnia*; especially troublesome in glasshouses. Widely distributed in Europe.

DESCRIPTION **Adult:** 5–7 mm long; robust-bodied with relatively short antennae; extremely variable in colour, varying from green, yellowish-brown or reddish-brown to black (Fig. **14**). **Egg:** 1.0×0.25 mm; creamish-white and flask-shaped. **Nymph:** green to brownish, with a pair of black dots dorsally on each thoracic segment.

LIFE HISTORY Adults overwinter in debris on the ground or in other suitable shelter, emerging in the following March or April. Eggs are laid during May, most commonly in the buds and stems of wild host plants such as dock and sorrel (*Rumex*), groundsel (*Senecio vulgaris*) and nettle (*Urtica*). Nymphs then feed on the youngest shoots, new adults appearing from July onwards. Except in more northerly districts, a larger second generation of nymphs is produced and these commonly cause damage to cultivated plants, new adults eventually appearing from September onwards.

DAMAGE Adults sometimes produce a localized yellowing of the leaves, with brown necrotic spots marking the position of the feeding punctures (Fig. **15**); attacks on young tissue may also result in the leaves or flowers becoming puckered and distorted. Nymphs cause noticeable injury to young shoots, which become twisted, swollen and often blind.

CONTROL As for *Lygocoris pabulinus* (p. 25).

15 Tarnished plant bug damage to leaf of *Primula*.

3. Family **CERCOPIDAE** (froghoppers)

Often called 'spittle-bugs'; the nymphs develop on plants within a mass of froth ('cuckoo-spit'), a secretion produced from the hind end of the body and through which air bubbles are forced from a special canal by abdominal contractions. The hind tibiae of adults bear just a few stout spines (cf. leafhoppers, p. 28).

16 Potato capsid, *Calocoris norvegicus*.

Calocoris norvegicus (Gmelin)
Potato capsid

A generally common capsid on herbaceous weeds; occasionally also a minor pest of cultivated *Chrysanthemum* and certain other ornamental Compositae. Widely distributed in Europe; also present in Canada.

DESCRIPTION **Adult:** 6–8mm long; fore wings green, sometimes tinged with reddish-brown; pronotum green and straight-sided, and often marked with a pair of black dots (Fig. **16**). **Nymph:** mainly green to yellowish-green, with black hairs.

LIFE HISTORY Eggs are laid in late July and August in cracks in woody or semi-woody stems of various plants. They hatch in the following May or early June. The nymphs then feed on the buds, growing points, flowers and foliage of herbaceous plants, including clover (*Trifolium*), nettle (*Urtica*) and various members of the family Compositae, including chamomile (*Anthemis*), mayweed (*Matricaria*) and thistles (*Carduus* and *Cirsium*). Nymphs reach maturity in late June or July. There is just one generation annually.

DAMAGE Adults and nymphs produce necrotic spots which develop into holes; young shoots

become distorted and may be killed. Attacks tend to occur in weedy situations.

CONTROL Good weed control will reduce the likelihood of attacks developing; beds should also be kept free of dead plant tissue and other potential shelter. If necessary, apply an insecticide as soon as damage is seen.

Aphrophora alni (Fallén)
Alder froghopper

A common pest of deciduous trees and shrubs, including alder (*Alnus*), ash (*Fraxinus excelsior*), poplar (*Populus*) and willow (*Salix*); also occurs on certain herbaceous plants, including *Viola*. Eurasiatic. Widespread in Europe.

DESCRIPTION **Adult:** 8.0–9.5 mm long; pale greyish-brown to dark olive-brown with deep blackish punctures; head and pronotum with a median keel; fore wings with a pair of pale patches (Fig. **17**). **Nymph:** mainly greyish to creamish-white, with a distinctive pair of dark spots on the head between the eyes.

LIFE HISTORY Eggs are laid from July to October, each being deposited close to the ground in the old tissue of host plants. They hatch in the following spring, nymphs then developing within distinctive, round accumulations of spittle, concentrated mainly on the lower parts of plants. Adults appear from late June onwards, ascending host plants to feed on the young tissue. They are far more often seen than the nymphs (cf. *Aphrophora salicina*).

DAMAGE The nymphs cause little direct damage but the young adults produce distinctive calloused rings on the shoots, often weakening the new growth.

CONTROL As for *Philaenus spumarius* (p. 28).

17 Alder froghopper, *Aphrophora alni*.

18 Nymphs of willow froghopper, *Aphrophora salicina*.

Aphrophora salicina (Goeze)
syn. *grisea* Haupt; *salicis* (Degeer)
Willow froghopper
A generally common pest of poplar (*Populus*), sallow and willow (*Salix*); often established on ornamental and amenity trees. Widespread throughout Europe; also present in North America.

DESCRIPTION **Adult:** 10.0–10.5 mm long; greyish-yellow or greenish to olive-brown, finely punctured with black; elytra sometimes with an indistinct whitish basal triangular spot; head and thorax distinctly keeled. **Nymph:** head and thorax reddish-brown; abdomen creamish-white (Fig. **18**).

LIFE HISTORY Eggs, deposited on host plants during the summer or early autumn, hatch in the following spring. The nymphs then feed within large, dripping accumulations of spittle, often established in groups on the young shoots (Fig. **19**). These nymphal feeding shelters are especially obvious in June, adults appearing shortly afterwards.

DAMAGE Feeding by the nymphs can cause considerable damage, affecting the vigour and quality of host plants. The presence of masses of spittle on ornamentals is also unsightly.

CONTROL As for *Philaenus spumarius* (p. 28)

19 Spittle of willow froghopper on *Salix*.

Philaenus spumarius (Linnaeus)
syn. *leucophthalmus* (Linnaeus)
Common froghopper
nymph = cuckoo-spit bug
Generally abundant on a wide variety of trees, shrubs and low-growing plants; often a minor pest of lavender (*Lavendula*) and various other ornamentals, including *Aster*, barberry (*Berberis*), *Campanula*, *Chrysanthemum*, *Coreopsis*, *Geum*, golden rod (*Solidago*), *Lychnis*, *Mahonia*, *Phlox*, rose (*Rosa*), *Rudbeckia* and many others. Holarctic. Present throughout Europe.

DESCRIPTION **Adult:** 5–7 mm long; colour varying from yellowish, greenish or brown to blackish; head bluntly wedge-shaped with large eyes; elytra convex and often with dark markings; each hind tibia with an apical ring of spines (Fig. **20**). **Egg:** 1 mm long; oval. **Nymph:** mainly pale and unicolourous, with dark eyes (Fig. **21**).

LIFE HISTORY The frog-like adults occur mainly from July onwards and are often seen at rest on plants. They are rather sluggish but jump violently into the air if disturbed. Eggs are laid in the stems of host plants during September, typically in batches of up to 30. Eggs hatch in the following May. The sedentary nymphs then feed on the sap of host plants, injecting their needle-like mouthparts into a shoot or leaf vein. Throughout their development, they cover themselves with a protective mass of spittle-like froth, passing through five nymphal instars and becoming mature in the summer.

DAMAGE Young shoots on susceptible nymph-infested plants may become distorted and wilted; in some cases flowers are malformed. Adults cause no obvious damage. Ornamentals are also disfigured by the presence of the spittle, and this can reduce the marketability of container-grown plants in nurseries.

CONTROL Spittle and the bugs can be dislodged from infested plants by spraying with a high-pressure hose. Insecticides can also be applied but chemical treatment is rarely justified.

20 Common froghopper, *Philaenus spumarius*.

21 Nymph of common froghopper, *Philaenus spumarius*.

4. Family **CICADELLIDAE** (leafhoppers)

Adults and nymphs are free-living on the foliage of various plants. The hind legs of the very active adults are adapted for jumping and bear two rows of fine spines along each tibia (cf. froghoppers, p. 26). Reliable identification of most species involves examination of genitalia and is a specialist task; some species, however, are distinguishable by characteristic markings on the body or the elytra. The following are representatives of those most frequently associated with ornamental plants, and include those often reaching true pest status.

Graphocephala fennahi Young
syn. *coccinea* (Förster)
Rhododendron hopper
Widely distributed on *Rhododendron* in England, especially in the south, having been introduced from America in the 1930s; also now widely distributed in continental Europe.

DESCRIPTION **Adult:** 8.4–9.4 mm long; head yellow; thorax bluish-green, marked with red and yellow; abdomen red above; fore wings bluish-green, striped with red; hind wings grey (Fig. 22). **Nymph:** whitish to yellowish-green.

LIFE HISTORY Eggs are inserted into slits formed in scales on the flower buds from August to October. The eggs hatch in the following spring, nymphs feeding on the underside of the leaves from April or May to July or August. Adults and nymphs often occur in considerable numbers on the tips of young shoots. There is just one generation each year.

DAMAGE Direct feeding has little adverse effect on host plants; however, infested plants are more susceptible to bud blast (a fungal disease caused by *Pycnostysanus azaleae*), the spores probably gaining easy entry into plant tissue through the egg-laying slits.

CONTROL Apply an insecticide at about two-week intervals from August to October, to coincide with the main egg-laying period. Such treatment is justified only in sites where bud blast is established.

Alebra albostriella (Fallén)
Infestations of this generally distributed, and often common, leafhopper occur on oak (*Quercus*), frequently causing damage to the foliage of young ornamental plants. In mainland Europe, damage is also caused to alder (*Alnus*). Adults are variable in colour but often have the pronotum yellowish with a pair of longitudinal red or orange streaks and the fore wings yellow to hyaline, marked with red or orange (Fig. 23). In *Alebra*, unlike all other European genera, the wing venation includes a distinct appendix (Fig. 24).

22 Rhododendron hopper, *Graphocephala fennahi*.

23 Adult of *Alebra albostriella*.

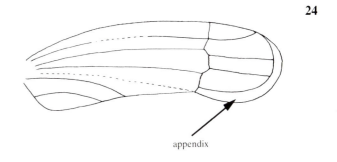

appendix

24 Venation of fore wing of *Alebra*.

Alebra wahlbergi (Boheman)
This generally common leafhopper is widely distributed in mainland Europe and in the southern half of Britain, causing noticeable damage to the foliage of various trees, especially elm (*Ulmus*), horse chestnut (*Aesculus hippocastanum*), lime (*Tilia*), maple (*Acer*) and sycamore (*Acer pseudoplatanus*). Adults, which occur from July to September, are relatively large (females up to 4.5 mm long); the fore wings are mainly pale yellow, sometimes streaked with yellow or orange, with the inner margin usually orange, pink or yellow.

25 Adult of *Eurhadina pulchella*.

26 Chrysanthemum leafhopper damage to leaves of *Salvia*.

Eurhadina pulchella (Fallén)

Although associated mainly with oak (*Quercus*), this species is also damaging to the foliage of other trees, including birch (*Betula*). It is widely distributed and generally common in Europe. Adults (*c.* 4mm long) are variable in colour but the fore wings are usually yellow with a distinct apical patch (Fig. **25**). In common with other members of the genus, individuals have a characteristically flattened appearance, the fore wings being noticeably broader medially and the head distinctly narrower than the pronotum.

Eupteryx melissae Curtis
Chrysanthemum leafhopper

A generally common species, attacking various members of the Labiatae (including ornamentals such as *Phlomis* and *Salvia*) and also various other plants, such as *Chrysanthemum*, hollyhock (*Althaea rosea*) and tree mallow (*Lavatera arborea*). Attacks are sometimes noted on garden or glasshouse-grown plants during the summer months, infested leaves developing a sickly, flecked appearance (Fig. **26**). The adults are 2.9–3.4mm long and mainly green, with distinctive black or greyish markings on the pronotum and scutellum (Fig. **27**).

27 Chrysanthemum leafhopper, *Eupteryx melissae*.

28 Cherry leafhopper, *Aguriahana stellulata*.

Aguriahana stellulata (Burmeister)
Cherry leafhopper

A distinctive, mainly white and generally common leafhopper (Fig. **28**), occurring on various ornamental trees, including cherry (*Prunus*) and lime (*Tilia*). Infestations are rarely important but leaf discoloration can be disfiguring, as on exotic hosts such as Indian horse chestnut (*Aesculus indica*).

Ribautiana ulmi (Linnaeus)
syn. *ocellata* (Curtis)

A generally common species, associated mainly with elm (*Ulmus*) and causing extensive discoloration of the expanded leaves; attacks may also occur on other trees, including hazel (*Corylus avellana*), hornbeam (*Carpinus betulus*), oak (*Quercus*), sallow (*Salix*) and whitebeam (*Sorbus aria*). Adults are present from May to November, breeding on the underside of the leaves; they often occur in considerable numbers. Individuals are 3.6–4.4 mm long; the fore wings are yellow, with greyish markings on the apical third; the head and pronotum are yellowish, the former with a pair of distinctive black spots and the latter usually with a single spot on the front margin (Fig. **29**). The active nymphs are yellowish to whitish.

29 Adult of *Ribautiana ulmi* alongside cast-off skin of final-instar nymph.

Typhlocyba bifasciata Boheman
Hornbeam leafhopper

This common yellow and black leafhopper breeds on elm (*Ulmus*) and hornbeam (*Carpinus betulus*), infestations occurring throughout the summer and leading to noticeable silvering of the leaves of garden hedges. The adults and nymphs often occur in large numbers on the underside of infested leaves, with the cast skins of earlier instars or generations clearly visible. Adults are 3.2–3.7 mm long, with pale yellow fore wings, each marked with a pair of black crossbands (Fig. **30**) or with the basal two-thirds mainly or entirely black. The characteristic pattern on the fore wings readily distinguishes this species from *Edwardsiana flavescens* (p. 33) and other leafhoppers associated with elm or hornbeam.

30 Hornbeam leafhopper, *Typhlocyba bifasciata*.

Typhlocyba quercus (Fabricius)
Although most often noted on oak (*Quercus*), causing extensive discoloration of the leaves, this generally common species is also associated with various other hosts, including birch (*Betula*), hornbeam (*Carpinus betulus*), smoke tree (*Cotinus coggygria*) and southern beech (*Nothofagus*). Adults are 3.0–3.5 mm long, with the fore wings mainly white, marked with greenish-yellow to yellowish-brown and with prominent brick-red to orange patches (Fig. **31**). They occur from July to October.

Fagocyba cruenta (Herrish-Schaeffer)
syn. *douglasi* (Edwards)
Beech leafhopper

Generally distributed and often abundant on beech (*Fagus sylvatica*), commonly disfiguring hedges and specimen trees, the foliage becoming extensively flecked with silver. Adults are 3.4–4.1 mm long and

31 Adult of *Typhlocyba quercus*.

32 Beech leafhopper, *Fagocyba cruenta*.

33 Rose leafhopper, *Edwardsiana rosae*.

usually pale yellowish, with the fore wings often somewhat darkened towards the inner margin (Fig. **32**); darker forms also occur. Damage can also occur on other trees, including oak (*Quercus*), sycamore (*Acer pseudoplatanus*) and whitebeam (*Sorbus aria*).

Edwardsiana rosae (Linnaeus)
Rose leafhopper
A generally abundant and often important pest of rose (*Rosa*); also harmful to various other rosaceous trees, including hawthorn (*Crataegus*), rowan (*Sorbus aucuparia*) and whitebeam (*Sorbus aria*). Widely distributed in Europe; also present in North America.

DESCRIPTION **Adult:** 3.4–4.0 mm long; mainly pale yellowish, with a yellow abdomen (Fig. **33**). **Nymph:** translucent to whitish (Fig. **34**).

34 Nymphs of rose leafhopper, *Edwardsiana rosae*.

LIFE HISTORY Eggs are deposited in the autumn under the epidermis of young shoots of wild and cultivated rose bushes. Eggs hatch in the following spring, nymphs then feeding on the underside of the leaves; the first adults appear in late May or early June. Most adults then migrate to summer hosts, such as fruit trees, but some remain on rose and eventually deposit eggs in the leaves. These eggs hatch in late June or early July and a second generation of nymphs develops, with the next generation of adults appearing from mid-August onwards. These, along with new individuals returning to rose from summer hosts, finally deposit the winter eggs.

DAMAGE Infested leaves become extensively flecked and blanched (Fig. **35**). Attacks are especially harmful in hot, dry weather, and are often especially severe on climbing roses. Severe attacks may cause leaves to turn brown and to drop prematurely.

35 Rose leafhopper damage to leaves of *Rosa*.

CONTROL Apply a contact or a systemic insecticide at the first signs of damage. Most insecticides applied against aphids will also kill leafhoppers.

Edwardsiana crataegi (Douglas)

syn. *froggatti* (Baker); *oxyacanthae* (Ribaut)
This widely distributed species occurs commonly on various rosaceous trees, and is sometimes damaging to the foliage of crab apple (*Malus*), hawthorn (*Crataegus*), ornamental cherry (*Prunus*), rowan (*Sorbus aucuparia*) and willow-leaved pear (*Pyrus salicifolia*). Adults are similar in appearance to those of *Edwardsiana rosae* (p. 32) but the fore wings are darkened along the inner margin of the clavus.

Edwardsiana flavescens (Fabricius)

syn. *fratercula* (Edwards)
Foliage damage caused by this pale yellow (Fig. 36), widely distributed species occurs most frequently on beech (*Fagus sylvatica*) and hornbeam (*Carpinus betulus*), the adults being readily distinguished from the most common other species associated with these two hosts (on beech, *Fagocyba cruenta*, p. 31–32; on hornbeam, *Typhlocyba bifasciata*, p. 31).

Edwardsiana nigriloba (Edwards)

Sycamore leafhopper
This widely distributed leafhopper, which is restricted to sycamore (*Acer pseudoplatanus*), frequently causes extensive damage to the foliage of ornamental trees, nymphs feeding in the spring and early summer, and adults appearing in July. The adults (3.6–4.0 mm long) are mainly pale yellowish (with a characteristic black tubercule ventrally on the genital segment); the older nymphs are especially noticeable, having a characteristic black pattern which contrasts with the pale background of the underside of the leaves (Fig. 37).

Hauptidia maroccana (Melichar)

syn. *pallidifrons* (Edwards); *tolosana* (Ribaut)
Glasshouse leafhopper
A tropical or subtropical species, now well established on glasshouse plants in northern Europe. Infestations often occur on ornamentals such as *Calceolaria*, *Chrysanthemum*, *Fuchsia*, *Gloxinia*, heliotrope (*Heliotropium*), *Pelargonium*, *Primula*, *Salvia*, sweet-scented verbena (*Aloysia citriodora*) and tobacco plant (*Nicotiana*); in favourable areas, attacks also occur on outdoor plants such as chickweed (*Stellaria*), foxglove (*Digitalis purpurea*) and primrose (*Primula vulgaris*).

DESCRIPTION **Adult:** 3.1–3.7 mm long; mainly pale yellow with greyish or brownish markings, the

36 Adults of *Edwardsiana flavescens*.

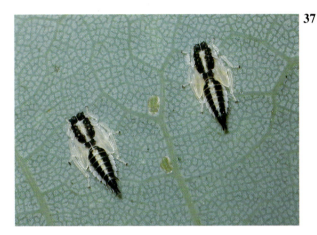

37 Final-instar nymphs of sycamore leafhopper, *Edwardsiana nigriloba*.

38 Glasshouse leafhopper, *Hauptidia maroccana*.

latter forming a pair of distinctive, chevron-like marks on the fore wings (Fig. 38). **Nymph:** whitish.

LIFE HISTORY Breeding is continuous throughout the year, all stages occurring on the underside of leaves of host plants. Eggs are deposited singly in the leaf veins, hatching in about a week at normal glasshouse temperatures. Nymphs feed for about a month, passing through five instars before reaching the adult stage, the underside of leaves becoming contaminated with the cast-off nymphal skins. Adults survive for up to three months, each female depositing up to 50 eggs. The duration of the various stages is extended in cool conditions, and greatly protracted during the winter when the eggs often take a month or more to hatch and nymphal development lasts for two or more months.

DAMAGE Growth of heavily infested plants is checked and seedlings killed, but on most hosts damage is limited to speckling, silvering or a blanched mottling of the foliage.

CONTROL If necessary, apply an insecticide or fumigate infested glasshouses. Good weed control within and around glasshouses will reduce the likelihood of attacks developing.

39 Ash leaf gall sucker, *Psyllopsis fraxini*.

5. Family **PSYLLIDAE** (psyllids or suckers)

Adults are very active, and have relatively large, membranous wings and strong hind legs adapted for jumping. The nymphs are flat, scale-like and slow-moving; they produce a white, waxy secretion and masses of sticky honeydew. Most species are free-living but some breed within leaf galls.

Psyllopsis fraxini (Linnaeus)
Ash leaf gall sucker
A common but minor pest of ash (*Fraxinus excelsior*). Eurasiatic. Widely distributed in Europe; introduced into North America.

DESCRIPTION **Adult:** fore wings 2.3–3.3 mm long, clear but with brown veins and a blackish basal pattern; body pale bluish-green to pale yellow, marked with black (Fig. **39**). **Nymph:** pale bluish-green; eyes red and prominent (Fig. **40**).

LIFE HISTORY The winter is passed in the egg stage on dormant shoots, eggs hatching at about bud-burst. Nymphs then feed on the young expanding leaves, clustered together within rolled leaf edges, secreting flocculent masses of white, waxy 'wool' and producing globules of honeydew. The

40 Nymphs of ash leaf gall sucker, *Psyllopsis fraxini*.

galled tissue turns yellow, changing through red and purplish to brown; such galls are especially conspicuous from summer until leaf-fall. Adults appear from June or July onwards, often sheltering within the galls along with any later-developing nymphs. In common with other psyllids, there are five nymphal instars; in this species there is a single generation annually.

DAMAGE The prominent leaf galls (Fig. **41**) disfigure infested plants, and can reduce the value of nursery stock, but are otherwise harmless.

CONTROL If necessary on nursery stock, galled leaves can be picked off and burnt.

41

42

41 Gall of ash leaf gall sucker on *Fraxinus*.

42 Alder sucker, *Psylla alni*.

Psyllopsis fraxinicola (Förster)
Ash leaf sucker

This widespread and often common species is associated with ash (*Fraxinus excelsior*), but does not produce leaf galls. Adults are greenish or yellowish-green, with clear, unpatterned wings; the nymphs are green (cf. *Psyllopsis fraxini*, p. 34).

Psylla alni (Linnaeus)
Alder sucker

A common pest of alder (*Alnus*). Present throughout northern Europe; also found in North America.

DESCRIPTION **Adult:** fore wings 3.8–4.8 mm long, clear but with dark veins; body bright green, later developing brown or red markings (Fig. **42**). **Nymph:** green, marked with black; antennae, legs and wing pads pale yellowish-brown to blackish (Fig. **43**).

LIFE HISTORY Eggs are laid in bark crevices during the autumn, hatching at bud-burst in the following spring. Nymphs then feed on the young shoots, developing within a sticky mass of white waxy 'wool' among which beads of honeydew accumulate (Fig. **44**). The nymphs are very active if disturbed; they will then scurry over the leaves or along the shoots and, if necessary, drop to the ground. Development is completed in the summer, adults occurring from June or July to October. There is just one generation annually.

DAMAGE The presence of flocculent masses of 'wool' on plants is unsightly, and infestations may also reduce the vigour of new shoots; damage, however, is rarely of significance.

43

43 Nymph of alder sucker, *Psylla alni*.

44

44 Nymphs of alder sucker on *Alnus*.

Psylla buxi (Linnaeus)
Box sucker

An often abundant pest of cultivated box (*Buxus sempervirens*). Widespread throughout Europe; introduced into North America.

DESCRIPTION **Adult:** fore-wings 2.9–3.7 mm long and shiny, with brownish-yellow veins; body pale green to yellowish-green (Fig. **45**). **Nymph:** mainly yellowish-green to pale green; tips of antennae blackish; eyes pale purplish-white (Fig. **46**).

LIFE HISTORY There is a single generation each year, overwintering eggs on the shoots hatching in the spring at about bud-burst. Nymphs then feed on the growing points, sheltered by a loose cabbage-like cluster of cupped terminal leaves which develops around them; these galls are very noticeable, each measuring about 10–20 mm across (Fig. **47**). One or a few individuals occur within each gall, the nymphs passing through five instars and becoming fully grown within about six weeks. The nymphs secrete conspicuous white waxen threads and considerable quantities of honeydew, the latter often spreading over infested foliage. Although adults appear from late April, May or June onwards, mating and egg laying do not occur until the late summer.

DAMAGE The galls check the growth of new shoots; they also spoil the appearance of hedges and remain on bushes long after the causal insects have disappeared. The sticky masses of honeydew, and sooty moulds which develop upon them, also disfigure plants. Although most damage is caused to the shoot tips, nymphal feeding may also extend to expanded leaves lower down the twigs, these becoming blistered, pallid and distorted.

CONTROL Specific recommendations are lacking but insecticides applied in late July to kill adults, or treatment in late April or early May to kill the young nymphs, are suggested.

Psylla melanoneura Förster
Hawthorn sucker

Infestations of this widely distributed and generally abundant psyllid are noted occasionally on cultivated hawthorn (*Crataegus*), the greenish (orange- to blackish-marked) nymphs feeding on the young shoots during May. Although sometimes present in large numbers, they cause little or no direct damage; however, by secreting considerable quantities of

45 Box sucker, *Psylla buxi.*

46 Nymph of box sucker, *Psylla buxi.*

47 Gall of box sucker on *Buxus sempervirens.*

honeydew and white waxen threads (Fig. **48**), this pest can attract attention on nursery stock, hedges or specimen trees, especially in dry conditions. Adults are reddish to brown or blackish (with characteristic whitish longitudinal lines on the thorax), and very active. They occur throughout most of the year, often inhabiting various non-host trees and shrubs (especially pines) upon which they also overwinter.

48 Nymphs of hawthorn sucker on *Crataegus*.

6. Family **TRIOZIDAE**

Trichochermes walkeri (Förster)
Buckthorn sucker

Widespread in mainland Europe, in association with common buckthorn (*Rhamnus cathartica*), and locally common in southern England. Infestations often occur on buckthorns in parks and gardens, the nymphs developing in distinctive, upward leaf-roll galls formed along part of the leaf margin (Fig. **49**). The galls are initiated in the spring and the nymphs complete their development in the summer, adults appearing from August onwards. The nymphs are pale greenish-yellow, elongate-oval and flattened, with an outer fringe of closely packed hairs, long, narrow wing pads and short antennae. Adults are 2.7–3.1mm long and brownish, with dark markings and relatively long, thin antennae; the fore wings (3.5–4.0mm long) are mottled with brown. Damage on infested bushes is often extensive and in severe cases most, if not all, leaves may be galled.

49 Buckthorn sucker galls on *Rhamnus cathartica*.

Trioza alacris Flor
Bay sucker

A generally common pest of bay laurel (*Laurus nobilis*) in mainland Europe, especially in Mediterranean areas. Commonly associated with container-grown plants, and often established on those adorning pavements in towns and cities; first introduced into Britain in the 1920s, and now widespread in the southern half of England. Also introduced into North and South America.

DESCRIPTION **Adult:** fore wings 2.4–3.1mm long, narrow, usually clear and with yellow veins; body pale whitish-yellowish, with brownish or blackish markings; antennae long and slender, with a black tip (Fig. **50**). **Nymph:** pale yellowish-white to greyish-

50 Bay sucker, *Trioza alacris*.

white, with greyish or blackish markings; antennae short and mainly pale, with the two apical segments black; body densely covered by tufts of white wax (Fig. **51**).

51

LIFE HISTORY Adults overwinter on bay laurel, sheltering amongst curled or dense clusters of leaves; they may also hide in leaf litter and other debris at the base of the plants. In spring, adults invade the young shoots to feed on the leaves. This causes the leaf edges to curl, and provides suitable sites within which small clusters of eggs are laid. Nymphs feed within these protective galls, secreting considerable quantities of honeydew; the leaf edges of galled leaves become thickened and further curled, changing to yellow, red or brown. Nymphs pass through five instars before attaining the adult stage, usually by mid-October. In favourable conditions there are two generations annually.

51 Nymphs of bay sucker, *Trioza alacris*.

DAMAGE Foliage damage on heavily infested plants is considerable and specimen plants are commonly disfigured by the prominent galls (Fig. **52**). Young plants are especially susceptible, the leaves often turning brown and dropping prematurely; shoot death may also occur. Infested plants are also affected by the masses of flocculent wax produced by the nymphs; they also become sticky with honeydew, upon which unsightly sooty moults develop.

52

CONTROL Few details are available on control measures but application of a systemic insecticide at the onset of leaf rolling has been suggested.

52 Galls of bay sucker on *Laurus nobilis*.

Trioza remota Förster
syn. *haematodes* Förster
Oak leaf sucker

A generally common but minor pest of oak (*Quercus*). Widely distributed in central and northern Europe; also found in Japan.

DESCRIPTION **Adult female:** fore wings 2.7–3.0 mm long, mainly clear with yellow or brownish veins; body brownish-red marked with cream. **Nymph:** 1.8–2.0 mm long; broad-bodied and scale-like, with a distinct fringe of setae; pale orange with a pair of deep orange stripes extending from head to tip of abdomen (Fig. **53**).

53

53 Nymph of oak leaf sucker, *Trioza remota*.

LIFE HISTORY Adults occur from September to May, overwintering in the shelter of evergreen plants and migrating to oak in the spring. Eggs are laid on the leaves during May. The sedentary nymphs then feed in small pits on the underside of the leaves during the summer, moulting into adults by about September. Adults remain on the host plants into October but then disperse to over-wintering sites.

DAMAGE Nymphal feeding causes a superficial pitting of the leaves and a noticeable pimpling on the upper surface (Fig. **54**); however, such damage is not important and does not result in distortion of the lamina.

54 Oak leaf sucker damage to leaf of *Quercus*.

7. Family **CARSIDARIDAE**

Psyllids with characteristically flattened, noticeably hairy antennae; mostly of tropical distribution.

Homotoma ficus (Linnaeus)
Fig sucker
This non-indigenous psyllid has been reported occasionally on common fig (*Ficus carica*) trees in various parts of Europe, including Austria, England, France, Jersey and Switzerland, but is local and uncommon. The winter is passed in the egg stage, with a single generation annually, brown-coloured adults (fore wings 3.4–4.3 mm long) occurring during the summer and pale brown nymphs develo-ping from early spring to about June; infestations are not of major importance.

55 Eucalyptus sucker, *Ctenarytaina eucalypti*.

8. Family **SPONDYLIASPIDAE**

A family of psyllids indigenous to Australia; one species has been introduced into Europe on impor-ted *Eucalyptus*.

Ctenarytaina eucalypti (Maskell)
Eucalyptus sucker
Established on *Eucalyptus* in various parts of main-land Europe, southern England and the Channel Islands, and often a pest in nurseries.

DESCRIPTION **Adult:** fore wings 1.3–1.9 mm long; body mainly dark brown, with the thorax partly orange; fore wings whitish with yellow veins and a waxy bloom; legs pale brown (Fig. **55**). **Egg:** shiny, pale yellow and pear-shaped. **Nymph:** greyish-yellow to orange-brown or blackish; first instar orange (Fig. **56**).

56 Nymphs of eucalyptus sucker, *Ctenarytaina eucalypti*.

LIFE HISTORY The relatively small, robust-bodied adults overwinter on host plants, depositing their first eggs in clusters on the young growth from February onwards. Eggs hatch in the early spring. Colonies then become established on the young, tender growth, the nymphs secreting honeydew and curly strands of whitish wax that are often blown about by the wind. There are two or more overlapping generations annually, but breeding may be continuous on plants growing in protected conditions, all stages of the pest often occurring together.

DAMAGE Infestations spoil the appearance of plants, contaminating the new growth with masses of honeydew upon which the cast nymphal skins accumulate. The shoots also become disfigured by loose masses of waxen wool. Feeding does not distort growth but heavy attacks will retard the growth of nursery plants.

CONTROL Little information is available but application of a contact insecticide, with additional wetter, might prove effective.

9. Family **ALEYRODIDAE** (whiteflies)

Small, moth-like insects coated with an opaque, white, waxy powder. The nymphs are flat and scale-like. Development includes a quiescent, non-feeding pupa-like stage.

Aleyrodes lonicerae Walker
syn. *fragariae* Walker; *rubi* Signoret
Honeysuckle whitefly
Sometimes noted on ornamental honeysuckle (*Lonicera*), snowberry (*Symphoricarpos rivularis*) and many other plants, contaminating the foliage with sticky honeydew upon which disfiguring sooty moulds also develop. Several overlapping generations occur each year, the pale yellow, scale-like nymphs and the yellow-bodied, white-winged adults occurring mainly on the underside of expanded leaves. The adults are whitish with a grey spot on each fore wing.

Asterobemisia carpini (Koch)
syn. *avellanae* (Signoret); *rubicola* (Douglas)
Hornbeam whitefly
Generally common in Europe on hornbeam (*Carpinus betulus*) but also associated with other hosts, including hazel (*Corylus avellana*) and *Rubus*; in mainland Europe various other hosts are also attacked, including beech (*Fagus sylvatica*), birch (*Betula*), false acacia (*Robinia pseudoacacia*), lime (*Tilia*) and maple (*Acer*). Eggs are laid on the underside of leaves in May and June. The scale-like, pale yellowish to yellowish-green nymphs then feed throughout the summer, completing their development in the autumn; they overwinter in the 'pupal' stage, adults (which are 1.0–1.1 mm long, yellow or orange, with clear white wings) emerging in the spring. Attacks on ornamental plants are usually unimportant, although heavy infestations cause foliage to become sticky with honeydew and blackened by accumulations of sooty moulds.

Pealius azaleae (Baker & Moles)
Azalea whitefly
Adults of this widely distributed but local pest of evergreen azaleas, especially *Rhododendron mucronatum* and *R. simsii*, appear during the early summer. They congregate on the underside of the leaves, where eggs are laid, making short flights if disturbed. Nymphs are pale green and scale-like. They develop on the underside of the leaves from late summer onwards, overwintering and completing their development in the following spring. There is one generation annually. The nymphs secrete considerable quantities of honeydew, the upper surface of leaves on infested plants commonly becoming coated and discoloured by sooty moulds. Infestations are reported from various parts of the world, including Australasia, Belgium, Britain, Japan and the Netherlands.

CONTROL As for *Siphoninus phillyreae*.

Siphoninus phillyreae (Haliday)
Phillyrea whitefly
A locally common but minor pest of mock privet (*Phillyrea latifolia*); also occurs on ash (*Fraxinus excelsior*). Widely distributed in mainland Europe but absent from northerly areas; in the British Isles confined mainly to southern England. Also occurs in parts of North Africa and Asia.

DESCRIPTION **Adult:** 1mm long; yellow, marked with grey; fore wings mainly white but somewhat smoky basally. **Egg:** 0.3×0.1mm; oblong, with a short stalk; yellow to brownish. **Nymph:** elliptical to oval; brownish, with a fringe of white waxy plates; profusely dusted with white, powdery wax. **Puparium:** 1.0×0.7mm; brown to brownish-black; elliptical, with a distinctive white waxen rim.

LIFE HISTORY Adult whiteflies first appear in late May and early June, depositing eggs in clusters on the underside of leaves. The eggs, which are liberally dusted with white waxen powder, hatch in 2–3 weeks. Nymphs then develop on the leaves, individuals pupating a few weeks later. A second generation of adults appears in August and September. Nymphs of this second generation overwinter, completing their development and eventually pupating in the following spring. In more favourable, southerly regions of mainland Europe there are three or more generations annually.

DAMAGE Infestations are usually unimportant, although large populations will contaminate foliage with honeydew, upon which sooty moulds develop (Fig. **57**).

CONTROL Spray with an insecticide as soon as adults are seen, and repeat at about two-week intervals as necessary.

Dialeurodes chittendeni Laing
Rhododendron whitefly

This locally distributed, generally uncommon European whitefly is associated with *Rhododendron*, and is sometimes a pest on *Rhododendron campylocarpum, R. catawbiense, R. caucasicum* and *R. ponticum*, and other species or cultivars where leaf cuticles are relatively thin or unprotected by hairs or scales. Adults are 1.2mm long, and pale yellow with pure white wings. They occur in June and July and are relatively sluggish, congregating on the underside of the leaves; if disturbed, they fly only a short distance before resettling. Eggs are laid singly on the underside of the leaves of host plants but without any accompanying waxy powder. Nymphs are flat and elliptical, greenish-yellow and semitransparent, without a waxy fringe. They feed from mid-July onwards, surviving throughout the winter; they pupate in the following spring, from mid-April onwards. The 'pupae' are oval (1.2×0.9mm), pale greenish-yellow to whitish-yellow, also without a waxy fringe. Because of their shape and colour, nymphs and 'pupae' are difficult

57 Colony of phillyrea whitefly on underside of leaf of *Phillyrea latifolia*.

to see; however, following the emergence of adults, the empty 'pupal' cases are more obvious. Light infestations are of little or no significance. However, if attacks are heavy, plant vigour is reduced, infested foliage becoming mottled with yellow, and covered with honeydew and sooty moulds.

CONTROL As for *Siphoninus phillyreae*.

Trialeurodes vaporariorum (Westwood)
Glasshouse whitefly

A tropical and subtropical species, introduced accidentally into northern Europe, where it is now a widespread and notorious glasshouse pest. Infestations occur on various ornamentals, including asparagus fern (*Asparagus plumosus*), *Begonia, Calceolaria, Cineraria, Coleus, Dahlia, Freesia, Fuchsia, Gerbera, Hibiscus, Pelargonium*, poinsettia (*Euphorbia heterophylla*), *Primula, Salvia, Solanum*, tobacco plant (*Nicotiana*) and many others; *Chrysanthemum* is a poor host and less commonly attacked. Infestations may also occur on seedling trees and shrubs growing under protection and on some, such as London plane (*Platanus × hispanica*), growing out-of-doors.

DESCRIPTION **Adult:** 1mm long; body pale yellow; wings pure white and held relatively flat when in repose. **Egg:** yellowish and broadly conical, and soon becoming grey. **Nymph:** pale green, oval, flat and scale-like. **'Pupa':** oval and whitish, with relatively short marginal wax processes and several pairs of long waxen dorsal tubes.

LIFE HISTORY Infestations occur on the underside of leaves of many cultivated glasshouse plants (Fig. **58**), and also on various weeds, the adults usually congregating on the foliage towards the tops of the plants and flying rapidly if disturbed. Each female is capable of depositing over 200 eggs during her lifetime, individuals usually surviving for from three to six weeks. The eggs are laid in distinctive circular groups on smooth leaves but tend to be more scattered on hairy ones. The eggs darken soon after laying; they hatch in about nine days at glasshouse temperatures of 21°C. The newly emerged nymphs crawl over the leaf surface for a short while but soon settle down to feed, inserting their mouthparts into the leaf tissue and remaining completely immobile throughout the rest of their development. They pass through three feeding stages and a non-feeding 'pupal' stage, before the new adults appear about 18 days later. Breeding is continuous under favourable conditions, with several overlapping generations annually, and is mainly parthenogenetic. The pest commonly survives between crops on glasshouse weeds or on outdoor plants in the near vicinity. Adults may also hibernate throughout the winter on outdoor weeds but they are unable to survive if weather conditions are severe.

DAMAGE Attacked plants are weakened and growth checked, the leaves sometimes becoming spotted with yellow or otherwise discoloured; heavy infestations may lead to premature leaf-fall. Plants also become contaminated with sticky honeydew and covered in sooty moulds.

CONTROL Various insecticidal programmes are recommended against this pest but choice of chemical is limited following the widespread development of resistance. Biological control, using the glasshouse whitefly parasite (*Encarsia formosa* Gahan), is also feasible.

Bemisia hancocki Corbett

A tropical and subtropical whitefly of uncertain status. It infests the underside of the leaves of host plants, including some ornamental species, and has been noted in several parts of Europe, including England where it was first found on bay laurel (*Laurus nobilis*) in the 1970s. Nowadays, the name *hancocki* Corbett is usually regarded merely as a synonym of *Bemisia afer* Priesner & Hosny, a polyphagous pest of cotton, peanut and other crops in Africa and Asia.

58 Colony of glasshouse whitefly, *Trialeurodes vaporariorum*.

Bemisia tabaci (Gennadius)
Tobacco whitefly

Infestations of this important tropical pest have occurred recently on imported ornamental plants in both England and the Netherlands. The insect has a very wide host range but in northern Europe is most likely to occur in glasshouses on ornamental plants such as *Gerbera*, *Gloxinia*, *Hibiscus* and poinsettia (*Euphorbia heterophylla*), and on vegetable crops such as sweet pepper and tomato. Adults and nymphs suck plant sap from the underside of the leaves of host plants, causing slight spotting of the tissue. They also secrete considerable quantities of honeydew. *Bemisia tabaci* is also a well-known virus vector. The life-cycle is similar to that of *Trialeurodes vaporariorum* (see above) but development is slower for any given temperature and may cease completely in the winter, even in warm houses. *Bemisia tabaci* is best recognised as scattered yellow scales (contrasting with the more clumped distribution and paler scales of *Trialeurodes vaporariorum*), the eggs being laid singly and randomly on the underside of the leaves. Also, the 'pupae' are slightly pointed posteriorly and lack horizontally directed waxen processes from the body wall; in repose, the wings of adults are held at a distinct roof-like angle.

CONTROL Suspected infestations should be reported to Plant Health authorities.

Aleurotuba jelinekii (von Frauenfeld)
Viburnum whitefly

This local, southerly distributed species occurs mainly on *Viburnum rotundifolia* and *V. tinus*; it does not invade other species of *Viburnum* but can occur on common myrtle (*Myrtus communis*) and on the strawberry tree (*Arbutus unedo*). Adults are 1.0–1.5 mm long, with whitish wings and a yellowish-orange body. They occur mainly during June and July, depositing eggs on the younger foliage. Nymphs feed during the summer on the underside of the leaves, completing their development in the autumn or following spring. The characteristic, scale-like 'pupae' (1.0–1.5 mm long) are black, with a whitish fringe and white waxen encrustations on the body (Fig. **59**). They occur throughout the winter and spring, and are very conspicuous when infested leaves are turned over. There is just one generation annually. Attacks are relatively harmless but heavily infested plants become contaminated with honeydew, upon which sooty moulds develop.

59 'Pupa' of viburnum whitefly, *Aleurotuba jelinekii*.

60 Narrow green pine aphid, *Eulachnus brevipilosus*.

10. Family **LACHNIDAE**

Aphids with terminal process of the antennae that are very short, the siphunculi are usually short, very hairy cones, and the cauda broadly rounded.

Eulachnus agilis (Kaltenbach)
Narrow spotted pine aphid

A local pest of Scots pine (*Pinus sylvestris*); infestations also occur on Corsican pine (*Pinus nigra* var. *maritima*). Widely distributed in mainland Europe; in the British Isles found mainly in southern England.

DESCRIPTION **Aptera:** 2.2–2.9 mm long; body elongate and spindle-shaped; bright green, with several reddish-brown dorsal platelets, most bearing a short, pointed seta; antennae short, pale greenish-brown, the third segment with long setae; siphuncular cones small; legs mainly pale green, the tibiae with long dark hairs.

LIFE HISTORY Colonies build up rapidly during the spring, with both winged and wingless forms present throughout much of the year. Aphid numbers decline during the summer but again reach a peak from late summer onwards. Sexual forms are produced in the autumn and eggs are deposited in the early winter, most placed singly in leaf scars. Individuals hatching from the eggs are very long-lived and more prolific than those produced parthenogenetically during the spring and summer.

DAMAGE Needles in the vicinity of colonies may become blackened, following the development of sooty moulds. Heavy infestations on nursery trees cause premature defoliation, with significant losses of the older needles.

CONTROL If required, apply a contact aphicide.

Eulachnus brevipilosus Börner
Narrow green pine aphid

A widespread but generally uncommon species, occurring mainly in pine-growing areas on the needles of Scots pine (*Pinus sylvestris*) but also on Austrian pine (*Pinus nigra*) and mountain pine (*Pinus mugo*). The aphids (1.7–2.2 mm long) are mainly green (Fig. **60**); they are distinguished from *Eulachnus agilis* by the inconspicuous dorsal platelets, the darker appendages and the short setae on the third antennal segment.

Eulachnus rileyi (Williams)
syn. *bluncki* Börner
Narrow brown pine aphid
This widely distributed species feeds mainly on Austrian pine (*Pinus nigra*), and is recognised by its brownish or orange-red body colour and associated waxy bloom.

Schizolachnus pineti (Fabricius)
Grey pine needle aphid
Generally common on young pines, especially Austrian pine (*Pinus nigra*), beach pine (*Pinus contorta*) and Scots pine (*Pinus sylvestris*). Widespread throughout Europe.

DESCRIPTION **Apterous female:** 1.2–3.0 mm long; body broadly oval, dark greyish-green, covered with long, fine hairs and coated in white, mealy wax; siphunculi reduced to small cones. **Alate:** 2–3 mm long; greyish-green.

LIFE HISTORY Overwintered eggs hatch in the early spring, and colonies of wingless aphids then develop in rows along the needles (Fig. **61**). Colonies persist throughout the summer and autumn, and are often attended by ants. Winged forms are produced from May to September. Sexual forms, including winged males, occur in the late autumn and early winter prior to the deposition of winter eggs. Under favourable conditions colonies will also survive the winter.

DAMAGE Infested needles may turn yellow and drop prematurely; this affects the vigour of young plants, and reduces shoot and needle length in the following season.

CONTROL If required, apply a contact aphicide.

Cinara pilicornis (Hartig)
Brown spruce aphid
An often common pest of spruces, including Norway spruce (*Picea abies*) grown as Christmas trees; present throughout western Europe.

DESCRIPTION **Apterous female:** 2.1–4.7 mm long; greyish-brown to brownish, and covered with fine hairs; rostrum short and dagger-like; siphunculi short, arising from pigmented cones; cauda inconspicuous and crescent-shaped.

LIFE HISTORY Overwintering eggs laid on spruce, hatch in the early spring, before bud-burst. Colonies of aphids then develop on the underside of the shoots (Fig. **62**), both winged and wingless forms

61 Colony of grey pine needle aphid on *Pinus sylvestris*.

62 Colony of brown spruce aphid on *Picea abies*.

being produced from May to July. Sexual forms appear from August onwards, the fertilized females finally depositing eggs on the twigs during the late summer and autumn.

DAMAGE Heavy infestations are not directly harmful, although they may cause slight discoloration of the foliage. Sooty moulds developing on honeydew produced in profusion by the aphids can affect the appearance and, hence, the marketability of nursery ornamentals and trees grown for the Christmas market.

CONTROL Spray with a contact aphicide as soon as aphids are seen.

Cinara costata (Zetterstedt)

Small colonies of this southerly-distributed species are noted occasionally on young trees of Norway spruce (*Picea abies*), coating infested parts of branches or twigs with whitish wax (Fig. **63**). Adults (2.7–3.8 mm long) are light brown with darker markings, but often appear greyish due to the presence of secreted wax; the siphunculi are borne on small, dark, widely spaced siphuncular cones (cf. *Cinara pruinosa*, p. 47). The aphids occur mainly from May to July, but also in the autumn. Colonies are usually attended by ants.

63 Colony of *Cinara costata* on *Picea abies*.

Cinara cuneomaculata (del Guercio)
Larch aphid

This large (2.4–4.6 mm long), dark brown species occurs on larch (*Larix*), including nursery stock, usually in small colonies on the shoots. Heavy infestations may cause discoloration of foliage, the aphids also producing considerable quantities of honeydew upon which sooty moulds develop. Wingless aphids occur from May to September, overwintering eggs being laid on young twigs and shoots in the autumn.

Cinara cupressi (Buckton)
Cypress aphid

This species is associated with *Thuja* but mainly with cypress trees, including Lawson cypress (*Chamaecyparis lawsoniana*), Leyland cypress (× *Cupressocyparis leylandii*) (especially the golden form 'Castlewellan') and Monterey cypress (*Cupressus macrocarpa*), sometimes causing damage to nursery stock; hedges and specimen trees in parks and gardens may also be affected. The aphids, which secrete considerable quantities of honeydew, feed on the foliage, shoots and branches, causing discoloration, die-back of affected tissue and reduced plant vigour. They occur from May onwards, often in only small numbers, winged forms appearing from June to August. Adults (1.8–3.9 mm long) are mainly orange-brown to yellowish-brown, with blackish markings diverging back from the thoracic region, pale grey transverse stripes on the abdomen and a black band between the prominent black siphuncular cones; the rostrum is relatively short (*c.* 1 mm long). The pest is widely distributed in continental Europe, especially in favourable southern areas; in Britain attacks are most often reported in the south and west of England.

64 Colony of American juniper aphid on *Juniperus*.

CONTROL Where significant damage is expected to occur, apply a contact aphicide in late May or early June.

Cinara fresai (Blanchard)
American juniper aphid

Dense colonies of this large (2.2–4.2 mm long), dark brownish-grey species occur locally on juniper (*Juniperus*) in southern England from June onwards. This species is readily distinguished from *Cinara juniperi* (p. 46) by the two broken, divergent black stripes which extend back from the head (Fig. **64**).

65 Protective earthen canopy around colony of American juniper aphid.

The aphids infest the foliage and stems of various ornamental junipers (including *Juniperus chinensis, J. sabina, J. squamata* and *J. virginiana*), protected by earthen shelters constructed by ants (Fig. **65**); at least in the Netherlands, *Chamaecyparis* and *Thuja*) are also attacked. Infestations lead to the development of sooty moulds and hence to the blackening of shoots (Fig. **66**), affecting the quality and appearance of plants. Attacks may also cause die-back of shoots and, sometimes, death of complete trees.

CONTROL As for *Cinara pilicornis* (p. 44).

66 American juniper aphid damage to foliage of *Juniperus*.

Cinara juniperi (Degeer)
Juniper aphid
A widely distributed species on junipers, especially *Juniperus communis*. The aphids are pinkish-brown, slightly smaller (2.2–3.4mm long) than *Cinara fresai* (p. 45), and have a relatively short rostrum. Also, unlike *Cinara fresai*, infestations are usually restricted to the young, green shoots and needles. Sooty moulds develop on honeydew secreted by the aphids, and this can disfigure established ornamentals and nursery stock.

CONTROL As for *Cinara pilicornis* (p. 44).

Cinara pinea (Mordwilko)
Large pine aphid
This common, widely distributed, large-bodied (3.1–5.2mm long) aphid feeds on Scots pine (*Pinus sylvestris*), infesting young shoots bearing new or one-year-old needles. Adults vary from grey to orange-brown or dark brown, and are present from the spring to the late autumn or early winter when oviparous forms finally deposit the overwintering eggs. These eggs (*c.* 1.6mm long) are black and shiny, and laid in rows along the needles (Fig. **67**).

67 Eggs of large pine aphid, *Cinara pinea*.

They hatch in the spring and colonies of pale aphids develop on the shoots, individuals clustering together on the wood between the bases of the needles (Fig. **68**). Breeding continues throughout the summer, including the production of winged forms, later-produced individuals tending to be darker in colour than the spring forms. Colonies are relatively small but produce considerable quantities of honeydew. Although damage to established trees is slight, attacks on nursery trees may cause considerable damage, leading to discoloration of the needles, premature loss of needles and reduced plant vigour.

CONTROL As for *Cinara pilicornis* (p. 44).

68 Colony of large pine aphid on *Pinus sylvestris*.

69 Colony of *Cinara pruinosa* on *Picea abies*.

Cinara pruinosa (Hartig)

Colonies of this large (2.4–5.0mm long), dark green to brown aphid occur on the shoots and branches of spruce trees, especially Norway spruce (*Picea abies*). The rostrum is noticeably long, and the siphuncular cones very large and prominent. In summer, colonies move to the base of the trunk and to the roots, the winter being passed as adults in the soil or as eggs on young twigs. Colonies are ant-attended (Fig. **69**) and are often sheltered by earthen canopies constructed by these insects (Fig. **70**).

Tuberolachnus salignus (Gmelin)

Large willow aphid

An often common pest of willow, especially osier (*Salix viminalis*). Widely distributed, especially in central and southern Europe; in the British Isles most numerous in southern England.

DESCRIPTION **Apterous female:** 4.0–5.4mm long; blackish-brown with a large tubercule arising from the fourth abdominal tergite; body clothed in fine, grey hairs.

LIFE HISTORY Dense colonies of this large, conspicuous, long-legged aphid (Fig. **71**) occur on the branches and stems of willow trees from late June onwards but not until much later in some areas; breeding is entirely parthenogenetic, with populations reaching their maximum in the autumn. Small numbers of apterae will survive the winter in favourable areas but, elsewhere, colonies must be re-established in the following year by winged migrants that have been reared further south. Colonies produce vast quantities of honeydew and are constantly visited by sugar-seeking insects such as ants, bees, wasps and various flies.

70 Protective earthen canopy around colony of *Cinara pruinosa*.

71 Colony of large willow aphid on *Salix viminalis*.

47

DAMAGE The aphids can weaken shoot growth, most significant damage occurring on the canes of willows grown for basket-making. Plants, including ornamentals, are also contaminated by masses of sticky honeydew and sooty moulds. If social wasps (*Vespula* spp.) (see p. 381) are attracted to infested willow windbreaks in orchards, they may divert their attention to nearby ripening fruits, and this can develop into a significant problem.

CONTROL Apply a contact or a systemic aphicide as soon as infestations are seen.

Maculolachnus submacula (Walker)
syn. *rosae* (Cholodkovsky)
Rose root aphid
An often common pest of rose (*Rosa*), especially in England, Germany and the Netherlands; also present in North America.

DESCRIPTION **Apterous female:** 2.7–3.8 mm long; yellowish-brown to dark chestnut-brown or blackish-brown; antennae short and blackish-brown; legs blackish-brown; abdomen with dorsal hairs arising from small plates; rostrum short; siphunculi short and conical; cauda rounded and inconspicuous. **Egg:** 1 mm long; elongate-oval, black and shiny.

LIFE HISTORY Clusters of overwintering eggs occur on the lower portions of rose stems (Fig. **72**), and these commonly attract attention when bushes are pruned during the winter or early spring. The eggs hatch in the early spring, nymphs then migrating to the underground parts of the stems to begin feeding. Ant-attended colonies develop during the spring and summer on the superficial roots and stem bases, often protected by earthen shelters (Fig. **73**), winged forms spreading infestations to other rose bushes. Aphid numbers decline in the autumn, when eggs are deposited.

DAMAGE The growth of infested bushes may be checked, and persistent infestations can cause the death of plants.

CONTROL Shoots infested with winter eggs can be cut out and burnt. If required, apply a contact or a systemic aphicide in the spring, soon after egg hatch.

Lachnus roboris (Linnaeus)
This locally common species is associated with oak, including evergreen oak (*Quercus ilex*), colonies developing on the shoots and smaller branches during the summer months; infestations sometimes attract attention and can cause minor damage to

72 Eggs of rose root aphid on stem of *Rosa*.

73 Protective earthen canopy around colony of rose root aphid.

garden or specimen trees. The aphids (3.2–3.5 mm long) are shiny black to dark brown, with very long, yellowish-brown legs, and the siphunculi borne on large, blackish, hairy cones; the wings of alates are clouded with brown patches. Colonies are typically ant-attended, and often very large; however, the individual aphids tend to remain relatively well separated and do not cluster tightly together.

Lachnus longirostris (Börner)
A relatively uncommon species on oak (*Quercus*) and although sometimes noted on ornamentals rarely, if ever, troublesome. The aphids are similar in appearance and habits to the previous species but are larger (4.5–6.4 mm long) and have inconspicuous siphunculi.

Trama troglodytes von Heyden
Jerusalem artichoke tuber aphid
Infestations of this large aphid sometimes occur on the roots of cultivated Compositae, including *Arnica*, *Aster*, *Chrysanthemum maximum*, sneezewort (*Helenium*) and sunflower (*Helianthus*); they also occur on weeds such as dandelion (*Taraxacum*) and sow-thistle (*Sonchus*). Colonies are ant-attended and occur on the roots throughout the year. Reproduction is entirely parthenogenetic, winged forms appearing during the summer months and then dispersing to new hosts. The aphids reduce the vigour of infested plants and, especially in dry conditions, can cause stressed plants to wilt. Adults (3–4 mm long) are white and lack siphunculi; characteristically, if disturbed, they vibrate their long hind tibiae.

74 Colony of osier leaf aphid on *Salix viminalis*.

11. Family CHAITOPHORIDAE

Aphids with body and legs bearing long hairs, the terminal process of the antennae is long, the siphunculi pore-like or stumpy, and the cauda knob-like or rounded.

Chaitophorus beuthami (Börner)
Osier leaf aphid
A generally common pest of osier (*Salix viminalis*) and other osier-type willows; often present on ornamental trees. Widely distributed in Europe.

DESCRIPTION **Apterous female:** 2.3 mm long; pale greenish, relatively elongate and noticeably hairy; siphunculi stumpy and slightly broadened at the tip; cauda knob-shaped; young individuals more-or-less colourless. **Alate:** similar to aptera but with dark abdominal tergites.

LIFE HISTORY Colonies develop from spring onwards, the aphids feeding and breeding on the underside of the leaves (Fig. **74**); they often occur in considerable numbers but usually do not infest the young shoots. Although the aphids secrete honeydew, upon which sooty moulds develop, they are not particularly attractive to ants. By late summer, colonies are often much affected by predators.

DAMAGE Direct damage is unimportant but heavily infested trees become unsightly following the development of sooty moulds on the aphid-secreted honeydew.

75 Colony of sallow leaf aphid on underside of leaf of *Salix*.

Chaitophorus capreae (Mosley)
Sallow leaf aphid
A widely distributed and very common species on eared sallow (*Salix aurita*), grey sallow (*Salix cinerea*), pussy willow (*Salix caprea*) and other broad-leaved willows. The aphids are relatively small (less than 2 mm long), with interrupted black abdominal markings; the nymphs are mainly whitish. Colonies develop throughout the summer on the underside of the leaves (Fig. **75**) but have little or no direct effect on their hosts.

76 Colony of poplar leaf aphid on *Populus*.

77 Colony of *Chaitophorus niger* on *Salix*.

Chaitophorus leucomelas Koch
Poplar leaf aphid
Infestations of this generally common species occur on the underside of leaves and on the young shoots of black poplar (*Populus nigra*) and Lombardy poplar (*Populus nigra* 'Italica') but are not important. Apterae (2.0–2.7 mm long) are dull greenish-yellow, with distinctive dark markings; nymphs are paler and often more brightly coloured (Fig. **76**).

Chaitophorus niger Mordvilko
Although local and generally uncommon, colonies of this species are sometimes found on ornamental willows, including crack willow (*Salix fragilis*), purple willow (*Salix purpurea*), weeping willow (*Salix vitellina* var. *pendula*) and white willow (*Salix alba*). The aphids (2.0–2.2 mm long) are distinguished by their overall dark appearance, the tergites forming a solid black carapace (Fig. **77**).

78 Colony of *Chaitophorus salicti* on *Salix*.

Chaitophorus salicti (Schrank)
This local species (Fig. **78**) occurs on broad-leaved willows, such as common sallow (*Salix atrocinerea*) and eared sallow (*Salix cinerea*), and is sometimes found infesting nursery trees. Individuals are slightly larger and also darker than those of *Chaitophorus capreae* (p. 49) but distinguishable with certainty only by microscopic examination.

Periphyllus acericola (Walker)
A generally common species on sycamore (*Acer pseudoplatanus*), infesting both young and mature trees. Widely distributed in Europe.

DESCRIPTION **Apterous female:** 3–4 mm long; body pale green, oval and hairy; siphunculi stumpy and tapered; cauda broadly rounded. **Nymph:** pale green with long body hairs. **Dimorphic summer nymph:** pale yellow with long body hairs (Fig. **79**). **Alate:** 3.0–3.5 mm long; dark-bodied.

79 Dimorphic summer nymphs of *Periphyllus acericola*.

LIFE HISTORY In the spring, colonies of aphids develop on the underside of expanding sycamore leaves, nymphs of the first generation usually clustering along the major veins towards the leaf base; later, colonies (including winged forms) also develop on the new shoots. Colonies die out in the summer, and the species survives by aestivating as first-instar dimorphic nymphs, which cluster together (cf. *Periphyllus testudinaceus*, see below) beneath fully expanded leaves. Activity is resumed in the autumn, and is completed with the production of sexual forms and the deposition of winter eggs.

DAMAGE Infestations have little effect on tree growth but spring colonies can be of minor significance on nursery trees.

CONTROL Apply a contact aphicide in the spring, soon after egg hatch.

Periphyllus californiensis (Shinji)
Californian maple aphid

This North American species occurs on various species of maple (*Acer*), including downy Japanese maple (*Acer japonicum*) and smooth Japanese maple (*Acer palmatum*), having been introduced into parts of Europe, including southern England, on infested nursery stock. The aphids are 2.3–3.5mm long and dark olive-green to brown (Fig. **80**); the dimorphic nymphs are pale green and similar in appearance to those of *Periphyllus testudinaceus* (q.v.) but with just two series of abdominal plates (spinal and marginal).

Periphyllus testudinaceus (Fernie)
This often abundant, dark brownish-green to blackish (2.0–3.5mm long) species (Fig. **81**) infests various species of *Acer*, including sycamore (*Acer pseudoplatanus*) but especially wild maple (*Acer campestre*) and various ornamental maples; it is also associated with horse chestnut (*Aesculus hippocastanum*). Heavy infestations often develop in the spring on the young growth; later, the aphids occur abundantly beneath the expanded leaves. Colonies are ant-attended, the aphids producing copious quantities of honeydew. The life-cycle follows that of *Periphyllus acericola* but the dimorphic summer nymphs, which are green and fringed with curious leaf-like hairs, aestivate singly along the major leaf veins. Unlike those of *Periphyllus californiensis*, the dimorphic nymphs possess spinal, marginal and, between these, a pleural series of abdominal plates.

80 Colony of Californian maple aphid on *Acer*.

81 Nymphs and adults of *Periphyllus testudinaceus*.

12. Family CALLAPHIDIDAE

Aphids with terminal process of the antennae of variable length, the siphunculi usually stumpy or broadly conical but sometimes pore-like or long and swollen, the anal plate often divided into two lobes, and the cauda knob-like or rounded.

Callaphis juglandis (Goeze)
Large walnut aphid

A generally common but minor pest of walnut (*Juglans*); colonies develop in the underside of the leaves, the aphids causing slight yellowing and premature leaf-fall. Heavy infestations can reduce the vigour of young trees but attacks on established plants are of little or no importance. The aphids (2–4mm long) are bright greenish-yellow to yellow, marked with brownish-black, the body bearing numerous fine hairs; the wing veins of alates are dusky bordered (Fig. 82).

Myzocallis coryli (Goeze)
Hazel aphid

Generally abundant on both wild and cultivated hazel (*Corylus*), and often a minor pest of nursery stock. Eurasiatic; also now found in North America and New Zealand. Present throughout Europe.

DESCRIPTION **Alate female:** 1.3–2.2 mm long; shiny, whitish to yellowish or pale green, with large red eyes; siphunculi stumpy; cauda knob-like and projecting beyond a bilobed subanal plate; wing veins terminating in dusky spots on the wing margins. **Nymph:** pale, with body hairs capitate.

LIFE HISTORY Overwintered eggs hatch in the spring, aphids then feeding on the underside of the foliage from May onwards. Breeding continues throughout the summer months, all of the aphids developing into winged forms. Sexual morphs arise in November and mated oviparous females finally deposit eggs on the shoots.

DAMAGE Although shoots of infested plants are not distorted the foliage is disfigured by the accumulation of sticky honeydew and development of sooty moulds.

CONTROL Apply a contact or a systemic aphicide if infestations develop on nursery stock.

Myzocallis boerneri Stroyan
This species, presumably of southern European origin, infests Chinese cork oak (*Quercus variabilis*), chestnut-leafed oak (*Quercus castaneifolia*), cork oak (*Quercus suber*), Lucombe oak (*Quercus × hispanica* var. *lucombeana*) and Turkish oak (*Quercus cerris*). It occurs locally in parks, botanic gardens and arboreta in northern Europe but does not occur on native oaks. The alatae are relatively small (1.3–2.2 mm long) and mainly yellow (cf. *Myzocallis castanicola*).

Myzocallis castanicola Baker
Widely distributed and locally common on oak (*Quercus*) and sweet chestnut (*Castanea sativa*) but not an important pest. Alatae (2.2–2.6 mm long) are deep yellow to brown, with black, bar-like markings and black siphunculi.

82 Colony of large walnut aphid, *Callaphis juglandis*.

Myzocallis schreiberi Hille Ris Lambers & Stroyan
Minor infestations of this local, southerly distributed species occur on evergreen oak (*Quercus ilex*). The aphids, which are entirely parthenogenetic, occur on the underside of the leaves; they may be found throughout the year but are not harmful.

Tuberculoides annulatus (Hartig)
syn. *quercus* (Kaltenbach)
Oak leaf aphid

Widespread and generally abundant on oak, especially English oak (*Quercus robur*), the aphids developing in scattered colonies on the underside of the expanded leaves. Infestations often occur on young trees in parks, gardens and nurseries but the aphids cause little or no damage. The aphids (1.4–2.9 mm long) have distinctively banded antennae, stumpy siphunculi and a prominent, knob-like cauda; they vary considerably in colour being either yellow, green, salmon-pink or grey. Viviparous forms of this species, which occur throughout the summer months, are all winged.

Eucallipterus tiliae (Linnaeus)
Lime leaf aphid

An abundant and widely distributed pest of lime (*Tilia*), but of importance mainly as a producer of contaminating honeydew. Present throughout Europe; also occurs in Central Asia, New Zealand and North America.

DESCRIPTION **Alate female:** 1.8–3.0 mm long; black and yellow; wing veins dusky-bordered; siphunculi dark and stumpy; cauda rounded and bearing a

83 Lime leaf aphid, *Eucallipterus tiliae*.

84 Nymphs of lime leaf aphid, *Eucallipterus tiliae*.

small, pigmented, dorsal tubercule (Fig. **83**). **Nymph:** greenish-yellow, marked with several blackish platelets (Fig. **84**).

LIFE HISTORY Eggs laid on lime trees in the autumn hatch in the following spring. Colonies of winged aphids then build up on the new shoots and expanding foliage, infestations readily spreading from tree to tree throughout the spring and early summer. Reproduction slows down in mid-summer but picks up with the initiation of sexual forms in the early autumn.

DAMAGE Although infestations have little direct effect on tree growth the vast quantities of honeydew produced by this species are commonly a nuisance, regularly contaminating cars, garden furniture, pavements and plants beneath infested trees; such contamination is especially significant in dry weather. Sooty moulds developing on the honeydew can also be a problem.

CONTROL Apply a contact or a systemic aphicide in the spring, soon after bud-burst. Treatment of mature trees is impractical and not worth while.

Takecallis arundicolens (Clarke)
Bamboo aphid

This eastern Asian species was accidentally introduced into Europe, where it is now widely distributed and sometimes locally abundant on bamboo (*Bambusa*). The aphids occur under the expanded leaves but do not appear to be harmful. Adults are medium-sized (1.8–2.8 mm long), whitish or pale yellowish, with pale, dark-ringed antennae, short siphunculi and a small, black, oval cauda (Fig. **85**). Nymphs are pale with long, capitate body hairs, red eyes and the joints of the antennal segments dark (Fig. **86**). Reproduction is entirely parthenogenetic and all adults are viviparous alates.

85 Bamboo aphid, *Takecallis arundicolens*.

86 Nymph of bamboo aphid, *Takecallis arundicolens*.

Phyllaphis fagi (Linnaeus)
Beech aphid

A generally common pest of beech (*Fagus sylvatica*), and frequently injurious to nursery trees and hedges; infestations also occur on Persian ironwood (*Parrotia persica*). Widespread in Europe; also present in North America.

DESCRIPTION **Apterous female:** 2.0–3.2 mm long (dwarf summer form half-size); yellowish-green, coated with white flocculent masses of wax; siphunculi pore-like; cauda small, rounded and often inconspicuous.

LIFE HISTORY Black, oval eggs are deposited on beech twigs during the autumn, hatching in the following spring. Colonies of aphids, which reach the peak of their development by early summer, then develop on the shoots and underside of leaves, amongst accumulations of white wax (Fig. **87**). Winged forms appear after two wingless generations, spreading infestations from place to place. In mid-summer, dwarf apterae are produced and these undergo a period of aestivation before giving rise to sexual forms. Colonies then die out as winter eggs are laid.

DAMAGE Infested foliage becomes coated with masses of sticky honeydew, which accumulates amongst the secreted wax and upon which sooty moulds develop. Attacked leaves also become curled; they may also turn brown around the edges (Fig. **88**) and die prematurely.

CONTROL Apply an aphicide during the spring, as soon as infestations are discovered; systemic insecticides are usually more effective than contact materials.

Betulaphis quadrituberculata (Kaltenbach)

A locally common pest of birch (*Betula*). Widely distributed in Europe.

DESCRIPTION **Apterous female:** 1.3–2.0 mm long; greenish or yellowish, sometimes marked with darker green or yellow; abdomen with four rows of hairs along the back and a series along either side, the longer hairs distinctly capitate; siphunculi conical and stumpy; cauda broadly rounded and subtriangular, projecting slightly beyond a deeply cleft subanal plate. **Alate:** 1.5–2.2 mm long; similar to aptera but capitate hairs restricted to the eighth abdominal segment; larger-bodied individuals with head, antennae, siphunculi and parts of thorax, legs and abdomen dusky.

87 Colony of beech aphid, *Phyllaphis fagi.*

88 Beech aphid damage to leaves of *Fagus.*

89 Colony of *Betulaphis quadrituberculata* on *Betula.*

LIFE HISTORY Eggs overwintering on host trees hatch in the spring. Aphids then feed on the underside of the leaves, individuals of the first generation developing into either winged or wingless forms. Nymphs of subsequent generations also occur on the underside of the leaves, either singly or in small groups (Fig. **89**), most developing into wingless adults. There is a succession of generations throughout the spring and summer months but dense colonies develop only if conditions are especially favourable.

DAMAGE Infestations are usually of only minor importance and any effect on tree growth is slight, even if noticeable colonies develop.

Callipterinella tuberculata (von Heyden)
syn. *betularia* (Kaltenbach)
Minor infestations of this widespread but locally distributed species occur on birch (*Betula*), the aphids occurring singly or in small groups on the leaves (Fig. **90**). Colonies are sometimes noticed on the new foliage of young trees but are not of importance. The aphids (1–2 mm long) are basically greenish, with the body variably patterned by dark sclerites, the darkened areas often being extensive; the siphunculi are small but conspicuous, the cauda very short and inconspicuous.

Euceraphis betulae (Koch)
Silver birch aphid
A very common pest of silver birch (*Betula pendula*), infestations frequently developing on trees in gardens and nurseries. Widely distributed in Europe.

DESCRIPTION **Alate female:** 3–4 mm long; body elongate, mainly pale green or yellowish-green and coated in a bluish-white waxy secretion; legs and antennae very long; siphunculi very short; cauda knobbed (Fig. **91**). **Nymph:** waxy-green, with distinctly dark legs and siphuncular rims (Fig. **92**).

LIFE HISTORY This species overwinters in the egg stage on the shoots of birch trees. The eggs hatch in the spring. Nymphs then develop on the leaves, the first adults appearing in late May or early June. Infestations occur throughout the spring and summer, the aphids occurring on the youngest, still unfurling leaves and also on the underside of fully expanded ones. The adult aphids are always winged and do not aggregate, typically occurring singly; they are also very active, immediately dropping from the foodplant if disturbed. The nymphs either occur singly or in small groups.

90 Colony of *Callipterinella tuberculata* on *Betula*.

91 Silver birch aphid, *Euceraphis betulae*.

92 Nymph of silver birch aphid, *Euceraphis betulae*.

DAMAGE Foliage is not distorted but does become sticky with honeydew; infested plants are also disfigured by sooty moulds.

CONTROL Apply a contact aphicide in the spring, as soon as aphids are seen. Treatment of mature trees is impractical and not worth while.

Euceraphis punctipennis (Zetterstedt)
Downy birch aphid
This aphid (Fig. **93**) is virtually identical in appearance and habit to the previous species but breeds on downy birch (*Betula pubescens*).

93 Colony of downy birch aphid, *Euceraphis punctipennis*.

Drepanosiphum platanoidis (Schrank)
Sycamore aphid
Generally abundant on sycamore (*Acer pseudoplatanus*) and, rarely, other species of *Acer*, but mainly a problem as a copious producer of contaminating honeydew. Widely distributed in Europe; also present in North America.

DESCRIPTION **Alate female:** 3.2–4.3 mm long; elongate, pale green or greyish-green, sometimes tinged with reddish, with darker markings dorsally; antennae with a long terminal process; siphunculi long and swollen; cauda small and rounded (Fig. **94**). **Nymph:** pale green to whitish-green, with red eyes and blackish markings on the antennae (Fig. **95**).

LIFE HISTORY Overwintered eggs hatch at budburst. Small numbers of alate females then develop rapidly on the new shoots and beneath the expanding leaves. In summer, breeding ceases but alate females continue to survive on the underside of expanded leaves where they may often be found in considerable numbers. Although mainly sedentary they readily become active if disturbed. Breeding recommences in the autumn, when leaves begin to senesce, with populations again increasing rapidly as sexual forms are produced; winter eggs are eventually laid in November or December.

94 Sycamore aphid, *Drepanosiphum platanoides*.

95 Nymph of sycamore aphid, *Drepanosiphum platanoides*.

13. Family **APHIDIDAE**

The main family of aphids. Alternation of generations between winter and summer hosts is commonplace and life-cycles of individual species are often complex. Many species show an alternation of generations, having a primary host upon which asexual and sexual reproduction occurs and eggs are laid, and a secondary host where development is entirely parthenogenetic and viviparous. Migration between these alternate hosts is usually achieved following the production of winged forms.

Pterocomma salicis (Linnaeus)
Black willow aphid
An often common pest of osier (*Salix viminalis*) and various other willows. Holarctic. Present throughout Europe.

DESCRIPTION **Apterous female:** 3.5–4.2 mm long; body dull greyish-black to black, with greyish or whitish markings, and very hairy; antennae, legs and siphunculi orange-red; antennae short; siphunculi flask-shaped. **Nymph:** pale orange-brown to greenish-black, with whitish markings.

LIFE HISTORY Dense, compact colonies often develop during the summer on the shoots and young stems of willow trees and osiers (Fig. **96**). The colonies are often very large and are commonly attended by ants. The winter is spent in the egg stage in crevices in the bark of host plants.

DAMAGE Colonies on windbreaks and specimen trees or shrubs are disfiguring but of little importance. However, attacks on osiers grown for basket-making can be serious; rods may be stunted and the wood stained, even when only lightly infested, and new growth may be killed off.

CONTROL Apply an aphicide as soon as infestations are seen.

Pterocomma populeum (Kaltenbach)
A widely distributed aphid, infesting poplar (*Populus*) and forming dense colonies on the young branches and twigs. Damage caused, however, is slight. Apterae are greyish-green to greyish-brown, with a pair of dark spots on each body segment and a light covering of whitish wax (Fig. **97**).

Pterocomma steinheili (Mordvilko)
Colonies of this species, which has a similar life-cycle to that of *Pterocomma salicis*, occur on the young wood of willow (*Salix*); the aphids are distinguished by their reddish-brown to grey body colour (Fig. **98**).

96 Colony of black willow aphid, *Pterocomma salicis*.

97 Colony of *Pterocomma populeum* on *Populus*.

98 Colony of *Pterocomma steinheili* on *Salix*.

Hyalopterus pruni (Geoffroy)
Mealy plum aphid

A cosmopolitan and generally common pest of damson and plum, but also inhabiting certain other kinds of *Prunus* (such as almond, *Prunus dulcis*) grown as ornamentals. The winter is passed in the egg stage on host trees. Eggs hatch in the early spring, usually by the white-bud stage, and dense colonies of the pale green to bluish-grey, mealy-coated aphids (1.5–2.6mm long) eventually develop on the leaves and shoots (Fig. **99**). The aphids produce vast quantities of honeydew and, although they do not cause leaf curl (cf. *Brachycaudus helichrysi*, p. 67), infested plants are disfigured by the colonies and by the development of sooty moulds. In summer, winged aphids migrate mainly to reeds (*Phragmites communis*), a return migration to winter hosts occurring in the autumn.

99 Colony of mealy plum aphid on *Prunus*.

CONTROL On suitable hosts, apply a winter wash during the dormant period; alternatively, apply a contact or a systemic aphicide at the white-bud stage or after flowering.

Rhopalosiphum nymphaeae (Linnaeus)
Water-lily aphid

A generally common pest of aquatic plants, especially water-lilies (*Nymphaea* and *Nuphar*). Virtually cosmopolitan. Widely distributed in Europe.

DESCRIPTION **Apterous female:** 1.6–2.6mm long; dark olive-green to brown, lightly dusted with whitish wax; siphunculi relatively long and swollen, mainly pale but dark apically. **Alate:** 1.6–2.6mm long; dark brown to shiny black.

100 Colony of water-lily aphid, *Rhopalosiphum nymphaeae*.

LIFE HISTORY The winter is passed in the egg stage on blackthorn (*Prunus spinosa*) and various other species of *Prunus*, the primary hosts. Eggs hatch in the following spring and small colonies develop of the shoots. Winged forms are produced in the early summer and these migrate to various aquatic plants, including arrow-head (*Sagittaria sagittifolia*), reed-mace (*Typha*) and water-plantain (*Alisma plantago-aquatica*); large colonies often develop on water-lilies, the aphids congregating along the leaf veins (Fig. **100**) and also invading the flowers. In the autumn, there is a return migration to the primary hosts where sexual reproduction occurs and eggs are laid.

DAMAGE **Primary hosts:** if attacks are heavy, young leaves at the tips of the new shoots become crinkled. **Secondary hosts:** heavy infestations cause considerable distortion of the stems and foliage, and also discolour the flowers.

CONTROL **Primary hosts:** specific control measures are rarely required but, if necessary, apply an aphicide in the early spring. **Secondary hosts:** aphids can be removed from water-lilies by hosing infested plants with water; insecticides should not be used as they are at least potentially harmful to fish and to other aquatic life.

Rhopalosiphum insertum (Walker)

syn. *crataegellum* (Theobald)

Apple-grass aphid

Minor infestations of this generally common species sometimes occur on *Cotoneaster*, crab apple (*Malus*), hawthorn (*Crataegus*), Japanese quince (*Chaenomeles japonica*), medlar (*Mespilus germanica*), rowan (*Sorbus aucuparia*) and other species of *Sorbus*, causing leaf distortion (Fig. **101**). The aphids over-winter as eggs on rosaceous hosts, with colonies developing on the shoots in the following spring. Eventually, winged forms are produced and these migrate to the roots of grasses, the secondary hosts. A return migration to primary hosts occurs in the autumn. Apterae are 2.1–2.6 mm long, plump, and yellowish-green with darker longitudinal stripes down the body, short, dark-tipped antennae, and short, pale green, distinctly flanged siphunculi. Attacks on ornamentals are usually insignificant but may be of some importance on young plants and nursery stock.

CONTROL Winter washes, applied during the dormant season, are effective; alternatively, apply a penetrative or a systemic aphicide in the spring before the flowering stage.

Rhopalosiphum rufiabdominalis (Saski)

Rice root aphid

Infestations of this widely distributed but mainly tropical species have been introduced recently into the Netherlands, the aphids breeding on the roots of various pot plants. Adults (1.2–2.2 mm long) are dark green or olive-green, typically with the posterior part of the abdomen reddish.

101

101 Apple-grass aphid damage to leaves of *Sorbus sargentiana*.

102

102 Colony of black bean aphid on *Clematis*.

Aphis fabae Scopoli

Black bean aphid

An often major pest of cultivated beans and various ornamental plants, including guelder-rose (*Viburnum opulus*), mock orange (*Philadelphus coronarius*), spindle (*Euonymus*) and herbaceous plants such as *Clematis*, *Dahlia*, nasturtium (*Tropaeolum*), poppy (*Papaver*), pot marigold (*Calendula*) and *Yucca*. Cosmopolitan. Present throughout Europe.

DESCRIPTION **Apterous female:** 1.5–2.9 mm long; black to blackish-brown or blackish-green, often with whitish patches of wax on the abdomen; antennae much shorter than body; siphunculi dark and of moderate length; cauda blunt and finger-shaped. **Alate:** 1.8–2.7 mm long; mainly black, often with noticeable spots of white wax. **Nymph:** black to blackish-brown or blackish-green, often with distinct patches of white wax.

LIFE HISTORY The winter is usually passed in the egg stage on wild spindle-bushes, but eggs may also be deposited on ornamental spindle and, sometimes, on other woody hosts such as guelder-rose; in favourable situations, colonies may also survive on herbaceous hosts. Eggs hatch in the early spring, and colonies of aphids develop on the young leaves and shoots. Winged forms appear in May or June and these disperse to various herbaceous hosts, colonies on the primary host then dying out. Breeding on these secondary hosts continues throughout the summer (Fig. **102**), with the frequent production of winged forms and further spread to other summer hosts. Colonies are ant-attended, and are often common in July and August on outdoor ornamentals; colonies may also develop in glasshouses on hosts such as *Chrysanthemum*. In the autumn, there is a return migration to the primary host where, following the production of sexual forms, winter eggs are eventually laid.

DAMAGE **Primary hosts:** spring infestations cause considerable curling of leaves on the new shoots (Fig. **103**), damage remaining to spoil the appearance of bushes long after colonies have died out. **Secondary hosts:** leaves may be curled, but infestations on the buds and flower stalks are usually more important, affecting both the quality, flowering potential and appearance of plants.

CONTROL **Primary hosts:** apply an aphicide shortly after bud-burst. **Secondary hosts:** apply an aphicide as soon as infestations are seen, and repeat as necessary; systemic insecticides, particularly soil-applied granules, will give longest protection.

103 Black bean aphid damage to leaves of *Euonymus japonica*.

Aphis frangulae gossypii Glover
Melon and cotton aphid

Generally common on glasshouse ornamentals such as *Begonia*, calla lily (*Richardia*), *Chrysanthemum* and *Cineraria*. Virtually cosmopolitan. Widely distributed in Europe. [Other sub-species of the *Aphis frangulae* Kaltenbach complex are of lesser importance.]

DESCRIPTION **Apterous female:** 1.4–2.0 mm long; very dark green or blue-green, sometimes mottled yellowish-green; siphunculi relatively short and dark. **Alate:** 1.1–2.1 mm long; head and thorax dark; abdomen marked with dark spots.

LIFE HISTORY This subspecies breeds continuously under protection (where it is often associated with members of the Cucurbitaceae, including vegetable crops) but is unlikely to survive the winter out-of-doors in northern Europe. [Other subspecies overwinter as eggs on outdoor hosts, e.g. alder buckthorn (*Frangula alnus*); they inhabit various herbaceous hosts during the summer, including ornamentals such as *Hypericum* (Fig. **104**).]

DAMAGE Attacked leaves turn yellow and may also wilt and die (Fig. **105**). The appearance of ornamentals is also spoilt by the accumulation of honeydew, development of sooty moulds and presence of cast aphid skins.

104 Colony of melon cotton aphid on *Hypericum*.

105 Melon cotton aphid damage to foliage of *Chrysanthemum*.

CONTROL Apply an aphicide, ideally before flowers open, and repeat as necessary; ensure that the product chosen is not phytotoxic and will not damage flower petals (particularly applicable to begonias). Insecticidal fogs and plant dips can be used as alternatives to sprays or granules.

106 Viburnum aphid damage to shoots of *Viburnum carlesii*.

Aphis viburni Scopoli
Viburnum aphid

A common pest of wild and ornamental species of *Viburnum*, including guelder-rose (*Viburnum opulus*). Widely distributed in Europe.

DESCRIPTION **Apterous female:** 1.8–3.0 mm long; dark greenish-black to brownish-black; siphunculi relatively short. **Nymph:** brownish, liberally coated with white wax.

LIFE HISTORY This species is restricted to *Viburnum*, overwintering in the egg stage on shoots and branches. Eggs hatch in the early spring, and aphids then feed on the underside of leaves on the young shoots from April onwards. [*Aphis fabae* (p. 59) also infests guelder-rose but colonies appear black rather than brown.] The sexual phase in the life-cycle includes the production of a generation of wingless males. Colonies, which are ant-attended, may also extend onto the inflorescences. They reach their maximum development relatively early in the season and then decline as winter eggs are laid, aphids having largely disappeared by mid-summer.

DAMAGE Spring infestations curl the leaves of new shoots, the deformed foliage remaining on the bushes long after the aphids have disappeared. The presence of damaged shoots spoils the appearance of ornamental shrubs; *Viburnum carlesii* is especially susceptible and may be severely disfigured (Fig. **106**).

CONTROL Apply an aphicide in the spring, shortly after bud-burst.

107 Colony of cowpea aphid on *Robinia*.

Aphis craccivora Koch
Cowpea aphid

Uncommon in southern England but more numerous in continental Europe, where it is associated with various hosts, especially *Leguminosae*. The aphids are 1.4–2.0 mm long and shiny black, the nymphs lightly dusted with wax (Fig. **107**). Dense colonies often develop during the summer on the young shoots of false acacia (*Robinia pseudoacacia*); infestations are often abundant on such trees in towns and cities. In southern Europe this aphid is an important pest of crops such as lucerne (*Medicago sativa*).

108

108 Colony of *Aphis cytisorum* on *Laburnum*.

10

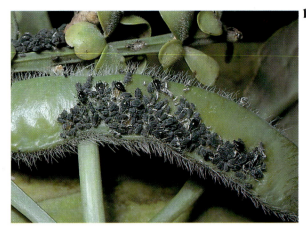

109 Colony of *Aphis cytisorum sarothamni*.

Aphis cytisorum Hartig
syn. *laburni* Kaltenbach
This greenish-black or blackish aphid infests the leaves and developing pods of *Laburnum* and Spanish broom (*Spartinum junceum*), colonies appearing somewhat greyish due to the presence of secreted wax (Fig. **108**). Infestations are often abundant on established hosts, but control measures are not worth while.

Aphis cytisorum sarothamni Franssen
This subspecies is very similar to *Aphis cytisorum* but occurs only on wild and cultivated broom (*Cytisus*), affecting the leaves and developing pods (Fig. **109**); cast skins accumulating amongst secreted honeydew on host plants are often more conspicuous on maturing pods than the live aphids (Fig. **110**). Although widespread, this aphid is not an important pest.

110 *Aphis cytisorum sarothamni* damage to *Cytisus*.

Aphis farinosa Gmelin
Small willow aphid
Dense ant-attended colonies of this widespread and often abundant aphid occur on the young shoots of sallow and willow (*Salix*), including ornamental cultivars (Fig. **111**). Breeding continues from spring to mid-summer, with sexual forms appearing from June or July onwards. The apterous viviparous females are 1.7–2.3 mm long, green and yellow mottled, with mainly pale, moderately long siphunculi (Fig. **112**). Males are characteristically reddish-orange; nymphs of this sex developing within colonies are sometimes mistaken for orange-coloured predatory midge larvae.

111 Colony of small willow aphid on *Salix*.

Aphis genistae Scopoli
Genista aphid
Widely distributed but local on cultivated Spanish gorse (*Genista hispanica*), forming dense, greyish colonies on the shoots. Heavy infestations, which may occur during the summer, weaken plants and spoil their appearance. Apterae are 1.4–2.6mm long, blackish-bodied (lightened by greyish wax) with relatively short siphunculi and an elongate cauda.

112 Small willow aphid, *Aphis farinosa*.

Aphis hederae Kaltenbach
Ivy aphid
This widespread and generally common, dark brown to blackish aphid is most often associated with ivy (*Hedera helix*) growing on old walls or tree trunks, infestations accumulating on the young leaves and shoots (Fig. **113**). Colonies also develop on ornamental ivy and on related house plants. Although commonly over-wintering in the egg stage, aphids may also survive the winter if conditions are favourable.

113 Colony of ivy aphid on *Hedera*.

Aphis ilicis Kaltenbach
Holly aphid
Infestations of this widely distributed, greenish-black to reddish-brown or greyish-brown aphid occur on new shoots and young foliage of holly (*Ilex*), causing significant leaf curling (Fig. **114**). Attacks tend to occur only on young plants and on severely pruned bushes or on hedges producing an abundance of new growth and, since colonies die out in the absence of young growth (mature foliage apparently being unsuitable for their continued survival), special control measures are rarely necessary. The winter is spent in the egg stage.

CONTROL If necessary, colonies on the young shoots of nursery plants can be eradicated by applying an aphicide; treatment of established plants is not worth while.

114 Colony of holly aphid on *Ilex*.

115 Colony of iris aphid, *Aphis newtoni*.

116 Earthen shelter around colony of iris aphid.

Aphis newtoni Theobald
Iris aphid

This dark green to brownish-black aphid is recorded only from various countries in north-western Europe, including England, Germany, the Netherlands and Scandinavia, where it is associated with *Iris*. Large numbers often build up on the lower parts of leaves (Fig. **115**) and, later, on the young inflorescences. Colonies are commonly attended by ants and are sometimes protected by ant-built earthen canopies (Fig. **116**).

CONTROL Infestations on irises growing in or near garden pools and ponds should be hosed off with water; where there is no risk of contaminating aquatic life, apply a contact aphicide.

117 Colony of green apple aphid on *Cotoneaster*.

Aphis pomi Degeer
Green apple aphid

Infestations of this generally common aphid often occur on *Cotoneaster*, crab apple (*Malus*), firethorn (*Pyracantha*), hawthorn (*Crataegus*), medlar (*Mespilus germanica*), rowan (*Sorbus aucuparia*) and other species of *Sorbus*, but they are of greater significance on fruiting apple trees. Heavy attacks on ornamentals occur most commonly from June or July onwards, dense ant-attended colonies developing on the underside of leaves and on the new shoots (Fig. **117**). The aphids are 1.3–2.3 mm long and bright green or yellowish-green, with short antennae and black or dark brown siphunculi. They secrete considerable quantities of honeydew, and foliage and shoots soon become blackened by the accumulation of sooty moulds. Heavy summer infestations cause leaf curl and check the growth of new shoots; although attacks on older shrubs are usually of little or no importance, damage to nursery stock may be considerable.

118 Colony of elder aphid, *Aphis sambuci*.

CONTROL Spraying with an aphicide in late May and again later in the season may be necessary on nursery stock and young plants; however, treatment of well-established shrubs is rarely justified.

Aphis sambuci Linnaeus
Elder aphid
Widespread and often abundant on elder (*Sambucus*), including ornamental forms. In spring, dense, ant-attended colonies develop on the young shoots (Fig. **118**), infestations being especially common on bushes growing in suburban areas. In July there is a migration of winged forms to the roots of various secondary hosts, including cultivated species of carnation (*Dianthus caryophyllus*) and saxifrage (*Saxifraga*); colonies on these summer hosts survive until the autumn, when a return migration to elder takes place. The winter is normally passed in the egg stage on elder but it is possible that a few aphids are also able to survive on the roots. Apterae on elder are 1.9–3.5mm long, greyish-green to yellowish-brown (coated with whitish or greyish wax), with dark, elongate and tapered siphunculi and a bluntly rounded cauda; apterae on summer hosts are smaller, and blue-green in colour.

Aphis schneideri (Börner)
Permanent currant aphid
Infestations of this local, but widely distributed, aphid are most numerous on blackcurrant (*Ribes nigrum*) but will also occur on the ornamental species *Ribes aureum*, causing distortion and tight bunching of the leaves. Adults are 1.2–2.2mm long and dark green, dusted with blue-grey wax (Fig. **119**). Colonies are typically ant-attended and especially damaging on young plants.

CONTROL Apply a winter wash during the dormant season; alternatively, spray with an aphicide in the spring, soon after bud-burst.

Aphis sedi Kaltenbach
Widely distributed on various members of the Crassulaceae, including cultivated species of *Kalanchoe* and *Sedum*. Infestations are often heavy on glasshouse and indoor plants, affected shoots becoming distorted and also coated in masses of sticky honeydew (Fig. **120**). The aphids are small (1.0–1.7mm long) and mainly dark green to blackish-green. They occur in large, ant-attended colonies, especially on the younger growth; infestations can also occur on aerial roots (Fig. **121**).

CONTROL Apply a suitable aphicide as soon as aphids are seen.

119 Colony of permanent currant aphid on *Ribes aureum*.

120 Colony of *Aphis sedi* on shoot of a cactus.

121 Colony of *Aphis sedi* on aerial roots of a cactus.

Dysaphis tulipae (Boyer de Fonscolombe)
Tulip bulb aphid
A common pest of stored bulbs and corms, including *Crocus*, lily (*Lilium*) and snowdrop (*Galanthus nivalis*), but especially *Gladiolus*, *Iris* and tulip (*Tulipa*); also a virus vector. Virtually cosmopolitan. Widely distributed in Europe.

DESCRIPTION **Apterous female:** 1.5–2.5 mm long; pale yellowish, greyish or pinkish-brown, dusted with white wax; siphunculi dark, tapered and relatively short; cauda small and triangular (Fig. **122**). **Alate:** 1.5–2.3 mm long; abdomen with a black dorsal mark and black siphunculi.

LIFE HISTORY This entirely parthenogenetic aphid occurs on bulbs and corms in store, breeding beneath the dried outer scales throughout the year if conditions are favourable. When infested bulbs or corms are planted out, aphid numbers build up rapidly on the young shoots; later, the flowers, flower spikes and developing seed pods may also be colonized. Colonies are sometimes ant-attended (cf. *Rhopalosiphoninus* spp., p. 74).

DAMAGE Infestations severely check and distort the growth of young shoots, and heavily infested bulbs and corms may fail to develop. A vector of lily symptomless virus and tulip breaking virus.

CONTROL Regular fumigation of bulbs and corms in store should keep this pest in check.

122 Colony of tulip bulb aphid, *Dysaphis tulipae*.

123 Colony of hawthorn-carrot aphid on *Crataegus*.

Dysaphis crataegi (Kaltenbach)
Hawthorn-carrot aphid
This aphid overwinters in the egg stage on hawthorn (*Crataegus*), where spring colonies of blackish aphids lightly dusted with wax (Fig. **123**) form conspicuous, deep-red pseudo-galls on the leaves (Fig. **124**). Winged aphids later migrate to wild and cultivated carrot (*Daucus carota*), where they initiate dense, ant-attended colonies on the tap root and leaf bases. Further winged forms then return to hawthorn, where winter eggs are laid.

124 Gall of hawthorn-carrot aphid on leaf of *Crataegus*.

Dysaphis sorbi (Kaltenbach)
syn. *brevirostris* (Börner)
Rowan aphid
Infestations of this aphid cause distortion of leaves of rowan (*Sorbus aucuparia*), infested shoots becoming contaminated with cast aphid skins, honey-dew and sooty moulds (Fig. **125**). Eggs overwinter, colonies then developing from April onwards, and culminating in the production of winged forms in late June or July. Apterae (*c.* 2mm long) on rowan are yellowish-brown to yellowish-green with pale siphunculi.

125 Rowan aphid damage to shoot of *Sorbus aucuparia*.

Brachycaudus helichrysi (Kaltenbach)
Leaf-curling plum aphid
An often serious pest of plum (*Prunus*), including various ornamental varieties, and a summer pest of herbaceous plants such as *Aster*, *Chrysanthemum*, *Cineraria*, forget-me-not (*Myosotis*) and periwinkle (*Vinca*). Cosmopolitan. Widely distributed in Europe.

DESCRIPTION **Apterous female:** 0.9–2.0mm long; brownish to yellowish-green; round and shiny; antennae short; siphunculi pale, short and flanged; cauda tongue-like.

LIFE HISTORY Eggs overwinter on *Prunus*, hatching in the early spring before bud-burst. At first, nymphs feed at the base of the buds but, eventually, as the buds open they move onto the new blossoms and foliage. Colonies develop throughout the spring, culminating in the production of winged forms in May. These aphids then migrate to summer herbaceous hosts, where colonies develop before a return migration to primary hosts in the autumn. Apterae on summer hosts are distinctly smaller and paler than those on primary hosts.

DAMAGE **Primary hosts:** infestations cause severe leaf curl, a yellow mottling of the foliage and distortion of the new growth and are especially harmful to nursery stock and young trees. **Secondary hosts:** plants are severely stunted and distorted; foliage also becomes yellow-mottled, affecting their overall appearance and quality.

CONTROL **Primary hosts:** on suitable hosts, apply a winter wash during the dormant period. Alternatively, spray with an aphicide at the white-bud stage; a post-flowering treatment may be necessary if reinfestation occurs. **Secondary hosts:** apply a contact or a systemic aphicide, preferably before flowers are open.

126 Colony of thistle aphid on *Senecio cineraria*.

Brachycaudus cardui (Linnaeus)
Thistle aphid
Minor infestations of this generally common aphid sometimes occur on ornamental plants, especially silver ragwort (*Senecio cineraria*) (Fig. **126**). The aphids are similar in appearance to *Brachycaudus helichrysi* but green to shiny brownish-black, with distinctive black siphunculi.

Hyadaphis passerinii (del Guercio)
syn. *lonicerae* Börner
Honeysuckle aphid
A generally common pest of honeysuckle (*Lonicera*). Virtually cosmopolitan. Widespread in Europe.

DESCRIPTION **Apterous female:** 1.3–2.3 mm long; dark bluish-green, with a waxy bloom; antennae and legs black; siphunculi black and swollen; cauda black and elongate (Fig. **127**). **Alate:** 1.3–2.3 long; abdomen green, mottled with darker green, with a dark patch at the base of the siphunculi.

LIFE HISTORY Colonies develop on the underside of honeysuckle leaves from the early spring onwards. Winged forms are eventually produced, and these migrate during the summer to umbelliferous hosts, especially wild hemlock (*Conium maculatum*), colonies on honeysuckle then dying out. A return migration to honeysuckle, the primary host, takes place in the autumn.

DAMAGE Heavy attacks can affect both shoot growth and flower development, infested hosts also becoming sticky with honeydew and fouled with sooty moulds (Fig. **128**).

CONTROL Apply a contact or a systemic aphicide as soon as infestations are seen.

Hyadaphis foeniculi (Passerini)
Fly-honeysuckle aphid
Virtually identical to the previous species, with which it is often confused, but associated with fly-honeysuckle (*Lonicera xylosteum*).

Coloradoa rufomaculata (Wilson)
Small chrysanthemum aphid
This very small, green aphid was introduced from America into northern Europe where it is now a widespread, but usually only minor, pest of glasshouse *Chrysanthemum*. The aphids infest the stems and underside of the leaves, including senescing foliage; they are also vectors of viruses, including chrysanthemum virus B. Apterae are 1.0–1.7 mm long with spatulate body hairs; the mainly dark siphunculi are very long (longer than the antennae) and swollen close to their apex; the cauda is triangular.

Longicaudus trirhodus (Walker)
Compact colonies of this pale yellowish-green (2.0–2.7 mm long) species develop on wild and cultivated columbine (*Aquilegia vulgaris*) during the summer months (Fig. **129**); meadow-rue (*Thalictrum*) is also a summer host. The aphids are unusual in possessing short siphunculi and a long cauda. Winged aphids, which have an irregular black mark

127 Colony of honeysuckle aphid, *Hyadaphis passerinii*.

128 Honeysuckle aphid damage to flower buds of *Lonicera*.

129 Colony of *Longicaudus trirhodus* on leaf of *Aquilegia*.

on the abdomen, migrate to rose (*Rosa*), the winter host; small colonies are produced on this woody host in the spring (Fig. **130**), before a return migration to summer hosts takes place in June.

Myzaphis rosarum (Kaltenbach)
Lesser rose aphid

A widely distributed, small, elongate-oval somewhat flattened, green to yellowish-green aphid, occurring on wild and cultivated rose (*Rosa*), especially climbers. The apterae are 1.0–2.2 mm long, with long, swollen, dark-tipped siphunculi and a long cauda (cf. *Chaetosiphon tetrarhodum*, p. 70). They feed mainly along the mid-rib on either side of the leaves, and are often hidden from view amongst leaflets that are still furled. Winged females occur during the summer; these are 1.2–2.0 mm long, and green or yellowish-green, with a darker mark on the abdomen. Infestations are usually noticed only when aphid numbers are unusually high. This species is not particularly damaging, and is usually kept in check by insecticides applied against more important rose-feeding aphids.

Elatobium abietinum (Walker)
Green spruce aphid

A major pest of ornamental and forest spruces (*Picea*), species of North American origin usually being more severely affected than Eurasiatic ones. Formerly restricted to North America but now widespread throughout Europe.

DESCRIPTION **Apterous female:** 1.5–1.8 mm long; green with red eyes; siphunculi long, cauda elongate.

LIFE HISTORY In the British Isles, and other areas with less hostile winters, this species is almost entirely parthenogenetic, wingless adults and nymphs occurring on spruce trees throughout the year. These aphids are most numerous from August to early June but details of their development can vary considerably from area to area. The aphids feed on the underside of the needles and are very active, often wandering along the shoots. Winged aphids appear from May onwards, and this brings an end to colony development. These winged aphids produce small numbers of nymphs, which aestivate for several weeks before feeding and completing their development. In many parts of mainland Europe a sexual phase occurs and the aphids overwinter in the egg stage.

DAMAGE Infestations do not occur on the new growth, but the aphids cause extensive yellowing of the older foliage (Fig. **131**), severe attacks also leading to a distinct bronzing (Fig. **132**) and premature loss of needles. Damage occurs from September onwards and, in mild conditions, can continue to

130 Colony of *Longicaudus trirhodus* on leaf of *Rosa*.

131 Green spruce aphid damage to needles of *Picea abies*.

132 Green spruce aphid damage to branch of *Picea abies*.

increase throughout the winter; in many cases, however, significant damage does not occur until the spring, reaching a peak in May.

CONTROL Spray with an aphicide when aphids are seen; in areas of high risk, best protection is afforded by spraying in late August or September.

Chaetosiphon tetrarhodum (Walker)
This small, pale green to yellowish-green aphid occurs on the young shoots and underside of leaves of rose (*Rosa*) but, unlike many rose-infesting species, the colonies are not ant-attended. Apterae are 1.0–2.6 mm long with capitate body hairs and a short cauda. Alatae are 1.2–2.4 mm long, greenish with a blackish head and thorax, and a black mark on the abdomen. In favourable conditions, colonies often persist throughout the year; they are especially damaging to climbing roses producing an abundance of new growth.

CONTROL Apply an aphicide as soon as infestations are seen, and repeat as necessary.

Liosomaphis berberidis (Kaltenbach)
Barberry aphid
This greenish-yellow, orange or pinkish aphid infests barberry (*Berberis*), colonies developing throughout the summer on the young shoots and underside of leaves (Fig. **133**). Apterae are 2.0–2.5 mm long with pale legs and antennae, and noticeably swollen siphunculi. Little or no damage is caused.

Cavariella aegopodii (Scopoli)
Willow-carrot aphid
This generally common aphid overwinters on willow, especially crack willow (*Salix fragilis*) and white willow (*Salix alba*). Infestations occur in the early spring on the young shoots (Fig. **134**). Winged forms are then produced, and these migrate in May to various umbelliferous hosts. Large colonies often develop on the summer hosts from late May to early July, before the production of winged forms and a return migration to the primary, winter hosts. Apterae (1.0–2.6 mm long) are green or yellowish-green, with swollen siphunculi and a pair of closely set hairs arising from a prominent tubercle on the eighth abdominal tergite; alatae (1.4–2.7 mm long) are green or yellowish-green, with a black patch on the abdomen.

Myzus cerasi (Fabricius)
Cherry blackfly
A generally common pest of cherry, including bird cherry (*Prunus padus*), wild cherry (*Prunus avium*) and various ornamental varieties. Eurasiatic; also found in Australasia and North America. Widely distributed in Europe.

133 Colony of barberry aphid on *Berberis*.

134 Colony of willow carrot aphid on *Salix triandra*.

135 Cherry blackfly damage to leaves of *Prunus avium*.

DESCRIPTION **Apterous female:** 1.5–2.6 mm long; shiny, dark brown to black; front of head emarginate; siphunculi black, moderately long and tapered. **Alate:** 1.4–2.1 mm long; blackish, but abdomen yellowish-brown with a large black dorsal patch. **Nymph:** purplish to blackish.

LIFE HISTORY Eggs, overwintering on the spurs and young shoots of cherry, hatch in March or April, before the white-bud stage. Colonies of wingless aphids then develop on the underside of the young leaves, populations building up rapidly. Winged forms are produced in June, at the climax of colony development, and these fly away to establish colonies on summer hosts such as bedstraw (*Galium*) and speedwell (*Veronica*). Colonies on cherry then decline and eventually die out, although breeding often continues throughout July. A return migration to cherry takes place in the autumn, where sexual reproduction takes place and, eventually, eggs are laid.

DAMAGE Colonies cause severe distortion of young shoots and leaves (Fig. **135**), affected foliage later turns brown or black; persistent infestations distort growth and may kill shoots. Attacks are especially serious on young trees and nursery stock.

CONTROL Apply a winter wash during the dormant period. Alternatively, spray with a contact aphicide during the spring, either at the white-bud stage or immediately after flowering; some systemic aphicides have proved phytotoxic to certain kinds of ornamental cherry.

Myzus ligustri (Mosley)
syn. *ligustri* (Kaltenbach)
Privet aphid
A locally common pest of privet (*Ligustrum vulgare*), and sometimes of importance on garden hedges in suburban areas. Widespread in mainland Europe; in the British Isles most numerous in southern England. Also present in North America.

DESCRIPTION **Apterous female:** 1.2–1.5 mm long; shiny yellow to yellowish-green; antennae long and green; siphunculi long, yellow to green, with dusky tips, and slightly swollen just beyond the mid-region; cauda yellowish and tapered. **Alate:** 1.5–1.8 mm long; yellowish to yellowish-green, with a brown head, brown thoracic lobes and brown abdominal crossbars, the latter often forming a distinct patch between the dark, moderately long, slightly swollen siphunculi.

136 Privet aphid damage to leaves of *Ligustrum*.

LIFE HISTORY Eggs are deposited on the shoots of privet in the early winter, hatching in the following spring. Colonies of aphids then develop on the foliage in tightly rolled leaves of the new shoots, reaching the peak of their development from June or July onwards. Winged forms occur at various times and those appearing on privet late in the season produce sexual forms which eventually deposit the overwintering eggs.

DAMAGE Infestations are especially numerous on regularly clipped hedges which provide an abundance of new growth, the individual leaves becoming discoloured and curling longitudinally to produce noticeable distortion on the young shoots (Fig. **136**). Attacks can be severe, affecting both the appearance and, if persistent, vigour of hosts; infestations may also lead to leaf death and premature defoliation.

CONTROL Apply an aphicide in the spring, as soon as infestations are seen.

Myzus ornatus Laing
Violet aphid
A generally common, polyphagous pest of herbaceous plants; often a persistent problem on protected ornamentals such as African violet (*Saintpaulia*), azalea (*Rhododendron*), *Begonia*, busy lizzie (*Impatiens sultani*), *Veronica* and violet (*Viola*). Cosmopolitan. Widely distributed in Europe.

DESCRIPTION **Apterous female:** 1.0–1.7 mm long; pale brownish-yellow or dull green, with paired, blackish markings on the thorax and abdomen; antennal tubercles distinctly convergent; siphunculi pale, cylindrical and moderately long. **Nymph:** pale green, with dark red eyes.

LIFE HISTORY This species is entirely partheno-genetic, occurring in small numbers on the leaves of various ornamental plants. Under protection, breeding continues throughout the year. The aphids produce considerable quantities of honeydew, which coats the foliage and upon which white, cast-off aphid skins often accumulate (Fig. **137**).

DAMAGE Direct feeding damage is of little importance, as the aphids do not form dense colonies. However, the vast quantity of honeydew produced is often a problem on pot plants, and the presence of cast-off nymphal skins on foliage and flowers can be unsightly.

CONTROL Apply an aphicide as soon as aphids are seen, and repeat as necessary; ensure that the product chosen is not phytotoxic, especially to flower petals. Insecticide fogs and plant dips can be used as alternatives to sprays or granules.

Myzus persicae (Sulzer)
Peach-potato aphid

An often abundant and very polyphagous pest of herbaceous plants, but of greatest significance as a vector of plant virus disease. Often present on ornamentals such as African violet (*Saintpaulia*), *Antirrhinum*, *Begonia*, *Calceolaria*, calla lily (*Richardia*), carnation and pink (*Dianthus*), *Chrysanthemum*, *Fuchsia*, hyacinth (*Hyacinthus*), morning glory (*Ipomaea*), nasturtium (*Tropaeolum*), *Phlox*, *Primula*, rose (*Rosa*), sweet pea (*Lathyrus odoratus*), tulip (*Tulipa*), violet (*Viola*), winter cherry (*Solanum capsicastrum*) and various cacti. Cosmopolitan. Widely distributed in Europe.

DESCRIPTION **Apterous female:** 1.2–2.5 mm long; pale green to yellowish-green; head with inwardly directed tubercules at base of antennae; siphunculi moderately long, with apical half slightly swollen, pale but dark tipped; cauda triangular. **Alate:** 1.4–2.3 mm long; head and thorax blackish-brown; abdomen green to yellowish-green and often pinkish, with a dark dorsal patch which includes a pale central mark.

LIFE HISTORY Although sometimes overwintering in the egg stage on nectarine and peach (*Prunus persica*), this aphid more frequently survives as adults and nymphs on herbaceous, secondary hosts such as outdoor brassicas and protected lettuce. The aphids breed parthenogenetically throughout much of the year, winged forms spreading infestations from host to host during the summer months. The aphids are very restless and frequently wander over the food plant, on flower crops often migrating

137 Violet aphid damage to leaves of *Begonia*.

138 Colony of peach-potato aphid, *Myzus persicae*.

upwards to invade the buds (Fig. **138**). Colonies reach their maximum development in July but are never very populous.

DAMAGE This aphid is a major vector of plant viruses, including carnation latent virus, chrysanthemum virus B and mosaic viruses of, for example, carnation, dahlia and orchids. Direct feeding is rarely significant, although the aphids may be damaging on certain protected crops, causing distortion of leaves, buds and flowers, and stunting of terminal shoots. Although populations are usually small, the mere presence at harvest of aphids on flower crops can affect marketability and is, therefore, unacceptable.

CONTROL Apply an aphicide as soon as infestations are seen, and repeat as necessary. Populations of this aphid have developed resistance to certain pesticides and this can restrict the choice of effective materials. Ensure that the product chosen is not phytotoxic.

Myzus ascalonicus Doncaster
Shallot aphid

A widely distributed, virtually cosmopolitan species; unknown before 1940 but now an often common pest of herbaceous plants, including ornamentals such as *Gladiolus*, lily (*Lilium*) and winter cherry (*Solanum capsicastrum*). The aphids breed asexually throughout the year, often surviving the winter on stored bulbs and corms, and on various glasshouse plants. In mild conditions colonies can also survive the winter out-of-doors. Winged migrants appear in the spring, and these spread infestations to various herbaceous summer hosts. Infestations lead to considerable distortion and malformation of the foliage or flower trusses of affected plants; subsequent growth from previously infested bulbs or corms in store may also be affected when these are planted out, the new shoots often being weak and noticeably distorted. Apterae (1.1–2.2mm long) are pale brown, greenish-brown or yellowish-brown, shiny and distinctly convex; the head is emarginate, with slightly convergent prominences, the siphunculi distinctly swollen towards the tip and the cauda bluntly triangular but barely visible from above.

CONTROL As for *Myzus ornatus* (pp. 71–72).

Cryptomyzus korschelti (Börner)

Infestations of this widely distributed species are associated with *Ribes alpinum*. The aphids cause noticeable discoloration and blistering of the foliage, and damage is often severe (Fig. **139**). The aphids occur on currant throughout the spring and early summer, having overwintered in the egg stage. Winged forms are produced during the summer, and these migrate to woundwort (*Stachys*), the secondary host, where breeding continues. There is a return migration to currant in the autumn. The aphids are delicate, shiny and rather plump, and whitish to pale orange (Fig. **140**); the siphunculi are relatively long and thin.

Pentalonia nigronervosa Coquerel
Banana aphid

A tropical and subtropical species, established in various parts of mainland Europe on hot-house palms; also occurs in hot-houses on certain other monocotyledonous hosts. The aphids are small (1.1–1.8mm long) and mainly dark brown, with stout siphunculi (Fig. **141**). Colonies develop on the underside of the leaves and are usually attended by ants.

139 Galls of *Cryptomyzus korschelti* on *Ribes alpinum*.

140 Colony of *Cryptomyzus korschelti* on *Ribes alpinum*.

141 Colony of banana aphid, *Pentalonia nigronervosa*.

Idiopterus nephrelepidis Davis
Fern aphid

An American species, now widely distributed in Europe on cultivated ferns, especially ladder fern (*Nephrolepis exalta*), maidenhair fern (*Adiantum capillus-veneris*), *Polypodium* and *Pteris* growing in heated conditions; infestations may also spread to African violet (*Saintpaulia*), *Cyclamen* and *Streptocarpus*.

DESCRIPTION **Apterous female:** 1.3 mm long; dark green to black, with capitate body hairs and whitish antennae; legs long, slender and mainly white; siphunculi white but blackish basally. **Alate:** 1.4–1.6 mm long; black to dark olive-green; wings mottled with dark brown.

LIFE HISTORY Aphids breed parthenogenetically throughout the year, producing winged forms in both spring and summer. Infestations are most common in glasshouses and hot-houses but can persist out-of-doors in mild districts.

DAMAGE Infested fern fronds are distorted, and may turn black and die.

CONTROL Aphicides should be used only if absolutely necessary, as ferns are often susceptible to chemicals; if treatment is required, spray with a contact, rather than with a systemic, product.

Rhopalosiphoninus latysiphon (Davidson)
Bulb and potato aphid

Infestations of this cosmopolitan species occur on plant roots and on tulip (*Tulipa*) bulbs in store, building up rapidly during forcing; they also occur on other stored bulbs, corms and tubers, including *Gladiolus*. When developing leaves are attacked they may turn brown and shrivel at the tips. The aphids (1.4–2.5 mm long) are plump and dark olive-green, with shiny black, strongly swollen siphunculi. Unlike *Dysaphis tulipae* (p. 66), which also attacks tulip bulbs, colonies are never attended by ants.

Rhopalosiphoninus staphyleae (Koch)
syn. *tulipaella* (Theobald)
Mangold aphid

Colonies of this widespread aphid also occur during the winter on stored bulbs and corms, including *Crocus*, day lily (*Hemerocallis*), lily (*Lilium*) and tulip (*Tulipa*). Aphids are 1.5–2.4 mm long and mainly olive-brown, with dark crossbands along the back and swollen siphunculi; the head and first two antennal segments bear numerous small spines.

142 Mottled arum aphid, *Aulacorthum circumflexum*.

Aulacorthum circumflexum (Buckton)
Mottled arum aphid

An often abundant, polyphagous pest of indoor and glasshouse ornamentals, including asparagus fern (*Asparagus plumosus*), *Begonia*, *Calceolaria*, calla lily (*Richardia*), *Cineraria*, *Chrysanthemum*, *Cyclamen*, *Freesia*, *Fuchsia*, lily (*Lilium*), *Petunia*, *Primula*, rose (*Rosa*), saxifrage (*Saxifraga*), tulip (*Tulipa*) (including dry bulbs in store), violet (*Viola*) and various orchids and ferns; also important as a virus vector. Virtually cosmopolitan. Widely distributed in Europe.

DESCRIPTION **Apterous female:** 1.2–2.6 mm long; shiny whitish to greenish-yellow, with blackish markings on the abdomen which often form a horseshoe-shaped pattern; antennae and legs mainly pale; siphunculi and cauda pale and elongate (Fig. **142**). **Alate:** 1.6–2.4 mm long; similar to aptera. **Nymph:** similar to adult but lacking the black body pattern.

LIFE HISTORY This species is entirely parthenogenetic. Breeding continues throughout the year if conditions are favourable but the aphids (Fig. **143**) are usually most numerous in the early spring. Winged forms occur occasionally, and these help to spread infestations from plant to plant; along with apterae, they are responsible for transmitting certain plant viruses. Infestations often occur on outdoor plants during the summer, especially in sheltered situations. This species secretes considerable quantities of honeydew.

DAMAGE Aphids occur on both foliage and flowers, and heavy infestations may prove directly harmful to many kinds of ornamental plants; attacks on cyclamen and certain lilies, e.g. calla lily, are especially damaging. Infested plants are often soiled by honeydew and sooty moulds; they may also be harmed by viruses, such as dahlia mosaic virus, primula mosaic virus and tulip breaking virus, which are commonly transmitted by this species.

CONTROL Apply a contact or a systemic aphicide as soon as infestations are discovered, and repeat as necessary. Ensure that the product chosen is suitable for use on the host plant and will not be phytotoxic.

Aulacorthum solani (Kaltenbach)
syn. *pseudosolani* (Theobald)
Glasshouse and potato aphid
In unheated situations, a generally common pest of glasshouse-grown crops, including ornamentals such as *Capsicum*, *Geranium*, *Pelargonium* and winter cherry (*Solanum capsicastrum*); also present in summer on various outdoor hosts. Virtually cosmopolitan. Widely distributed in Europe.

DESCRIPTION **Apterous female:** 1.8–3.0 mm long; pear-shaped, shiny greenish-yellow, with darker patches at base of siphunculi; antennae about as long as body; siphunculi long, slender and tapered, pale with dark tips and distinctly flanged (Fig. **144**). **Alate:** 1.8–3.0 mm long; head and thorax dark brown to black, abdomen yellowish-green marked with dark brown spots and crossbars. **Nymph:** yellowish-green and shiny, with dusky legs and antennae.

LIFE HISTORY All stages may overwinter, including eggs, but the aphids will also continue to breed parthenogenetically if conditions are favourable, producing both wingless and winged forms. In spring, infestations spread to various hosts, both indoors and outside, colonies reaching the peak of development in July but largely dying out by the autumn, except in favourable, protected habitats.

DAMAGE Infestations weaken plants; they can also spoil their appearance and affect quality.

CONTROL As for *Aulacorthum circumflexum*.

Acyrthosiphon pisum (Harris)
syn. *onobrychis* (Boyer de Fonscolombe); *pisi* (Kaltenbach)
Pea aphid
A generally common pest of certain cultivated legumes, including sweet pea (*Lathyrus odoratus*). Virtually cosmopolitan. Widespread in Europe.

DESCRIPTION **Apterous female:** 2.5–4.4 mm long; elongate, pale green to yellowish or pinkish, with

143 Colony of mottled arum aphid, *Aulacorthum circumflexum*.

144 Apterous females of glasshouse and potato aphid, *Aulacorthum solani*.

long antennae, legs and cauda, and very long siphunculi. **Alate:** 2.3–4.3 mm long; similar to aptera. **Nymph:** similar to adult but lightly dusted with wax.

LIFE HISTORY Adults and eggs overwinter on perennial plants such as clover (*Trifolium*), lucerne (*Medicago sativa*), sainfoin (*Onobrychis vicifolia*) and trefoil (*Lotus*). Colonies develop on these plants in the spring, winged forms migrating to peas and beans from May onwards. Immigrants infest the growing points, and infestations are often overlooked until aphid numbers build up. The aphids, which readily drop to the ground if disturbed, often form significant infestations in June and July, colonies usually persisting on such hosts into the autumn. The aphids are commonly preyed upon by various natural enemies, including coccinellids, predatory midge larvae and syrphid larvae, and these help to keep minor infestations in check.

DAMAGE Most damage is caused to the growing points during June and July, leaves becoming distorted and turning yellow. Pea enation mosaic virus, pea leaf roll virus and pea mosaic virus are also introduced and spread by this insect.

CONTROL Apply a contact or a systemic aphicide but only if significant infestations develop. Ensure that the chosen product will not be phytotoxic.

Acyrthosiphon pisum spartii (Koch)
This subspecies occurs on wild and cultivated broom (*Cytisus*), infestations often building up on the shoots and developing pods (Fig. **145**). Attacks, however, rarely if ever justify the application of an insecticide.

Acyrthosiphon malvae (Mosley)
syn. *pelargonii* (Kaltenbach)
Pelargonium aphid
Infestations of this medium-sized, pale green or yellowish aphid sometimes occur on glasshouse-grown *Pelargonium*, often in company with other species such as *Aulacorthum circumflexum* (p. 74). They also infest other ornamentals, including *Cineraria*, but are rarely important. Apterae are 2.5–3.0 mm long with very long antennae, divergent antennal tubercules, and elongate, tapered siphunculi and cauda; unlike *Macrosiphoniella* and *Macrosiphum* (q.v.) the siphunculi are not reticulated at the apex.

Microlophium primulae (Theobald)
Colonies of this pale yellow to pale yellowish-green aphid sometimes occur on cultivated *Primula*. The aphids congregate on the underside of the leaves amongst cast-off nymphal skins, which (like those of leafhoppers (see p. 28 *et seq.*) commonly remain attached to the hairy leaf surface (Fig. **146**). Although aphid numbers are sometimes large, foliage is not distorted and attacks are of little or no importance. Apterae are 2.2–2.5 mm long, with long, tapered siphunculi and relatively long legs; alatae are pale yellow with black abdominal markings.

Metopolophium dirhodum (Walker)
Rose-grain aphid
Widely distributed on rose (*Rosa*) from autumn to late spring, overwintering in the egg stage. Unlike the main rose-inhabiting species *Macrosiphum rosae* (p. 78), spring-formed colonies do not persist on rose beyond June, the aphids then migrating to cereals and grasses. *Metopolophium dirhodum* is usually unimportant on rose but in some years it is present in great profusion, and vast clouds of winged migrants are produced; in such circumstances, rose leaves and flower buds often become

145 Colony of *Acyrthosiphon pisum spartii* on *Cytisus*.

146 Colony of *Microphium primulae* on *Primula*.

147 Colony of rose-grain aphid on *Rosa*.

covered in white, cast-off nymphal skins, giving the superficial appearance of an outbreak of disease (Fig. **147**). Apterae are elongate (2–3 mm long), and mainly shiny yellowish-green, with a darker, mid-dorsal longitudinal stripe; the antennae, legs, siphunculi and cauda are mainly pale and relatively long. Alatae (1.6–3.3 mm long) have a uniformly green abdomen.

CONTROL Specific treatment against this species is rarely necessary, infestations usually being unimportant or kept in check by aphicides applied in the spring against *Macrosiphum rosae* (p. 78). When accumulations of cast nymphal skins are discovered, it is usually too late to justify application of an insecticide.

148

148 Colony of lupin aphid, *Macrosiphum albifrons.*

Macrosiphum albifrons Essig
Lupin aphid
An important North American pest of lupin (*Lupinus*). In Europe first reported in southern England in 1981; now more widely distributed and also present in mainland Europe, including Germany and the Netherlands.

DESCRIPTION **Apterous female:** 3.2–4.5 mm long; large, pale blue-green, with a white, waxy coating; antennae, legs, siphunculi and cauda all long; siphunculi pale brown with darker tips (Fig. **148**). **Alate:** 3.2–4.5 mm long; similar to aptera but the siphunculi entirely dark. **Nymph:** similar to adult but pre-aptera with uniformly dark siphunculi (Fig. **148**).

LIFE HISTORY Breeding colonies develop on all parts of host plants, including flower spikes (Fig. **149**), pods and senescing tissue; the aphids may also invade the roots. The reproductive rate is high and infestations build up quickly. Aphids overwinter in the crowns of mature plants, infestations spreading rapidly during the summer, following the production of winged forms.

149

149 Colony of lupin aphid on *Lupinus.*

DAMAGE Severest damage is caused to young flowers, with whole spikelets being aborted. Plants may be killed if infestations are heavy.

CONTROL Apply a contact or a systemic aphicide as soon as infestations are seen.

Macrosiphum euphorbiae (Thomas)
syn. *solanifolii* Ashmead
Potato aphid
A polyphagous North American species, first introduced into Europe in about 1917 and now cosmopolitan. Infestations occur commonly on ornamental plants, including calla lily (*Richardia*), carnation and pink (*Dianthus*), *Cineraria*, columbine (*Aquilegia vulgaris*), *Dahlia*, *Eucalyptus*, *Freesia*, *Gladiolus*, hollyhock (*Althaea rosea*), *Iris*, pot marigold (*Calendula*), snapdragon (*Antirrhinum*), sweet pea (*Lathyrus odoratus*), tulip (*Tulipa*) and valerian (*Valeriana*). Present throughout Europe.

DESCRIPTION **Apterous female:** 1.7–3.6 mm long; greyish-green to pink, and often shiny; spindle-shaped, with long antennae and legs; siphunculi very long, slender and curved; cauda long and finger-shaped. **Alate:** 1.7–3.4 mm long; similarly coloured to aptera but antennae, head, thorax and siphunculi usually yellowish-brown. **Nymph:** long-bodied and pale, with a dark, central, longitudinal stripe and a slight wax coating (Fig. **150**).

LIFE HISTORY Although sometimes overwintering in the egg stage on rose and related rosaceous hosts, most populations survive parthenogenetically in protected situations as adults or nymphs. Aphid numbers increase rapidly on various hosts from early spring onwards, infestations spreading to other plants in May and June as winged forms are produced. Attacks are especially common in unheated glasshouses and often develop on flower stalks (Fig. 150).

DAMAGE Infested plants may become stunted and distorted; the aphids are also capable of transmitting certain plant virus diseases, including freesia mosaic virus and pea enation mosaic virus, but are not important vectors.

CONTROL Apply a contact or a systemic aphicide as soon as infestations are discovered, and repeat as necessary. Ensure that the product chosen is suitable for use on the host plant and will not be phytotoxic.

Macrosiphum rosae (Linnaeus)
Rose aphid
An important and often abundant pest of rose (*Rosa*). Virtually cosmopolitan. Widely distributed in Europe.

DESCRIPTION **Apterous female:** 1.7–3.6 mm long; green to pink or reddish-brown; spindle-shaped with long antennae and legs; siphunculi long, black (N.B. pale in all other rose-infesting species) and tapered; cauda pale and elongate (Fig. 151). **Alate:** 2.2–3.4 mm long; green to pinkish-brown, distinctly marked with black along the sides of the abdomen.

LIFE HISTORY Overwintered eggs often occur on rose bushes, although adult aphids may also survive the winter if conditions are favourable. Substantial colonies build up during the spring and summer (Fig. 152), infested shoots often becoming smothered in the aphids. In summer, winged forms spread infestations to other rose bushes; they may also migrate to secondary hosts such as holly (*Ilex aquifolium*), scabious (*Knautia*) and teasel (*Dipsacus fullonum*). Rose bushes remain liable to invasion throughout the summer months, and new colonies are often present through to the autumn until their development is curtailed by the onset of cold weather.

DAMAGE Infestations check the growth of buds and new shoots, and commonly contaminate plants with sticky honeydew and sooty moulds. Foliage and flowers may also be disfigured.

150 Colony of potato aphid on *Calendula*.

151 Rose aphid, *Macrosiphum rosae*.

152 Colony of rose aphid on *Rosa*.

CONTROL Frequent applications of contact or systemic aphicidal sprays are necessary throughout the growing season to keep rose bushes free of infestation. Treatments should begin in April and continue as required at appropriate intervals during the summer. A systemic granular aphicide can give extended protection, and is often the ideal treatment for potted glasshouse plants.

Illinoia azaleae (Mason)

A common pest of container-grown azalea (*Rhodo-dendron*) in Europe, having been introduced from North America on infested plants. Heavy infestations reduce the vigour of host plants and will also cause defoliation. Attacks may also occur on open-bedded bushes growing in parks and gardens. Reproduction is typically parthenogenetic and viviparous, and is continuous so long as conditions remain favourable. Adults are medium-sized, spindle-shaped, deep green and shiny, with long, mainly pale, slightly swollen siphunculi (Fig. **153**).

Illinoia lambersi (MacGillivray)

A North American species, now well established on cultivated azalea (*Rhododendron*) in parts of Europe, including Denmark, England and the Netherlands, colonies developing upon the new shoots and flower buds (Fig. **154**). The aphids are similar in appearance to *Illinoia azaleae*, but are larger and either green, pink or yellow, the various colour forms often occurring together.

Dactynotus tanaceti (Linnaeus)

This uncommon aphid occurs occasionally on glass-house-grown *Chrysanthemum*, forming colonies on the underside of leaves. Apterae are 2.0–2.5 mm long, dark reddish-brown, with long antennae and long, black siphunculi and capitate body hairs.

Macrosiphoniella sanborni (Gillette)
Chrysanthemum aphid

A generally common pest of *Chrysanthemum*. Of eastern Asian origin but now cosmopolitan. Widely distributed in Europe.

DESCRIPTION **Apterous female:** 1.0–2.3 mm long; shiny, dark reddish-brown to blackish-brown; siphunculi black, short and stout; cauda black and elongate, slightly longer than siphunculi. **Alate:** 1.8–2.6 mm long; similar to aptera.

LIFE HISTORY This species occurs on cultivated chrysanthemum in or near greenhouses, reproducing parthenogenetically throughout the year. The aphids colonize the underside of leaves, and also infest the buds and flower stems. Winged forms are produced during the summer months, and these help to

153 Colony of *Illinoia azaleae* on *Rhododendron*.

154 Colony of *Illinoia lambersi* on *Rhododendron*.

spread infestations from place to place, but there are no sexual stages in the life-cycle.

DAMAGE Infested buds and flowers are malformed. The aphids can also transmit viruses such as chrysanthemum vein mottle virus and chrysanthemum virus B.

CONTROL Apply a contact or a systemic aphicide as soon as infestations are seen, and repeat as necessary; certain sprays can cause damage to the flowers.

Macrosiphoniella absinthii (Linnaeus)
syn. *artemisiae* (Buckton); *fasciata* Del Guercio

Infestations of this widely distributed species are sometimes reported on wormwood (*Artemisia*). The aphids are dark reddish-brown, partly coated in white wax to form an attractive and characteristic pattern; the siphunculi and cauda are black (Fig. **155**). The dark appearance of the aphids contrasts with the pale silvery foliage, so colonies are very noticeable; colonies are also ant-attended.

Macrosiphoniella oblonga (Mordvilko)

This widely distributed species occurs throughout the year on protected *Chrysanthemum*, but may also overwinter in the egg stage at the base of plant stems. The aphids feed singly on the underside of older leaves, and on the flower stalks. They are easily dislodged from host plants, dropping to the ground if disturbed. In addition to *Chrysanthemum*, infestations occur commonly out-of-doors on wild chamomiles (*Anthemis* and *Chamaemelum nobile*) and mayweed (*Tripleurospermum*). The aphids are 5 mm long, slender-bodied, green to pale green, with a darker dorsal longitudinal stripe; the legs are very long and thin, the siphunculi broad and moderately long but without an apical flange, and the cauda blunt and about as long as the siphunculi. Apterous males are reddish-brown, with a relatively small body and long legs.

155 Colony of *Macrosiphoniella absinthii* on *Artemisia*.

14. Family **HORMAPHIDIDAE**

A small group of often small, scale-like aphids with siphunculi much reduced or absent.

Cerataphis orchidearum (Westwood)

Orchid aphid

This small, reddish-brown to black aphid breeds parthenogenetically on glasshouse orchids. Apterae are 1.0–1.6 mm long, flattened and scale-like, with very short legs and antennae, short, stubby siphunculi, a knob-like cauda and a marginal fringe of wax. They occur on various species of orchid, including tropical climbing orchids (*Vanilla*); some hosts, including *Cattleya*, *Cymbidium*, *Cypripedium*, *Dendrobium* and *Odonotoglossum*, are especially liable to be damaged but the aphid is usually not of major significance.

15. Family **MINDARIDAE**

A small group of aphids with reduced, ring-like siphunculi and a rather long cauda; associated with conifers (cf. family Adelgidae, p. 87).

Mindarus abietinus Koch

Balsam twig aphid

This pale grey species forms small colonies during the summer on the succulent new growth of fir (*Abies*) trees. They secrete considerable quantities of waxen 'wool' and cause considerable distortion of the foliage. Colonies are sometimes noted on ornamentals but are of only minor importance.

Mindarus obliquus (Cholodkovsky)

Spruce twig aphid

Individuals of this species are small, elongate and white, and occur throughout the summer on the tips of new shoots of spruce (*Picea*), producing masses of bluish-white waxen 'wool'.

16. Family **PEMPHIGIDAE**

Aphids with terminal process of the antennae short, the eyes reduced to three facets, the siphunculi stumpy cones, pore-like or absent, and the cauda broadly rounded; body often with groups of well-developed wax glands. Associated with trees and shrubs, often forming galls and sometimes migrating in summer to herbaceous plants or grasses.

Eriosoma lanigerum (Hausmann)

Woolly aphid

Common throughout the world on apple (*Malus*) but also a pest of ornamentals such as *Cotoneaster*, firethorn (*Pyracantha*), hawthorn (*Crataegus*), Japanese quince (*Chaenomeles japonica*) and *Sorbus*.

DESCRIPTION **Apterous female:** 1.2–2.6 mm long; purplish-brown, covered with masses of white, mealy wax; body with numerous wax plates; antennae short; siphunculi pore-like.

LIFE HISTORY Nymphs, devoid of a wax coating, overwinter in crevices or under loose bark of suitable host plants, becoming active in March or April. By the end of May breeding colonies, now coated in masses of waxen 'wool', develop on the branches and spurs; they are often common around wounds and splits in the bark. Breeding continues until the autumn, with the production of a small number of winged forms in July; although a few eggs may be deposited in the autumn these fail to develop, the life-cycle being dependent upon the production of nymphs by viviparous females.

DAMAGE The masses of flocculent wax accumulating on heavily infested plants (Fig. **156**) are unsightly and can be a nuisance if infested branches overhang or lie alongside garden paths. The aphids also induce the development of disfiguring galls; these cause new growth to be malformed and the wood of older branches to split.

CONTROL Winter washes, applied during the dormant period, will give some control; alternatively, apply a drenching aphicidal spray in June but, because of the risk to bees, do not treat when plants are in flower.

Eriosoma ulmi (Linnaeus)
Currant root aphid

This generally common Eurasiatic species overwinters on elm (*Ulmus*), eggs hatching in the spring. Colonies of greyish aphids then develop on the underside of curled leaves, protected by flocculent masses of bluish or whitish wax (Fig. **157**). In June and July, winged forms emerge; they then migrate to currant and gooseberry (*Ribes*), where colonies of greyish or pinkish-grey aphids develop on the roots amongst masses of whitish wax (Fig. **158**). Winged aphids return to elm in the autumn where, after production of sexual forms, eggs are eventually laid. Infested elm foliage is severely curled and shoot growth checked; however, attacks, although often common in gardens, do not harm established trees. Summer infestations on the roots of ornamental currant (e.g. *Ribes sanguineum*) can affect growth of young plants and are of particular importance on containerized nursery stock.

CONTROL **Currant:** apply an insecticidal drench to the roots of affected plants as soon as infestations are found.

Kaltenbachiella pallida (Haliday)

This widespread species occurs during the summer on various Labiatae, including mint (*Mentha*), marjoram (*Origanum vulgare*), thyme (*Thymus*) and woundwort (*Stachys*), colonies of very small (0.9–1.3 mm long), creamish-white aphids developing on the roots amongst flocculent masses of white wax. Winged forms appear in the late summer and early autumn, and then migrate to elm (*Ulmus*) where, eventually, eggs are laid. The eggs hatch in the following spring, wingless aphids invading the unfurling foliage to initate large, pale green galls

156 Colony of woolly aphid, *Eriosoma lanigerum*.

157 Gall of currant root aphid on leaf of *Ulmus*.

158 Colony of currant root aphid on roots of *Ribes sanguinium*.

on the leaves. Each gall is coated in short whitish hairs, and develops at the base of the mid-rib as a conspicuous swelling (15–20 mm across) which protrudes both above and below the lamina (Fig. **159**). These galls mature in the early summer, winged aphids escaping through a stellate opening and eventually establishing colonies on summer herbaceous hosts. Galls on primary hosts, although disfiguring the leaves of ornamental specimen trees, do not affect plant growth.

Tetraneura ulmi (Linnaeus)
Elm leaf gall aphid
Widespread in north-western Europe, forming large, conspicuous bean-like galls on the expanded leaves of various kinds of elm (*Ulmus*). The galls develop from the upper surface of the leaf lamina, each attached by a narrow stalk (Fig. **160**). The galls, which may exceed 15 mm in height, are initially green but later turn cream and finally brown (Fig. **161**). Pale yellowish or yellowish-white aphids develop within the galls, amongst whitish masses of flocculent wax, with winged forms finally escaping in the summer through a conspicuous basal aperture (Fig. **162**). These aphids migrate to various grasses where subterranean, ant-attended colonies occur on the roots, with a return migration to the primary host taking place in the autumn. The elm galls are very conspicuous, and are often numerous on infested trees; however, they do not cause significant damage.

Pachypappa tremulae (Linnaeus)
syn. *populi* (Fabricius)
Spruce root aphid
A generally common and sometimes abundant pest of spruce, including Norway spruce (*Picea abies*) grown as Christmas trees. Often present on the roots of nursery trees, including container-grown stock, but also invading the superficial roots of older trees. Present throughout central and northern Europe.

DESCRIPTION **Apterous female [on spruce]:** 1.5–2.0 mm long; white to creamish-white, with a brownish head, antennae and legs; wax glands are prominent but there are no siphunculi (cf. *Rhizomaria piceae*, p. 83).

LIFE HISTORY This species overwinters in the egg stage on aspen (*Populus tremula*); grey poplar (*Populus canescens*) is also suitable as a primary host. In spring, eggs hatch and wingless aphids (fundatrices) develop on the smaller twigs; their offspring develop on the young shoots, causing a slight bending of the petioles. In June, winged forms appear and these migrate to plants in the

159 Gall of *Kaltenbachiella pallida* on *Ulmus*.

160 Young gall of elm leaf gall aphid on *Ulmus*.

161 Old galls of elm leaf gall aphid on *Ulmus*.

shade of spruces; their progeny enter the soil to initiate ant-attended colonies on the roots, the aphids developing amongst flocculent masses of white wax. In the early autumn, after one or two wingless generations, winged aphids are produced; these then fly to primary hosts, where sexual forms are produced. These hide beneath woolly, waxen coverings in crevices in the bark; after mating, females eventually deposit the winter eggs. Colonies of wingless aphids on the roots of spruces may also persist throughout the year.

DAMAGE **Aspen:** aphids cause slight leaf curl but this is of no importance. **Spruce:** heavy attacks cause death of plants.

CONTROL **Container-grown spruces:** apply a suitable aphicide as a soil drench. **Open-bedded spruces:** infested plants should be dug up and burnt, and a soil insecticide incorporated into the soil before replanting.

Stagona pini (Burmeister)
syn. *crataegi* Tullgren
Pine root aphid
Small colonies of this generally common and widely distributed pest occur throughout the year on the roots of pine (*Pinus*) trees. Infested parts of the root system become covered in masses of bluish-white waxen 'wool', amongst which may be found the small (1.5–2.0mm long) greyish-white aphids; individuals possess numerous wax glands but no siphunculi. Attacks are especially important on container-grown plants, nursery stock and transplants, severe infestations causing yellowing of the needles, wilting and death of plants. Hawthorn (*Crataegus*) is the primary host of this pest.

CONTROL **Container-grown pines:** apply a suitable aphicide as a soil drench. **Open-bedded pines:** infested plants should be dug up and burnt, and a soil insecticide incorporated into the soil before re-planting.

Rhizomaria piceae Hartig
syn. *vesicalis* (Koch)
This species, which overwinters on white poplar (*Populus alba*), breeds during the summer on the roots of spruce (*Picea*). It often occurs in company with *Pachypappa tremulae* (p. 82), from which it may be distinguished by the presence of pore-like siphunculi, but is usually present in lesser numbers. Infestations in spruce nurseries are sometimes damaging.

CONTROL As for *Pachypappa tremulae* (see above).

162 Mature galls of elm leaf gall aphid on *Ulmus*.

163 Poplar-buttercup aphid, *Thecabius affinis*.

Thecabius affinis (Kaltenbach)
syn. *ranunculi* (Kaltenbach)
Poplar-buttercup aphid
Generally common on black poplar (*Populus nigra*) and, less frequently, white poplar (*Populus alba* 'Pyramidalis'), the summer forms infesting the roots of various species of *Ranunculus*, especially creeping buttercup (*Ranunculus repens*). Eurasiatic. Widespread in Europe.

DESCRIPTION **Apterous female [on poplar]:** 4.0–4.5mm long; pale green to greyish-green, with dark brown head, antennae and legs, and coated with white waxy powder; legs short and robust (Fig. **163**). **Alate [ex: poplar]:** 2.6–3.0mm long; mainly blackish-brown, with a green to purplish-green abdomen; legs long and thin.

LIFE HISTORY Eggs overwintering on poplar trees hatch in the spring. Wingless aphids then develop within pale green, yellowish-green, pouch-like galls, each formed from a folded leaf (Fig. **164**). After a further generation of wingless aphids, winged forms are produced; these migrate from July onwards to the roots of buttercups. Subterranean colonies of wingless aphids then develop on these summer hosts, with winged forms appearing in the autumn and returning to poplar in October and November.

DAMAGE **Poplar:** galls affect the development of new shoots, disfiguring host plants; however, attacks on established trees are of little or no consequence. **Ranunculus:** heavy infestations on the roots produce unsightly masses of white wax and reduce the vigour of host plants; however, attacks on cultivated species are uncommon and rarely important.

Thecabius auriculae (Murray)
Auricula root aphid

This species infests the roots of auricula (*Primula auricula*) and certain other related plants. The aphids feed on the roots, and are especially damaging on potted plants and on those growing in glasshouses or under other protection. The aphids, which occur on the roots throughout the year, are small (1.3–1.5 mm long), pale yellowish-white or pale greenish-white, with brownish legs, and are coated with white mealy wax. Winged forms, which are larger (2.5 mm long) and mainly brown to green, also occur. The leaves of infested plants turn yellow, and may also become mottled and distorted, with considerable quantities of 'wool' accumulating around the collar and amongst the root system. Severely affected plants may wilt and die.

CONTROL Apply an insecticidal drench treatment; alternatively, dip the roots of affected plants in a dilute solution of insecticide and repot into clean compost.

Pemphigus bursarius (Linnaeus)
Lettuce root aphid

A generally common gall-forming aphid on Lombardy poplar (*Populus nigra* 'Italica') but most important as a pest of cultivated chicory and lettuce. Present throughout Europe; also found in North America and Australia.

DESCRIPTION **Apterous female:** 1.6–2.5 mm long; elongate-oval, yellowish-white, with a tuft of white wax posteriorly; small, dark abdominal wax plates clearly visible; siphunculi reduced to mere rings (Fig. **165**). **Alate:** 1.7–2.2 mm long; abdomen brownish-orange.

LIFE HISTORY This species overwinters in the egg stage on Lombardy poplar. In the spring, wingless

164 Gall of poplar-buttercup aphid on *Populus*.

165 Colony of lettuce root aphid in gall on *Populus*.

166 Mature gall of lettuce root aphid on *Populus*.

females initiate characteristic pouch-like galls by feeding on the petioles of leaves. Small colonies of aphids develop within these structures until, in the early summer, winged forms eventually escape through a beak-like opening (Fig. **166**) and migrate to lettuce and other Compositae, including wild lettuce (*Lactuca*) and sow-thistle (*Sonchus*). Colonies of wingless aphids then build up on the roots until the autumn, when winged forms return to poplar. These give rise to a generation of sexual forms which eventually deposit winter eggs.

DAMAGE **Poplar:** autumn populations distort and discolour the foliage but, as with the spring galling, such damage is not of significance.

Pemphigus phenax Börner & Blunck
syn. *dauci* (Goureau)
Carrot root aphid
This species overwinters as eggs laid on the bark of Lombardy poplar (*Populus nigra* 'Italica'). The eggs hatch in the spring; young aphids then move to the unfurling leaves where they induce the formation of mid-rib galls. Each gall becomes an elongate, somewhat wrinkled, reddish swelling (often tinged with yellow laterally), packed with numerous wax-secreting aphids. In summer, winged forms migrate to carrot (*Daucus carota*) and establish dense colonies on the roots, the aphids producing copious amounts of white waxen wool. There is a return migration to poplar in the autumn. Although the leaf galls on poplar trees may attract attention, infestations are not of importance.

Pemphigus populinigrae (Schrank)
syn. *filaginis* (Boyer & Fonscolombe)
Poplar-cudweed aphid
Widely distributed in association with black poplar (*Populus nigra*), eggs overwintering and hatching in the spring. Young aphids then form bulbous, pouch-like galls on the mid-rib of the leaves (Fig. **167**). Aphids (fundatrices) initiating the galls are 2.6–2.8mm long and mainly green to greyish-green, characteristically with four-segmented antennae; they develop amongst a mass of mealy wax (Fig. **168**), producing mainly brown-bodied alates, which eventually escape from the galls and fly to secondary hosts such as cudweed (*Gnaphalium*). There is a return migration to poplar in the autumn.

Pemphigus spyrothecae Passerini
Poplar spiral-gall aphid
Widely distributed and often common on poplar, especially Lombardy poplar (*Populus nigra* 'Italica'). Overwintered eggs hatch in the spring, and aphids feed on the leaf petioles which then curl into characteristic, distinctly spiral, pouch-like galls about 10mm long (Fig. **169**); the galls change from green

167 Gall of poplar-cudweed aphid on *Populus*.

168 Colony of poplar-cudweed aphid in gall on *Populus*.

169 Gall of poplar spiral-gall aphid on *Populus*.

through red to brown as they mature, and are often abundant on ornamental trees. Breeding continues within the galls until wingless adults emerge in August. These aphids then crawl to the bark, where small numbers of so-called pseudo-eggs are deposited. These hatch to produce sexual forms which eventually mate, the fertilized females then depositing winter eggs. The growth of shoots with a high proportion of galled leaves may be affected adversely but it is unlikely that treatment to control the pest would prove worth while.

Pemphiginus vesicarius (Passerini)
This species occurs mainly in the Apennines and parts of Asia but is also present in the South Tyrol, from northern Italy into Austria, forming distinctive galls on black poplar (*Populus nigra*). Galls develop at the base of the young shoots as irregular, brownish growths several centimetres across (Fig. **170**). Greyish-bodied nymphs develop within the galls amongst flocculent masses of whitish wax, winged adults eventually escaping through small openings which eventually appear in the wall and dispersing to unknown summer hosts. Vacated galls persist on host trees as distinctive, shrunken, black, woody structures (Fig. **171**). Such galls are sometimes found on young amenity trees but damage caused is of little or no importance.

17. Family THELAXIDAE
Aphids with the body flattened and the eyes reduced to three facets; antennae short, with a short terminal process, the siphunculi broad and cone-like, and the cauda knob-like or broadly rounded.

Glyphina betulae (Linnaeus)
Associated with birch (*Betula*), and often common throughout the summer and autumn months. The dark green, white-marked aphids cluster on the shoot tips and young leaves (Fig. **172**), often forming dense colonies. Wingless forms are 1.4–1.8 mm long, with short antennae, stout legs, a rounded cauda and small, cone-like siphunculi, and the body coated with short spine-like hairs; they occur most commonly from June to the end of August and again in October and November. Alates, which appear from mid-June to late July, are 1.3–1.8 mm long, and dark green with a dark brown to blackish head and thorax; the abdomen bears numerous tubercules and short, truncated siphunculi.

170 Young gall of *Pemphiginus vesicarius* on *Populus*.

171 Old gall of *Pemphiginus vesicarius* on *Populus*.

172 Colony of *Glyphina betulae* on *Betula*.

Thelaxes dryophila (Schrank)

syn. *quercicola* Westwood

Generally common on oak (*Quercus*), and often abundant on young trees. Eurasiatic. Widely distributed in Europe.

DESCRIPTION **Apterous female:** 2.2–2.8 mm long; reddish-brown to green, with marked segmentation and a pale dorsal line, lightly dusted with whitish wax; body flattened and hairy; eyes small, antennae short, siphunculi reduced to flat cones; cauda knob-like.

LIFE HISTORY Overwintered eggs hatch in the spring, dense colonies of wingless aphids developing on the young shoots and underside of oak leaves (Fig. 173) from May onwards. Winged forms are produced in the summer, and these spread infestations from tree to tree. Sexual forms appear in August, mated females eventually depositing winter eggs. The aphids produce copious quantities of honeydew and the colonies are strongly ant-attended.

DAMAGE Vigour of heavily infested shoots is reduced, persistent infestations on young trees being of particular significance.

CONTROL If infestations appear on nursery stock, apply a contact or a systemic aphicide.

173 Colony of *Thelaxes dryophila* on *Quercus*.

174 Gall of spruce pineapple-gall adelges on *Picea abies*.

18. Family ADELGIDAE

Conifer-feeding, aphid-like insects but, unlike true aphids, with short antennal segments, reduced wing venation, no siphunculi and entirely oviparous females; alatae with five antennal segments. Life-cycles are very complex, often involving a variety of different morphs and alternation of host plants.

Adelges abietis (Linnaeus)

A spruce pineapple-gall adelges

An often abundant pest of spruces, and commonly damaging to ornamentals, including Norway spruce (*Picea abies*) grown as Christmas trees. Present throughout the natural range of spruces and widely introduced with such plants to many other areas.

DESCRIPTION **Apterous female (pseudo-fundatrix):** yellowish-green to light green, with five-segmented antennae and five pairs of abdominal spiracles. **Alate female (gallicola):** 1.8–2.2 mm long; yellow with five-segmented antennae; fore wings 2.5–2.8 mm long.

LIFE HISTORY This adelgid completes its complete life-cycle on one host, forming characteristic pineapple-like galls on the young shoots of Norway spruce (Fig. 174). The galls vary considerably in size, those on the vigorous young shoots usually being the largest, but are most frequently about 15–20 mm long. In this species the number of adult morphs is restricted to just two: wingless pseudo-fundatrices and winged gallicolae. Gallicolae escape from mature galls during the late summer, usually during August and September, and then deposit eggs. These hatch into nymphs which overwinter close to the buds and eventually mature into the pseudo-fundatrices in the following spring. They then deposit pale yellow eggs, usually in batches of about 50, partly covering them with white waxen threads. Nymphs hatching from these eggs are the gall-inducing, gall-inhabiting forms. They feed at the base of the needles, causing a localized swelling which eventually develops into the familiar pine-apple-like gall, each gall containing numerous chambers within which groups of the pale pinkish-orange nymphs develop (Fig. **175**). This species exhibits only a limited migration, so that heavy populations tend to build up locally on affected trees.

DAMAGE The galls cause considerable distortion of the shoots but, unlike those formed by *Adelges viridis* (p. 90), they usually fail to encircle the shoot and do not, therefore, stop further growth. However, galling on trees no more than a few years old may be extensive, rendering young spruces grown as Christmas trees unmarketable.

CONTROL To prevent gall formation on young trees, spray with an appropriate insecticide during mild weather from November to late February.

Adelges cooleyi (Gillette)
Douglas fir adelges
A common pest of conifers, alternating between spruce (*Picea*) and Douglas fir (*Pseudotsuga menziesii*). Originally restricted to western North America but introduced into Europe in the 1930s, where it is now widely distributed.

DESCRIPTION **Alate female (gallicola on spruce):** 1.7–2.5 mm long; reddish-brown to purplish-black. **Alate female [ex: Douglas fir]:** 1.2–1.7 mm long; reddish-brown to purplish-black.

LIFE HISTORY Eggs deposited on Douglas fir by the overwintering forms hatch at bud-burst. The insects then develop on the needles, amongst conspicuous masses of white, waxen 'wool'. Both winged and wingless forms occur, the former departing in June for primary hosts, including Sitka spruce (*Picea sitchensis*). On spruce, sexual forms are produced. These eventually give rise to overwintering morphs which finally mature in the spring and produce a gall-forming generation. The latter initiate characteristically elongate galls on the shoots, each

175 Colony of spruce pineapple-gall adelges, *Adelges abietis*.

176 Douglas fir adelges damage to foliage of *Pseudotsuga menziesii*.

177 Douglas fir adelges on *Pseudotsuga menziesii*.

gall maturing in the summer. Winged forms (galli-colae) then migrate back to Douglas fir from late summer onwards where, eventually, wingless females settle down to overwinter on the underside of the leaves.

DAMAGE **Spruce:** the shoot galls cause a twisting of the new growth. **Douglas fir:** foliage is yellow-mottled (Fig. **176**) and tree growth is retarded; the foliage also becomes heavily encrusted with masses of white, fluffy wax and discoloured by sooty moulds which develop on the copious quantities of honey-dew excreted by the adelgids (Fig. **177**).

Adelges laricis (Vallot)
syn. *coccineus* (Ratzeburg)
Larch adelges

An often common pest of larch (*Larix*), including ornamentals, but mainly present on older trees; also gall-forming on spruce (*Picea*). Originally restricted to the central Alps but now widespread throughout Europe; also introduced into North America.

DESCRIPTION **Alate female [ex: larch]:** 1.0–1.5 mm long; dark green with a greyish-green head and thorax; fore wings *c.* 0.5 mm long. **Alate female [ex: spruce]:** 1.9–2.0 mm long; greyish to blackish; fore wings *c.* 0.6-0.7 mm long. **Egg:** greyish-black, coated in whitish wax.

LIFE HISTORY Blackish or purplish-grey nymphs overwinter on the bark of one-year-old larch shoots, often settled close to a bud. They mature in April, appearing blackish-grey, and then deposit unpro-tected clusters of eggs at the base of the leaf spurs. Nymphs developing from these eggs feed on the new foliage, turning into either winged or wingless adults about a month later. The former then disperse to primary hosts, including Sitka spruce, while the latter continue to breed on the larch needles, producing vast quantities of waxen 'wool' and sticky honeydew. This honeydew commonly accu-mulates as large globules amongst the protective canopy of waxen fibres (Fig. **178**). Individuals migrating to spruces give rise to wingless sexual forms and, in the following spring, gall-forming nymphs develop. Galls formed by this species on spruce are characteristically waxy, creamish and relatively small (*c.* 10–15 mm long) (Fig. **179**). They mature in the summer, and the fully winged galli-colae which emerge, fly back to larches, to settle on the needles and then deposit eggs.

DAMAGE **Larch:** infested foliage is disfigured with masses of white waxen 'wool' (Fig. **180**). The

178 Larch adelges on *Larix*.

179 Galls of larch adelges on *Picea*.

180 Larch adelges on *Larix*.

needles become discoloured and distorted, heavily infested trees often appearing blue; attacks may also lead to premature loss of needles and to die-back of shoots. **Spruce:** the galls sometimes prevent further shoot growth.

Adelges nordmannianae (Eckstein)
syn. *neusslini* (Börner); *schneideri* (Börner)
Silver fir migratory adelges

This dark brown to black adelgid is an important pest of young silver fir (*Abies alba*) trees; heavy infestations also occur on other kinds of firs, including Cilician fir (*Abies cilicica*) and Caucasian fir (*Abies nordmanniana*). Nymphs overwinter on the shoots, and eventually mature and deposit clusters of brownish-orange eggs. These hatch at about bud-burst, the nymphs feeding on the needles and commonly causing significant discoloration, stunting and distortion; severely infested trees may be killed. Infestations may also occur on the stems of host trees but, unlike *Adelges piceae* (see below), with which this species is commonly confused, very little waxen 'wool' is produced. Several generations may persist on fir-tree hosts but, in summer, some winged migrants also appear. These disperse to spruce (*Picea*) trees where, in the following year, small, rounded terminal galls (about 10 mm across) are eventually produced. Colonies of this species produce considerable quantities of honeydew.

Adelges piceae (Ratzeburg)
Silver fir adelges

Restricted to silver fir trees and most commonly present on species of North American origin (e.g. *Abies balsamea*, *A. grandis*, *A. lasiocarpa* and *A. procera*). Colonies occur mainly on the stems, the dark brown or blackish individuals producing noticeable quantities of waxen 'wool' (unlike *Adelges nordmannianae*, with which — when on *Abies alba* — this insect is commonly confused). Infested tissue often becomes distorted and swollen; in severe cases, growth is significantly affected and host trees may be killed.

Adelges viridis (Ratzeburg)
syn. *laricis* (Hartig)
A spruce pineapple-gall adelges

Generally common on spruces, especially Norway spruce (*Picea abies*) and Sitka spruce (*Picea sitchensis*), forming pineapple-like galls on the shoots; these are similar to those produced by *Adelges abietis* (p. 87) but tend to be more elongated and usually completely encircle the shoots

181 Gall of *Adelges viridis* on terminal shoot of *Picea*.

182 Mature galls of *Adelges viridis* on *Picea*.

(Fig. **181**), thereby stopping further growth (Fig. **182**). Winged individuals emerge from the galls in July (earlier than *Adelges abietis*), and then migrate to larch where eggs, the overwintering stage, are laid. On larch, the secondary host, pale green, wingless forms develop from mid-April onwards (earlier than other larch-infesting adelgids), causing a characteristic kinking of the needles (Fig. **183**). Eventually, pale green or yellow, winged forms are produced and these fly back to spruces, where their eggs hatch into summer, sexual forms. Their progeny overwinter and eventually produce nymphs which initiate the next generation of pineapple galls.

Pineus pini (Macquart)
Scots pine adelges

A widespread and common, but minor, pest of Scots pine (*Pinus sylvestris*); other two-needled pines are also hosts. Present throughout Europe; also introduced with the host plant to all other parts of the world.

DESCRIPTION **Apterous female (sisten):** 1.0–1.2 mm long; dark brown to dark red, and almost spherical; head and prothorax heavily chitinized; antennae three-segmented; abdomen with four distinct pairs of spiracles and an ovipositor. **Alate female (sexupara):** 1.0–1.2 mm long; mainly reddish-grey, antennae five-segmented; fore wings 1.4–1.7 mm long, hyaline with the veins often tinged with red. **Egg:** oval and orange.

LIFE HISTORY Nymphs, which hatch from eggs deposited in the autumn, overwinter and mature in the following March. They then deposit clusters of eggs, the progeny invading the new shoots and eventually developing into either winged or wingless forms. The former migrate to other sites from the end of May to late June, while the latter initiate a further wingless generation on the 'parent' plant; all females are oviparous. There may be three or more wingless generations each year, colonies persisting mainly on the bark of the shoots and youngest stems and producing masses of white, fluffy 'wool'.

DAMAGE The presence of waxen 'wool' disfigures host plants and is especially unsightly on young garden or nursery trees (Fig. **184**).

Pineus strobi (Hartig)
syn. *strobus* (Ratzeburg)
Weymouth pine adelges

Essentially similar to *Pineus pini* but slightly smaller and of North American origin, having been introduced into Europe in the mid-1800s. It infests Weymouth pine (*Pinus strobus*) and certain other five-needled pines (Fig. **185**), and is widely distributed on such hosts in parks and gardens. Overwintering nymphs mature and deposit eggs about a month later than *Pineus pini* and, in consequence, winged forms do not appear until mid-June. A second, wingless generation develops during the summer and further egg clusters are produced during August. Nymphs appearing in August and September eventually overwinter on the shoots. Considerable amounts of white, waxen 'wool' are produced on the stems and younger branches.

183 *Adelges viridis* damage to foliage of *Larix.*

184 Colony of Scots pine adelges on *Pinus sylvestris.*

185 Colony of Weymouth pine adelges, *Pineus strobi.*

19. Family PHYLLOXERIDAE

A small group of insects, structurally similar to the Adelgidae but alates with just three antennal segments.

Phylloxera glabra (von Heyden)
Oak leaf phylloxera
A generally common but minor pest of English oak (*Quercus robur*). Widely distributed in Europe.

DESCRIPTION **Apterous female:** 0.70–0.85 mm long; yellowish, marked with orange; oval-bodied and scale-like (Fig. **186**). **Egg:** 0.25 mm long; elongate-oval, yellow and shiny (Fig. **187**). **Nymph:** similar to adult but smaller.

LIFE HISTORY The winter is passed in the egg stage in crevices in the bark of oak trees. In spring, nymphs invade the new growth, to feed on the underside of the leaves. Once maturity is reached, eggs are deposited in small circles, all female individuals being oviparous. There are several asexual generations throughout the summer. However, in the autumn, smaller-bodied, winged sexual forms are produced, the mated females eventually depositing the winter eggs.

DAMAGE Infestations cause yellow and brown spotting of leaves (Fig. **188**); this sometimes leads to a general browning of the foliage, followed by premature leaf-fall. Infestations on mature trees are of little or no importance but persistent attacks on young trees can reduce plant vigour.

CONTROL Young trees should be treated in spring with a systemic or a contact aphicide, in the latter case ensuring that the spray reaches the underside of the leaves.

20. Family DIASPIDAE (armoured scales)

Body of female protected by a hard, scale-like covering formed from cast-off nymphal skins and wax.

Abgrallaspis cyanophylli (Signoret)
A tropical, polyphagous species. Often established as a persistent glasshouse pest in temperate regions, including northern Europe where it is most often noted on cacti; infestations also occur on orchids and palms. The scales are elongate-oval, 1–3 mm long, and mainly yellow to pale brown.

186 Colony of oak leaf phylloxera on *Quercus*.

187 Eggs of oak leaf phylloxera, *Phylloxera glabra*.

188 Oak leaf phylloxera damage to leaf of *Quercus*.

Aspidiotus nerii Bouché
Oleander scale

This widely distributed scale insect is a common glasshouse pest in northern Europe, infesting various ornamentals such as *Acacia*, asparagus fern (*Asparagus plumosus*), azalea (*Rhododendron*), *Cyclamen*, *Dracaena*, oleander (*Nerium oleander*) and palms. In favourable districts, infestations also occur out-of-doors on hosts such as Japanese laurel (*Aucuba japonica*); in continental Europe, the pest is often present on decorative container-grown oleanders. The female scales are flat, rounded (1–2 mm across) and whitish, with a central yellow spot; male scales are similar but smaller (cf. *Carulaspis*, p. 94). The scales often occur on the stems of host plants but are usually most abundant on the leaves (Fig. **189**). *Aspidiotus nerii* occurs throughout southern Europe, North Africa, the United States of America and Australasia, but the exact distribution and host range are uncertain due to frequent confusion with the closely related species *Aspidiotus hederae* (Vallot).

189 Colony of oleander scale on *Nerium oleander*.

190 Colony of rose scale on *Rosa*.

Aulacaspis rosae (Bouché)
Rose scale

This widespread species occurs most commonly on non-hybrid rose (*Rosa*), but will also attack blackberry (*Rubus fruticosa*), scales of both sexes developing on the stems (Fig. **190**). Female scales are 2.0–2.5 mm across, whitish to greyish and more-or-less oval, with the brownish-yellow nymphal exuviae usually placed excentrically at the margin; male scales are elongate (0.8 × 0.3 mm) and white. The scales spoil the appearance of host plants, infested wood developing a scurfy appearance; heavy attacks, which occur most commonly in sheltered sites, also reduce plant vigour. Eggs are laid under fertilized female scales in July or early August, orange-coloured nymphs appearing shortly afterwards. These nymphs wander over host plants but eventually become sedentary, male nymphs appear in the autumn but female nymphs usually not until the following spring. Individuals reach maturity by May or June, when mating occurs.

CONTROL Spray with an insecticide in late summer or early autumn, when mobile nymphs appear.

Diaspis boisduvalii Signoret
Orchid scale

A common and virtually world-wide pest of glasshouse-grown orchids, especially *Calanthe*, *Cattleya*, *Cymbidium* and *Epidendrum*; infestations also occur on palms. Widespread in Europe.

DESCRIPTION **Female scale:** 2 mm long; flat and oval, yellowish to brownish, and translucent. **Male scale:** 0.8–1.0 mm long; elongate, with three distinct longitudinal ribs, and coated with white waxen threads. **Adult female:** yellow to orange-yellow. **Adult male:** orange-yellow, with pale legs and antennae (the latter very long) and a moderately long caudal spine. **Egg:** minute, oval and yellow. **Nymph:** yellow.

LIFE HISTORY Under suitable conditions, breeding is continuous and infestations can build up rapidly, especially on the underside of the leaves of both young and established monocotyledonous pot-plants. Scales of the different sexes are often clustered separately, dense groupings imparting a distinctive waxen appearance to infested parts of plants.

DAMAGE Infestations disfigure and weaken host plants; the thick masses of white wax associated with male scales are especially unsightly.

CONTROL Treat with a suitable insecticide, applying two or three sprays at two- to three-week intervals.

Carulaspis juniperi (Bouché)
Juniper scale
A common pest of ornamental cypresses (*Chamaecyparis* and *Cupressus*), juniper (*Juniperus*) and *Thuja*; also associated with giant sequoia (*Sequoiadendron giganteum*). Widely distributed in Europe, including Belgium, France, Germany and the British Isles, but detailed distribution uncertain due to confusion with related species.

DESCRIPTION **Female scale:** 1.0–1.5mm across; rounded, slightly convex and whitish with an excentric yellow spot (Fig. **191**). **Male scale:** 0.5–1.0mm long; narrow with a distinct longitudinal rib; mainly white (Fig. **192**). **Egg:** oval; pale yellowish-white. **Nymph:** pale greenish-yellow.

LIFE HISTORY This species infests the foliage, shoots and fruits of various hosts, and is often present in considerable numbers. Scales are mature in the late autumn or early winter, when males appear and mating takes place. The scales are readily dislodged from the host plant, and the overwintering mated females are frequently taken by insectivorous birds. Eggs are deposited beneath the protective scales in the following May, the females then dying. The eggs hatch in June. Nymphs then swarm over the host plants before settling down to feed where they will eventually reach maturity.

DAMAGE Heavy infestations disfigure ornamentals and also cause considerable discoloration of the foliage, affected shoots and branches looking dull and distinctly unthrifty.

CONTROL Effective treatment requires the application of an insecticide in the summer, to coincide with the appearance of the active first-instar nymphs.

Carulaspis minima (Targioni-Tozzetti)
syn. *carueli* (Signoret)
This more southerly distributed species is virtually

191 Female scale of juniper scale, *Carulaspis juniperi*.

192 Male scale of juniper scale, *Carulaspis juniperi*.

indistinguishable from *Carulaspis juniperi*, but slight microscopical strucural differences are apparent in adults. It is recorded from various parts of Europe; confirmed hosts include *Chamaecyparis lawsoniana*, *Cupressus*, *Juniperus virginiana* and *Thuja*.

Chionaspis salicis (Linnaeus)
syn. *alni* Signoret; *populi* (Baerensprung)
Willow scale
Locally common on alder (*Alnus glutinosa*), ash (*Fraxinus excelsior*), sallow and willow (*Salix*); infestations also occur on broom (*Cytisus*), *Ceanothus*, elm (*Ulmus*), flowering currant (*Ribes sanguineum*), lilac (*Syringa vulgaris*), lime (*Tilia*), maple (*Acer*), poplar (*Populus*), privet (*Ligustrum vulgare*), spindle (*Euonymus*) and winter jasmine (*Jasminum grandiflorum*). Widely distributed in Europe; also present in North America.

DESCRIPTION **Female scale:** 1.5–2.3mm long; whitish to waxy yellowish-white, and irregularly pear-shaped. **Male scale:** 0.5–1.0mm long; white, elongate and very narrow, with a central and two lateral longitudinal ribs. **Adult male:** 1 mm long; orange to reddish-orange, with bright yellow legs and antennae; alate and apterous forms occur. **Nymph:** rusty-red, oval and flattened.

LIFE HISTORY Eggs are laid beneath the scales of fertilized females in August. They remain *in situ* throughout the winter, protected by the dead maternal scale. The eggs hatch in May, the newly emerged red-coloured nymphs often clustering in large, conspicuous groups on the bark of host plants (Fig. **193**). They soon disperse and settle down to feed, developing to maturity a few weeks later. Adult males occur from late June to mid-July, apterous forms considerably outnumbering those with wings.

DAMAGE Heavy encrustations impart a whitish appearance to the bark of host plants (Fig. **194**), and have a deleterious effect on the growth of young trees and shrubs.

CONTROL Apply an insecticide in May, immediately after egg hatch. On suitable deciduous hosts, application of a winter wash during the dormant period is also recommended.

Pinnaspis aspidistrae (Signoret)
Fern scale
Infestations of this scale occur commonly on the stems and fronds of glasshouse-grown *Aspidistra*, ferns and palms. Female scales are 2.5–3.0 mm long, reddish-brown and mussel-shaped; male scales are *c*. 1 mm long, white, ribbed and elongate.

Unaspis euonymi (Comstock)
Euonymus scale
A local pest of ornamental spindle, particularly *Euonymus japonica*. Widely distributed in southern Europe; also found further north, including certain coastal areas in southern England.

DESCRIPTION **Female scale:** 2–3 mm long; brown and mussel-shaped, oyster-shaped. **Male scale:** 2.0–2.5 mm long; pale yellow to white and elongate, with three longitudinal ridges.

LIFE HISTORY There are two generations annually, eggs hatching and nymphs swarming on host plants in June and in early September. The nymphs eventually settle down to feed on the foliage and stems of host plants, and finally mature (Fig. **195**). Adult males then emerge and mating takes place with young female scales.

193 Nymphs of willow scale on *Salix*.

194 Colony of willow scale on *Fraxinus*.

195 Colony of euonymus scale, *Unaspis euonymi*.

196 Euonymus scale damage to leaves of *Euonymus*.

197 Fig mussel scale damage to leaf of *Ficus carica*.

DAMAGE Attacks often cause leaf discoloration (Fig. **196**) and lack of vigour, heavy infestations causing severe decline and, eventually, death of plants.

CONTROL Spray with an insecticide in early summer or autumn as soon as mobile nymphs are seen.

Lepidosaphes conchyformis (Gmelin in Linnaeus)
syn. *ficus* (Signoret)
Fig mussel scale

This species occurs out-of-doors in southern Europe, the scales often covering the underside of the leaves of common fig (*Ficus carica*) and causing a noticeable pale mottling of the foliage (Fig. **197**). The scales are about 2 mm long, relatively narrow, straight or slightly curved and slightly expanded posteriorly (Fig. **198**). Infested plants are also reported in northern Europe but here the insect is able to survive only under artificial conditions.

198 Colony of fig mussel scale, *Lepidosaphes conchyformis*.

Lepidosaphes machili (Maskell)
Cymbidium scale

In northern Europe, infestations of this scale insect are sometimes noted on the leaves and stems of glasshouse-grown *Cymbidium*. The mature scales are about 2 mm long, and mussel-shaped but rather narrow. Although disfiguring host plants, they are rarely numerous and are of only minor importance.

Lepidosaphes ulmi (Linnaeus)
syn. *pomorum* (Bouché)
Mussel scale

Generally abundant throughout most of Europe on various trees and shrubs, including fruit trees and ornamentals such as box (*Buxus sempervirens*), *Ceanothus*, *Cotoneaster*, crab apple (*Malus*), hawthorn (*Crataegus*), heather (*Erica*), Japanese quince

199 Mussel scale, *Lepidosaphes ulmi*.

(*Chaenomeles japonica*) and rose (*Rosa*). The mature female scales are 2.0–3.0 mm long, elongate mussel-shaped and grey to yellowish-brown (Fig. **199**). They often encrust the bark of mature host plants but cause little or no damage. Eggs overwinter, protected by the dead remains of the maternal scale, and hatch in the following late May or June. Nymphs then crawl over the branches and trunks before settling down to feed. They reach maturity by the end of July, eggs being laid in late August and September.

21. Family **COCCIDAE** (soft scales, wax scales)

Body of female forming a hard, smooth or wax-covered, often tortoise-shaped, scale.

Chloropulvinaria floccifera (Westwood)
Cushion scale
An often common pest of glasshouse-grown *Camellia japonica* and orchids; in favourable areas, infestations also occur out-of-doors. Probably of southern European origin but nowadays found in many parts of the world. Widespread in Europe.

DESCRIPTION **Female scale:** 2.5 × 2.0 mm; oval and yellowish, surmounting a long (10–15 mm), thin, white egg-sac. **Egg:** minute, oval and pinkish.

LIFE HISTORY Infestations occur on the foliage and shoots throughout the year with, in suitably warm conditions, a succession of generations each year. Elsewhere, there is just one generation annually, immature scales overwintering and the characteristic egg-sacs being secreted in April and May soon after mating has occurred. Mature females perish soon after completing their egg-sacs, falling away from the host plant before the eggs hatch.

DAMAGE Attacked plants are often disfigured by accumulations of flocculent waxen 'wool'.

CONTROL Egg-sacs can be picked off and destroyed. If required, apply an insecticide immediately after egg hatch.

Pulvinaria regalis Canard
Horse chestnut scale
An American species, unknown in Europe before the 1960s. Now widely distributed in southern England and north-west France on urban deciduous trees and shrubs, including bay laurel (*Laurus nobilis*), dogwood (*Cornus*), elm (*Ulmus*), horse chestnut (*Aesculus hippocastanum*), lime (*Tilia*), *Magnolia*, maple (*Acer*), *Skimmia japonica* and sycamore (*Acer pseudoplatanus*).

200 Horse chestnut scale, *Pulvinaria regalis*.

201 Colony of horse chestnut scale, *Pulvinaria regalis*.

DESCRIPTION **Female scale:** 7 mm long; brown; roundedly triangular with a distinct posterior cleft (Fig. **200**). **Male scale:** 3 mm long; similar to female scale but smaller, narrower and lighter brown. **Egg:** minute, oval and whitish. **Nymph:** oval, flat and brownish, with a distinct posterior cleft.

LIFE HISTORY Unlike most scale insects, this species displays considerable mobility during its developmental stages. Eggs hatch in June to July, and the first-instar nymphs, or crawlers, then migrate to the underside of leaves to begin feeding. In September, when about 2 mm long, the nymphs move onto the twigs where feeding continues throughout the winter, but growth is slow. After bud-burst, however, nymphal development is rapid and size differences soon become obvious between the smaller males (if present) and larger females, with dense colonies developing on suitable hosts (Fig. **201**). Individuals become mature by May. Before egg laying, the mature females migrate from the twigs to the main branches and trunks. At this time, female scales may accidentally drop from host trees, and may then occur on various other nearby plants or objects. Eggs, up to 2,000 per female, are deposited beneath the scales, mainly on the branches and trunks of host plants but also on walls, pavements and elsewhere. Although males occur, females are capable of reproducing parthenogenetically and usually form the bulk of populations.

DAMAGE Although some hosts, especially limes and maples, are colonized extensively, the scales appear to have little or no effect on tree growth. Their presence on amenity or ornamental trees, however, is unsightly and commonly causes concern, such trees often bearing whitish marks on the bark long after the dead scales have fallen away.

CONTROL On deciduous trees, apply a winter wash during the dormant season; alternatively, if practical, apply an insecticide in the summer to kill the young nymphs.

Pulvinaria ribesiae Signoret
Woolly currant scale
Although associated mainly with fruiting currant and gooseberry, infestations of this widely distributed but local species also occur occasionally on flowering currant (*Ribes sanguineum*). Infested bushes are disfigured, especially during May and June when large, white egg-sacs are present beneath the mature female scales (Fig. **202**) and strands of wind-blown 'wool' become distributed over the shoots and branches; the foliage is also blackened by sooty moulds which develop throughout the summer on honeydew secreted by the insects. The nymphal scales are yellowish-brown, elongate-oval and slightly convex; after a brief wandering stage in the early summer, they settle on one-year-old wood to complete their development in the following spring. The mature female scales are dark brown to blackish, 4–6 mm long and oval with a distinct posterior cleft.

CONTROL Apply a winter wash during the dormant period or spray with an insecticide in the summer, when first-instar nymphs are active.

Pulvinaria vitis (Linnaeus)
Woolly vine scale
Very similar to *Pulvinaria ribesiae* but chestnut-brown and slightly larger; occurs most commonly on birch (*Betula*), but can also develop on other trees and shrubs such as alder (*Alnus*), *Cotoneaster*, hawthorn (*Crataegus*) and willow (*Salix*). Although sometimes of significance on glasshouse-grown hosts, attacks on outdoor ornamentals are rarely important.

202 Woolly currant scale, *Pulvinaria ribesiae*.

203 Hydrangea scale, *Eupulvinaria hydrangeae*.

Eupulvinaria hydrangeae (Steinweden)
Hydrangea scale
A recently recognized pest of ornamental trees and shrubs in parts of northern Europe, including Belgium, France, Germany and the Netherlands, apparently having spread northwards from southern Europe. Host plants include, field maple (*Acer campestre*), *Hydrangea*, lime (*Tilia*), plane (*Platanus*), *Viburnum* and various rosaceous trees and shrubs. The brownish female scales occur on the shoots, branches and trunks, their white egg-sacs developing during the summer (Fig. **203**) and active first-instar nymphs later migrating over the trees before settling down to feed and develop.

Coccus hesperidum Linnaeus
Brown soft scale
A very common, polyphagous and virtually worldwide pest of glasshouse ornamentals; also attacks outdoor plants growing in favourable situations, including many parts of mainland Europe and southern England. Commonly infested hosts include azalea (*Rhododendron*), bay laurel (*Laurus nobilis*),

Camellia, Citrus, Clematis, Escallonia, Ficus, Geranium, Hibiscus, holly (*Ilex*), ivy (*Hedera*), oleander (*Nerium oleander*), poinsettia (*Euphorbia heterophylla*), *Stephanotis*, *Viburnum* and various ferns.

DESCRIPTION **Female scale:** 3.5–5.0 mm long; very flat and oval; translucent-yellow to brown, with an often distinct median longitudinal ridge and rib-like markings.

LIFE HISTORY This species is viviparous, and usually parthenogenetic, each female producing about a thousand nymphs over a period of 2–3 months. The young nymphs wander over host plants for a few days before settling down to feed on the leaves, individuals clustering along the mid-rib and other major veins; the scales commonly overlapping one another, forming dense colonies (Fig. **204**). Breeding is continuous, so long as conditions remain favourable, the complete life-cycle from birth to maturity occupying about two months at average glasshouse temperatures.

DAMAGE The scales secrete considerable quantities of honeydew, foliage beneath infested leaves often becoming severely blackened by sooty moulds, which spoils the appearance of ornamentals and can also check growth; the honeydew-secreting scales are also attended by ants.

CONTROL Apply an insecticide to coincide with the appearance of young nymphs, and repeat as necessary.

Lichtensia viburni Signoret

A locally distributed species, occurring mainly on ivy (*Hedera helix*) but also associated with *Viburnum tinus*; minor infestations sometimes occur on ornamental plants and hedges in England, Wales and parts of mainland Europe. The female scales are 3–5 mm long, brown and somewhat fleshy (Fig. **205**). In May, the females mature and produce large numbers of whitish to pale brownish-white eggs; the eggs are contained in prominent (5–6 mm long) egg-sacs, composed of sticky masses of white waxen 'wool', which almost completely cover the maternal scales (Fig. **206**). Eggs hatch in late June, the emerged nymphs soon attaching themselves to both sides of expanded leaves to begin feeding. Development is relatively slow, with the scales maturing in the following spring. Adult males usually appear in late April and May. The scales from which they emerge are 2.0–2.5 mm long, white and elongate-oval, with distinct longitudinal ridges. This pest produces considerable quantities of honeydew upon which disfiguring sooty moulds develop.

204 Colony of brown soft scale, *Coccus hesperidum*.

205 Female scale and egg mass of *Lichtensia viburni*.

206 Female scales of *Lichtensia viburni*.

Eulecanium tiliae (Linnaeus)

syn. *capreae* (Linnaeus); *coryli* (Linnaeus)
Nut scale
A locally common pest of ornamental trees and shrubs, including alder (*Alnus*), *Ceanothus*, Cotoneaster, elm (*Ulmus*), firethorn (*Pyracantha*), hawthorn (*Crataegus*), hornbeam (*Carpinus betulus*), horse chestnut (*Aesculus hippocastanum*), lime (*Tilia*), oak (*Quercus*), rose (*Rosa*), spindle (*Euonymus*) and sycamore (*Acer pseudoplatanus*). Widely distributed in Europe; also present in North America.

DESCRIPTION **Female scale:** 5–6 mm across; dark chestnut-brown to light brown or greyish-brown, and strongly convex (Fig. **207**). **Male scale:** 2.0–2.5 mm long; greyish, elongate-oval (Fig. **207**). **Adult male:** reddish-crimson, with a relatively short caudal spine and a pair of long caudal filaments. **Egg:** pale yellowish-white. **Nymph:** pinkish to orange-yellow.

LIFE HISTORY Adult males emerge in late April and early May, eggs eventually being deposited by fertilized females about a month later. Eggs hatch towards the end of the summer, the nymphs wandering over host plants before settling down. Young overwintering male and female scales are similar in appearance, and about 1.5–2.0 mm long, but the more elongate appearance of the former becomes obvious in the spring as development recommences. This species may also breed parthenogenetically.

DAMAGE Heavy infestations retard growth, and host plants may be killed. Minor attacks are of little or no significance.

CONTROL On deciduous trees, apply a winter wash during the dormant period; alternatively, spray with an insecticide immediately after egg hatch.

Parthenolecanium corni (Bouché)

Brown scale
A generally common pest of trees and shrubs, including ornamentals such as *Ceanothus*, Cotoneaster, crab apple (*Malus*), *Elaeagnus*, *Escallonia*, firethorn (*Pyracantha*), flowering currant (*Ribes sanguineum*), honeysuckle (*Lonicera*), Japanese quince (*Chaenomeles japonica*), *Magnolia*, ornamental cherry (*Prunus*), rose (*Rosa*) and *Wisteria*. Present throughout Europe and many other parts of the world.

DESCRIPTION **Adult female scale:** 4–6 mm long; more-or-less oval, very convex and roughened; chestnut-brown and often shiny (Fig. **208**). **Egg:** minute, oval, whitish and shiny. **Nymph:** oval, flat, pale greenish to orange or brownish.

207 Colony of nut scale on *Acer pseudoplatanus*.

208 Female scales of brown scale, *Parthenolecanium corni*.

LIFE HISTORY In most situations, eggs are laid beneath the female scales in May or June. They hatch from mid-June onwards, the young nymphs (often called 'crawlers') wandering to the young leaves and shoots to begin feeding. Second-stage nymphs, which are also mobile, appear in August. They continue feeding on the young growth but in the autumn they migrate to the branches and twigs to settle down for the winter, gradually changing colour from greenish to brownish. Activity is resumed in the following spring, individuals settling permanently and growing rapidly, the back becoming increasingly hardened and convex to form the familiar protective scale. Although most races of this species are parthenogenetic, males occur on some hosts in certain areas. Also, details of the life-cycle vary according to conditions, two or three generations sometimes occurring annually under glass.

DAMAGE Infestations disfigure and weaken host plants, and often cause premature leaf-fall.

CONTROL As for *Eulecanium tiliae* (p. 100).

Parthenolecanium persicae (Fabricius)
syn. *crudum* (Green)
Peach scale
Infestations of this scale insect occur on various ornamental plants, including *Citrus*, *Ficus*, honeysuckle (*Lonicera*), false acacia (*Robinia pseudoacacia*), ornamental cherry (*Prunus*) and rose (*Rosa*). In northern Europe, attacks are usually heaviest under glass or on hosts growing against sheltered, sunny walls. Adult female scales (5–6 mm long) are shiny brown, with a distinct anal cleft and a slight longitudinal keel. The nymphs are elongate-oval, yellowish-brown to brownish-orange, marked with dark brown, and rather flat; long, straight glass-like strands of silk radiate outwards from the body. Infested stems become coated in scales (Fig. **209**), the insects secreting considerable quantities of honeydew and the foliage eventually becoming disfigured by accumulations of sooty moulds.

CONTROL Apply an insecticide immediately after egg hatch. Winter washes, applied during the dormant season, are also recommended on deciduous outdoor hosts.

Parthenolecanium pomeranicum (Kawecki)
Yew scale
Similar in appearance and biology to *Parthenolecanium corni* (see above) but present only on yew (*Taxus baccata*), the scales occur mainly on the

209 Colony of peach scale, *Parthenolecanium persicae*.

210 Yew scale on *Taxus baccata*.

leaves (Fig. **210**). Infestations are unsightly, sooty moulds developing on honeydew secreted by the scales; the pest is also directly harmful, heavy attacks leading to defoliation of hedges and specimen trees. The insect is widely distributed in mainland Europe; in Britain it is most often noted in southern England.

Saissetia coffeae (Walker)
syn. *hemisphaericum* (Targioni-Tozzetti)
Hemispherical scale
An often abundant glasshouse scale, infesting a wide variety of hosts, including asparagus fern (*Asparagus plumosus*), *Begonia*, carnation (*Dianthus caryophyllus*), *Ficus*, oleander (*Nerium oleander*), *Stephanotis* and various ferns and orchids. Infestations rarely cause significant damage, especially if plants are otherwise healthy and well-tended, but hosts are often contaminated by honeydew upon which sooty moulds develop. Eggs and nymphs

occur at all times of the year, even in unheated glasshouses; reproduction is entirely parthenogenetic, each female depositing up to 2,000 eggs in a tight cluster beneath the protection of the scale and then dying. The rounded to oval female scales are 2–3 mm across, strongly convex and reddish-brown to blackish (Fig. 211); oval forms are slightly larger, sometimes exceeding 4 mm in length.

CONTROL The scales are very noticeable and, if necessary, can be removed from host plants by hand without causing damage; eggs clustered beneath the scales should also be destroyed.

22. Family ERIOCOCCIDAE

Cryptococcus fagisuga Lindinger
syn. *fagi* (Baerensprung)
Beech scale
A generally abundant pest of beech (*Fagus sylvatica*). Widely distributed in Europe; accidentally introduced into North America in the late 1800s, where it has since become a significant pest.

DESCRIPTION **Adult female:** 0.75–1.0 mm long; hemispherical and pale lemon-yellow, coated with white flocculent threads of wax (Fig. 212). **Egg:** 0.15 mm long; pale yellow (Fig. 212). **First-instar nymph:** 0.25 mm long; lemon-yellow, with three pairs of legs and five-segmented antennae.

LIFE HISTORY This species is parthenogenetic and has a single generation. Eggs are laid from June to August, hidden under the protective waxen wool in small string-like clusters of up to eight. They hatch 6–7 weeks later. The very active first-instar nymphs ('crawlers') either burrow beneath the remains of adjacent dead scales or swarm over the trunk and branches before settling down in suitable bark crevices. They then overwinter before moulting to the entirely sedentary second nymphal stage. These second-instar nymphs eventually moult to the adult stage, individuals usually reaching maturity in early June. The scales secrete considerable quantities of white, waxen 'wool', and often cover considerable areas of the main trunks and branches (Fig. 213). On heavily infested sections of bark, the scales commonly overlap the shrivelled, black remains of previous generations.

DAMAGE Scale-encrusted bark disfigures ornamental trees, but infestations usually have little or no direct effect on the growth or well-being of their hosts. However, in some regions (e.g. Denmark and North America) attacks are associated with pathogenic fungi (*Nectria* spp.) and lead to the decline and eventual death of affected trees.

211 Hemispherical scale on *Encephalartos ferox*.

212 Colony of beech scale, *Cryptococcus fagisuga*.

213 Colony of beech scale on *Fagus*.

Gossyparia spuria (Modeer)
syn. *ulmi* (Linnaeus) [in part]
Elm scale

A widely distributed species, associated with elm (*Ulmus*), sometimes forming noticeable encrustations on amenity trees in towns and cities. The adult females (*c.* 2.5 mm long) are dark red and oval, with a short pair of anal papillae. They occur during the summer on the bark of the trunks and main branches amongst flocculent masses of whitish wax, each insect depositing about 250 eggs before dying. The eggs hatch from late August onwards, first-instar nymphs migrating onto the leaves to feed. Second-instar nymphs eventually overwinter, resting in sheltered places on the bark of host trees. In the following spring, the nymphs invade the young shoots and foliage. Later, they move onto the trunk and larger branches where they settle down to mature. There is one generation annually.

Pseudochermes fraxini (Kaltenbach)
syn. *fraxini* (Newstead)
Ash scale

A local insect, restricted to ash (*Fraxinus*), colonies occurring beneath white, felt-like masses of wax on the stems of small trees or on the branches of larger ones (Fig. **214**). Adult males appear from October to early November, crawling over the bark in search of females; they are wingless, 0.9 mm long and mainly orange to orange-yellow, with black eyes. The orange-red females reach maturity in the spring; they then deposit eggs and die. These eggs hatch in mid-June, reddish first-stage nymphs then swarming over the bark before settling down to feed. Infestations develop rapidly on trees placed under stress by the removal of nearby shelter, and often occur on isolated amenity trees in towns and cities.

23. Family PSEUDOCOCCIDAE (mealybugs)

Typically elongate-oval insects, with poorly developed antennae but well-developed legs; body covered in a flocculent or mealy, waxen secretion.

Pseudococcus affinis (Maskell)
syn. *latipes* Green; *obscurus* Essig
Glasshouse mealybug

A widely distributed tropical or subtropical species, generally common in Europe on a wide variety of glasshouse ornamentals and house plants.

214 Colony of ash scale on *Fraxinus*.

215 Colony of glasshouse mealybug on *Cyperus*.

DESCRIPTION **Adult:** 4 mm long; body pinkish, covered with white mealy wax; caudal processes about half as long as body.

LIFE HISTORY Eggs are laid in batches within a white, waxy sac (Fig. **215**). After egg hatch, nymphs eventually disperse over infested plants, often congregating in the axils of buds, at the base of leaves and within leaf sheaths and curled leaves. There are several generations each year and, under favourable conditions, breeding is continuous.

DAMAGE Adults and nymphs suck the sap of host plants, impairing growth and, sometimes, leading to defoliation. Infested plants are also disfigured by the presence of clumps of white wax and by accumulations of honeydew and sooty moulds.

CONTROL Mealybugs can be controlled by applying insecticides but the protective masses of 'wool' often prevent penetration of sprays. On a small scale, remove the pests by hand or dab them with a paint brush charged with an insecticide. Biological control of mealybugs on ornamentals in greenhouses is possible, using the subtropical predatory coccinellid *Cryptolaemus montrouzieri* Mulsant.

Pseudococcus calceolariae (Maskell)
syn. *fragilis* Brain; *gahani* Green
Citrophilus mealybug
Unlike most mealybugs, this species is indigenous to northern Europe. In favourable districts, infestations occur out-of-doors on various plants, including *Ceanothus*, currant (*Ribes*), false acacia (*Robinia pseudoacacia*), *Forsythia*, juniper (*Juniperus*) and *Laburnum*; glasshouse ornamentals are also attacked. Adults are 3–4 mm long and broadly oval, with the body partly coated with white, mealy wax; the caudal filaments are relatively thick (Fig. **216**).

216 Citrophilus mealybug, *Pseudococcus calceolariae*.

Pseudococcus longispinus (Targioni-Tozzetti)
Long-tailed mealybug
Infestations of this common glasshouse mealybug often occur on cacti, lilies, orchids, vines and many other ornamentals. Individuals are relatively small (adults, *c.* 2.5 mm long), with very long caudal processes. Colonies secrete considerable quantities of honeydew (Fig. **217**).

Planococcus citri (Risso)
Citrus mealybug
An often abundant pest on glasshouse ornamentals, developing rapidly under conditions of high humidity and temperature, with up to eight generations annually. Heavy infestations commonly develop on *Amaryllis*, cacti, ferns, orchids, vines and many other plants. Adults are 3–4 mm long and pinkish, coated with whitish wax; the peripheral and caudal waxen processes are characteristically short and stout (Fig. **218**).

217 Long-tailed mealybug, *Pseudococcus longispinus*.

Nipaecoccus nipae (Maskell)
Palm mealybug
Unsightly infestations of this tropical mealybug sometimes occur on glasshouse-grown palms, individuals tending to settle on the leaf bases. Adults are about 3.5 mm long, salmon-pink in colour, with discrete creamish-white waxen cones distributed over the rather convex body.

Rhizoecus falcifer Künckel de Herculais
syn. *terrestris* (Newstead)
Root mealybug
Dense subterranean colonies of this and certain

218 Citrus mealybug, *Planococcus citri*.

other closely related species often occur on the roots of (mainly) glasshouse-grown ornamentals, surrounded by fragile masses of white wax (Fig. **219**). In common with container-grown plants infested with root aphids, the colonies often occur at the periphery of the root ball. The mealybugs are elongate (1.0–2.3 mm long), with distinct, five-segmented, geniculated antennae and a pair of very short, waxy anal appendages; although greenish-yellow, they appear white due to the mealy coating of wax. Reproduction is parthenogenetic, the mature females depositing their white, translucent eggs in batches inside cottonwool-like ovisacs; there is a succession of generations throughout the year. The foliage of attacked plants becomes dull, heavily infested plants making poor growth and eventually wilting.

CONTROL Dip the roots of infested plants in a dilute solution of insecticide and then repot into clean compost. Application of an insecticidal drench is also recommended but often less effective.

Trionymus diminutus Leonardi
syn. *calceolariae* (Maskell)
New Zealand flax mealybug
An Australasian species, associated mainly with New Zealand flax (*Phormium tenax*); also occurs on *Cordyline australis*. Introduced on such plants to other parts of the world, including northern Europe where it is most often noted in south-western England.

DESCRIPTION **Adult female:** 4–5 mm long; grey to purplish-grey or dark red, dusted with a white, waxen secretion that forms long filaments. **Nymph:** light grey, dusted with white wax.

LIFE HISTORY Eggs of this parthenogenetic species are laid in masses within protective ovisacs formed mainly on the leaf blades. The first-instar nymphs are very mobile, especially when hatching from eggs deposited in unfavourable situations; later instars and adults are more-or-less sedentary, the insects often clustering together at the leaf bases. There are several generations annually, all stages often occurring together.

DAMAGE Infested plants are contaminated by vast quantities of honeydew; masses of flocculent wax can also spoil the appearance of plants. Attacked plants are weakened and can be killed.

CONTROL As for *Pseudococcus affinis* (p. 103).

219

219 Colony of root mealybug on roots of *Lavendula*.

220

220 Fluted scale, *Icerya purchasi.*

24. Family **MARGARODIDAE**

Icerya purchasi Maskell
Fluted scale
Originally an Australian species, now a serious pest of citrus plants in various parts of the world; infestations also occur on ornamentals, especially *Acacia*, broom (*Cytisus*), *Mimosa* and rose (*Rosa*). Attacks are common in citrus-growing regions of southern Europe, and may also occur occasionally on glasshouse-grown ornamentals and house plants further north. Female scales are about 3 mm long, brown and oval, each surmounting a large (*c.* 10 mm long), white, distinctly grooved egg-sac which contains several hundred eggs (Fig. **220**).

Order **THYSANOPTERA** (thrips)

1. Family **THRIPIDAE**

Thrips with a flattened body, narrow and pointed wings, and a saw-like ovipositor.

Heliothrips haemorrhoidalis (Bouché)
syn. *adonidum* Haliday
Glasshouse thrips

This tropical or subtropical thrips is a frequent glasshouse pest in temperate regions, infesting many ornamental plants including azalea (*Rhododendron*), *Begonia*, calla lily (*Richardia*), *Chrysanthemum*, ferns, *Fuchsia*, orchids, palms, rose (*Rosa*) and vine (*Vitis*). Under favourable conditions, breeding may continue throughout the year, but this insect cannot survive northern European winters in unprotected situations. Adults are 1.2–1.8mm long and dark brown, with an orange tip to the abdomen and pale yellow antennae, legs and wings; the fore wings are narrow, with few, very short setae on the veins; the antennae are eight-segmented, with the terminal segment needle-like; the tarsi are one-segmented. The yellowish-brown nymphs secrete a reddish fluid, that is deposited in large drops on the surface of host plants and upon which brown fungal growths develop. In addition, adult and nymphal feeding causes a brownish, silvery or whitish spotting of leaves and flowers.

CONTROL Spray with an insecticide as soon as damage is seen, and repeat as necessary; fogs and smokes are also recommended. Ensure that the produce chosen is not phytotoxic to the plants requiring treatment.

Parthenothrips dracaenae (Heeger)
Banded-winged palm thrips

This subtropical species is widely distributed in glasshouses in northern Europe, infesting various plants such as *Citrus*, *Croton*, *Cycas*, *Dracaena*, *Ficus*, *Kentia*, *Stephanotis* and *Tradescantia*. Infestations are also common on house plants. Affected foliage often becomes extensively damaged, both the upper (Fig. **221**) and the lower (Fig. **222**) surfaces developing distinctive silvery patches. Adults (*c.* 1.3mm long) are yellow to brown, with a strongly reticulated head and thorax; the antennae are seven-segmented, with the terminal segment needle-like; the fore wings, which lack a costal fringe of hairs, are very broad, banded with brown but mainly hyaline with a reticulated pattern; the legs are brown, with yellow tibiae and tarsi; the tarsi are one-segmented.

221 Thrips damage to leaf of *Ficus elastica*, viewed from above.

222 Thrips damage to leaf of *Ficus elastica*, viewed from below.

CONTROL As for *Heliothrips haemorrhoidalis* (p. 106). Regular spraying of house plants with water is also effective.

Dendrothrips ornatus (Jablonowski)
Privet thrips

Infestations of this locally common thrips occur on the upper surface of leaves of lilac (*Syringa vulgaris*) and privet (*Ligustrum vulgare*), causing noticeable silvering and distortion (Fig. **223**); in continental Europe, this species is also reported on alder (*Alnus*) and lime (*Tilia*). There are two or more generations each year, thrips occurring on host plants from April to November. Adults (0.9–1.1 mm long) are brown, with short, six-segmented antennae and three white crossbands on each fore wing.

CONTROL Apply a contact or a systemic insecticide as soon as damage is seen.

Anaphrothrips orchidaceus Bagnall
Yellow orchid thrips

An introduced thrips, sometimes infests glasshouse-grown orchids in northern Europe, producing brown patches on the green tuberous tissue. Adults (1.3 mm long) are yellow-bodied, with a pair of brown stripes behind the eyes; the antennae are eight-segmented, and mainly brown; the fore wings are light brown, but pale basally, and the legs yellow.

CONTROL As for *Heliothrips haemorrhoidalis* (p. 106).

Frankliniella occidentalis (Pergande)
Western flower thrips

An important, polyphagous thrips, occurring widely in North America and also now found on various ornamental herbaceous plants in Denmark, England, France, the Netherlands, Sweden and West Germany, originally having been introduced into glasshouses in Europe on imported *Chrysanthemum* cuttings. Ornamentals commonly infested in northern Europe include *Achimenes*, African violet (*Saintpaulia*), busy lizzie (*Impatiens sultani*), cape primrose (*Streptocarpus*), *Chrysanthemum, Cineraria, Cyclamen, Gerbera, Gloxinia, Pelargonium* and vervain (*Verbena*). Also an important vector of plant virus diseases.

DESCRIPTION **Adult:** 2mm long; pale yellow to brownish-yellow; antennae eight-segmented; distinguishable with certainty from closely related species only by microscopical examination. **Nymph:** translucent to golden-yellow.

223 Privet thrips damage to leaf of *Ligustrum*.

224 Western flower thrips damage to flower of *Chrysanthemum*.

LIFE HISTORY Although most frequently associated with chrysanthemums, attacks can also become established on various other glasshouse plants. The thrips infest both surfaces of the leaves but tend to occur more commonly on the underside, causing a noticeable scarring and excreting tell-tale specks of black frass, a typical symptom of thrips attack; they also occur commonly beneath bud scales and between the calyx and petals of open blooms. Breeding is continuous under suitable conditions, the complete life-cycle (including egg, two nymphal and two pseudopupal stages) occupying about 2–3 weeks at temperatures of 20–30°C.

DAMAGE Thrips feeding causes a silvering and spotting of tissue (Fig. **224**), and also noticeable blanching (Fig. 225), especially to petals, significant damage often being caused by a relatively small number of individuals; infestations may also lead to noticeable distortion of host plants. The effects of tomato spotted wilt virus, transmitted by this pest to various hosts (including chrysanthemum and gloxinia), are often severe.

CONTROL This insect is, apparently, tolerant to many insecticides; it is also difficult to eradicate due to its secretive habits. In some countries, it is a notifiable pest and subject to statutory control. If infestations are suspected, or if thrips populations in glasshouses prove difficult to control, expert help should be sought.

Frankliniella intonsa (Trybom)
Flower thrips
This yellowish to reddish-brown, 1.2–2.8 mm long, flat-bodied thrips infests a wide variety of ornamentals, causing minor damage to the foliage and petals. Infestations are most often reported on heather (*Erica*). Unlike the following species, the front projection of the head is triangular and there are eight pronotal bristles.

CONTROL Apply an insecticide as soon as damage is seen; ensure that the product chosen will not be phytotoxic.

Frankliniella iridis (Watson)
Iris thrips
Infestations of this widespread species sometimes occur on *Amaryllis* and *Iris*, nymphs and adults causing a silvering of the foliage and a white speckling of the flower petals. Female thrips occur throughout the year, hibernating during the winter months; males and nymphs are present during the summer and autumn. There are several generations each year, individuals developing through an egg, two nymphal, a prepupal and a pseudopupal stage. Adult females are 1.5–1.7 mm long, dark brown and flattened, with pale yellow (usually reduced) wings; the front projection of the head is rounded, and there are six strong pronotal bristles.

CONTROL Apply an insecticide as soon as thrips are found, ideally before the flowering stage.

225 Western flower thrips damage to petals of *Chrysanthemum*.

Thrips fuscipennis Haliday
syn. *menyanthidis* Bagnell
Rose thrips
A common outdoor and glasshouse pest of various flowering plants, including rose (*Rosa*) and many other ornamentals; infestations also occur on the young leaves of various trees. Widespread in Europe.

DESCRIPTION **Adult female:** 1.2–1.6 mm long; yellowish-brown to dark brown; legs brown; fore wings dark grey-brown, paler basally; antennae seven-segmented; comb of setae on hind margin of the eighth abdominal tergite incomplete centrally. **Nymph:** white to pale yellow.

LIFE HISTORY Outdoors, adult females hibernate and become active in the spring, depositing eggs from May onwards. Nymphs then occur on host plants from May to August or September and males from June to October, with up to four generations annually. In glasshouses, there are a greater number of generations; however, even under heated conditions, there is an obligatory period of hibernation from November onwards, with adult females sheltering in cracks and crevices, in debris or in the soil; the thrips reappear in the late winter or early spring, eggs being deposited from late February onwards.

DAMAGE Infested tissue becomes discoloured and distorted (Fig. **226**); on rose, attacks on the developing flowers lead to malformed, brown-streaked petals.

CONTROL Spray with an insecticide as soon as damage is seen.

226 Rose thrips damage to foliage of *Ostrya carpinifolia*.

Kakothrips pisivorus (Westwood)
syn. *robustus* (Uzel)
Pea thrips
A widespread and common pest of legumes, including sweet pea (*Lathyrus odoratus*), but rarely of significance on ornamentals. Adults occur from May to July or August, infesting and causing a silvering of the foliage, flowers and pods. There is a single generation each year, nymphs developing during June, July or August. When fully fed, the nymphs enter the soil, where they overwinter, adults appearing in the following spring. Adults (1.5–2.0mm long) are blackish-brown and flat-bodied, with dark brown, basally clear, fore wings; the antennae are eight-segmented, with the third segment yellow.

Thrips simplex (Morison)
syn. *gladioli* (Moulton & Steinweden)
Gladiolus thrips
An important pest of glasshouse-grown *Gladiolus* and, to a lesser extent, *Crocus*; less often, *Freesia*, *Iris* and lilies are also attacked. Probably introduced into northern Europe from southern Africa.

DESCRIPTION **Adult:** 1.5mm long; dark brown; antennae eight-segmented and mainly dark, with third and basal part of fourth and fifth segments pale; fore wings brown but pale basally; legs dark brown with paler tibial apices and tarsi. **Nymph:** yellow to orange.

LIFE HISTORY In northern Europe, this species generally survives the winter between the scales of stored corms where, if temperatures remain above 10°C, breeding can continue throughout the year. In the spring, when corms are planted, the thrips will be carried upwards on the enlarging stems and leaves; subsequently, they may also invade the developing flowers. Gladioli growing out-of-doors are sometimes infested during the summer, especially in hot, dry conditions.

DAMAGE **Crocus:** yellowish-brown areas develop beneath the skin of stored corms; new growth from damaged corms is poor and cream-coloured rather than white. **Gladiolus:** infested corms develop rough, greyish-brown surface patches; infestations on developing plants produce yellowish or silvery streaks on the foliage and stems, which may subsequently turn brown; when feeding extends to the flowers, the petals develop silvery flecks and, if attacks are severe, they may turn brown and die.

CONTROL Routine fumigation of stored corms is recommended, several treatments often being advocated to ensure effective control. Care is necessary to avoid introducing contaminated corms into clean stores. In the field, apply an insecticide as soon as damaging thrips are discovered.

Thrips tabaci Lindeman
syn. *debilis* Bagnell
Onion thrips
An often common pest of cultivated plants, including ornamentals such as asparagus fern (*Asparagus plumosus*), *Begonia*, calla lily (*Richardia*), carnation (*Dianthus caryophyllus*), *Chrysanthemum*, *Cineraria*, *Cyclamen*, *Dahlia*, *Gerbera* and orchids; especially damaging in glasshouses, and also a virus vector. Cosmopolitan. Widely distributed in Europe.

DESCRIPTION **Adult female:** 1.0–1.3 mm long; greyish-yellow to brown; antennae seven-segmented and yellowish-brown; fore wings pale brownish-yellow; comb of setae on hind margin of eighth abdominal tergite complete. **Nymph:** whitish to pale yellowish-orange.

LIFE HISTORY This entirely parthenogenetic species breeds continuously in favourable conditions. Eggs are laid in plant tissue and these hatch within 1–2 weeks. Two nymphal stages then occur on host plants, fully fed nymphs dropping to the ground about a week later. They then enter the soil, where the transformation to the adult stage takes place, again in about a week. Considerable populations may develop on glasshouse-grown plants; glasshouses are also subject to frequent invasion from outdoor populations. In unprotected situations, females usually overwinter in the soil and various generations of nymphs occur from May through to the winter.

DAMAGE Affected leaves become flecked extensively with silver, damage typically being associated with black grains of frass (Fig. **227**); flower petals may also become discoloured and distorted (Fig. **228**). Attacks are especially severe in hot, dry conditions.

CONTROL Apply an insecticide at the first signs of damage and repeat at regular intervals as necessary, preferably restricting treatments to the pre-flowering stages. Care is necessary to avoid using any product likely to cause damage to host plants.

227 Onion thrips damage to leaf of *Dianthus*.

228 Onion thrips damage to flower of *Cyclamen*.

Thrips atratus Haliday
Carnation thrips
This generally abundant species occurs throughout the year, breeding on various outdoor plants from May onwards. Adults are 1.3–1.8 mm long and mainly dark brown, with eight-segmented antennae and narrow, brown (basally paler) fore wings. They commonly enter glasshouses from June to September, and may then cause damage to flowers of carnation and pink (*Dianthus*).

CONTROL As for *Thrips tabaci*.

Thrips nigropilosis Uzel
Chrysanthemum thrips
Generally common and sometimes troublesome on glasshouse-grown *Chrysanthemum*, producing characteristic brownish-red marks at the base of petioles and brownish warts on the foliage. Adults occur throughout the year, with several generations of nymphs developing from May to November. The adults are 1.2 mm long, and mainly yellow with brownish markings, pale yellow fore wings and seven-segmented antennae; some adults may have reduced wings. The main outdoor hosts are plantains, especially ribwort plantain (*Plantago lanceolata*).

CONTROL As for *Thrips tabaci* (p. 110).

Limothrips cerealium Haliday
syn. *avenae* Hinds
Grain thrips
This mainly black-bodied thrips often appears in vast numbers during July or August, at about the time that cereal crops are ripening; the adult females may then invade glasshouses, to appear on crops such as *Chrysanthemum*. Damage due to other thrips (especially *Thrips tabaci*, p.110) is often present on glasshouse-grown ornamentals at the time of invasion by *Limothrips cerealium*, and the latter (which breeds only on cereals and grasses) is sometimes assumed to be the harmful species; unlike *Thrips tabaci*, *Limothrips cerealium* has eight-segmented antennae, the head longer than broad and a pair of stout spine-like setae on the tenth abdominal tergite. It is this and other species of *Limothrips* which commonly gives rise to the term 'thunder-flies'.

2. Family
PHLAEOTHRIPIDAE

Thrips without an ovipositor, the fore wings with just one vein; eggs hard-shelled, often sculptured, and deposited on plant tissue or in plant debris; development includes two (often brightly coloured or banded) nymphal stages, one prepupal and two pseudopupal stages.

Liothrips vaneeckei Priesner
Lily thrips
A potentially serious pest of lily bulbs in store, especially in the Netherlands; also present in North America.

DESCRIPTION **Adult:** 2.0–2.5 mm long; dark reddish-brown; antennae yellow to brown and eight-segmented; legs orange-yellow, long and slender; abdomen without an ovipositor but terminating in a distinct tube; fore wings strap-like, pale brown, with a pale base and a dark median band.

LIFE HISTORY Attacks are limited to stored bulbs, infestations developing close to the base plate between the scales, after eggs (which are hard-shelled) are placed on the plant tissue by mated females. There are several generations each year, adults or final-instar nymphs overwintering.

DAMAGE Infestations are often persistent and lead to the development of sunken rust-coloured spots towards the base of the outer scales. Damaged scales become soft and the outer ones papery; the latter may also drop off. If infested bulbs are planted out, they produce fewer new scales but plant development is little affected.

CONTROL Fumigation will check infestations on stored bulbs; in some countries, dip treatment of infested bulbs is recommended.

Order **COLEOPTERA** (beetles and weevils)

1. Family **SCARABAEIDAE** (chafers)

Chafers are large to very large, often brightly coloured beetles, with lamellated tips to the antennae. The larvae are large and fleshy, with a distinct head, powerful mouthparts, three pairs of strong legs and a swollen tip to the abdomen; they feed in the soil on decaying matter and plant roots, and are particularly abundant in light, well-drained sites near heathlands or woodlands.

Melolontha melolontha (Linnaeus)

syn. *vulgaris* Fabricius
Cockchafer

An often common pest of corms, roots, tubers and underground parts of the stems of various plants, including alpines, herbaceous ornamentals and nursery trees and shrubs. Eurasiatic. Widely distributed in Europe.

DESCRIPTION **Adult:** 20–30 mm long; chestnut-brown, the head and thorax darker, and partly coated in whitish hairs which often rub off; elytra each with five longitudinal lines; abdomen terminating in a blunt, downwardly directed spine; antennae with six (female) or seven (male) lamellae (Fig. **229**). **Egg:** oval, whitish or yellowish. **Larva:** 30–35 mm long; body white, with the last segment somewhat translucent and darkened by the underlying gut contents; head and legs brown and shiny; anal slit transverse and wavy, surmounted by two more-or-less parallel longitudinal rows of spines. **Pupa:** 25–35 mm long; whitish to brown.

LIFE HISTORY Adults occur in May or June. They are nocturnal, feeding at night on the buds, flowers or foliage of various trees and shrubs, and are frequently attracted to light. After a few weeks, the females burrow into the soil to a depth of 15–20 cm and then lay the eggs, typically in small batches of 12–30. The eggs swell considerably during an extended period of incubation, usually hatching after about four weeks. The larvae then attack plant roots, feeding for up to three years and passing through three clearly defined instars. Pupation takes place during the third summer in an earthen cell 60 cm or more below the surface.

22

229 Cockchafer, *Melolontha melolontha*.

Adults are produced about six weeks later but they do not emerge from the soil until the following spring.

DAMAGE Adults chew holes into the leaves of various trees and shrubs, but significant damage is usually restricted to buds of roses, deep holes being formed in the sides. The larvae destroy much of the root system of host plants, seriously restricting growth; in severe cases, plants wilt and die. Larval damage is especially liable to occur on plants in recently broken-up grassland or pasture.

CONTROL Thorough cultivation and good weed control can keep chafer grubs in check. Land known to be infested can be treated with a soil insecticide prior to planting but treatment of growing crops is impractical.

Amphimallon solstitialis (Linnaeus)
Summer chafer

This southerly distributed species is sometimes a pest in gardens and nurseries, the larvae feeding on the roots of various herbaceous plants, including ornamentals; roots of nursery trees are also attacked. The yellowish-brown adults (14–18 mm long) occur in June and July. They are active on warm evenings, the mated females depositing several white, oval eggs in the soil, and the eggs hatching about a month later. The larvae are white and elongate (up to 30 mm long), with a pale brown head and legs, and two divergent rows of spines surmounting the anal slit (Fig. 230); they feed from August onwards, usually completing their development within two years. Although capable of causing considerable damage, especially in their second summer, the larvae are usually present in only small numbers.

230 Larva of summer chafer, *Amphimallon solstitialis*.

Serica brunnea (Linnaeus)
Brown chafer

A locally common woodland insect, sometimes causing damage to nursery trees, including Norway spruce (*Picea abies*) grown as Christmas trees, but not an important pest of ornamentals. The larvae (up to 18 mm long) are creamish-white, with a yellowish-brown head, numerous reddish body hairs and the anal slit surmounted by a horizontal arc of spines (Fig. 231). They feed on plant roots, usually completing their development in two years. The adults (7–11 mm long) are mainly reddish-brown; they occur from June to August.

CONTROL As for *Melolontha melolontha* (p. 112).

231 Larva of brown chafer, *Serica brunnea*

Hoplia philanthus (Fuessley)
Welsh chafer

A widely distributed and locally common species, especially in light soils, the larvae sometimes causing damage to lawns and sports turf. The larvae are similar to those of the previous species but have characteristically long, stout claws on the front tarsi. Adults (7–11 mm long) are mainly black with reddish-brown elytra. They occur from May to July.

CONTROL **Lawns and turf**: if practical, apply an insecticide to damaged areas at the onset of adult emergence. Attempts to kill larvae with insecticides are rarely effective.

Phyllopertha horticola (Linnaeus)
Garden chafer

The colourful, metallic bluish-green and reddish-brown adults (7–11mm long) (Fig. **232**) of this widely distributed chafer are especially common in light-soiled grassland areas. They occur mainly in May and June, often flying during the daytime in warm, sunny weather. The larvae are relatively small (15mm long when fully grown) and characterized by the two parallel rows of spines above the anal slit; also, in repose they rest with the head close to the anal segment (Fig. **233**). They feed on plant roots, especially grasses, from June or July onwards, becoming fully grown by the autumn and eventually pupating in the following spring. Although the adults graze on leaves, flowers and fruits of many garden plants, and the larvae may occasionally damage the roots of herbaceous plants, this species is not an important pest of ornamentals. However, larval damage to lawns and sports turf is often extensive.

CONTROL As for *Hoplia philanthus* (p. 113).

Cetonia aurata (Linnaeus)
Rose chafer

This locally common, southerly distributed chafer is rarely important as a pest, but the adults do sometimes browse on the flowers of ornamental plants, especially honeysuckle (*Lonicera*), *Viburnum* and various members of the family Rosaceae, damage being reported most frequently on rose (*Rosa*). Adults (14–20mm long) are metallic golden-green above, purplish-red below, with a few irregular silvery markings on the elytra (Fig. **234**). They occur from late May onwards, depositing eggs in the soil during the early summer. The larvae (up to 30mm long) feed on rotting wood and decaying plant material for up to two, or sometimes three, years; they are distinguished from other chafer larvae by the reddish head and legs, and by the transverse rows of reddish hairs on the body.

2. Family **ELATERIDAE** (click beetles)

A small group of elongate beetles which, when lying on their backs, are capable of propelling themselves into the air, with an audible click. The soil-inhabiting larvae (commonly known as 'wireworms') are long, thin, more-or-less cylindrical and tough-skinned, with small thoracic legs and powerful jaws.

232 Garden chafer, *Phyllopertha horticola.*

233 Larva of garden chafer, *Phyllopertha horticola.*

234 Rose chafer, *Cetonia aurata.*

Athous haemorrhoidalis (Fabricius)
Garden click beetle
A generally common pest of herbaceous plants, including ornamentals such as *Anemone*, carnation (*Dianthus caryophyllus*), *Chrysanthemum*, *Dahlia*, *Gladiolus* and *Primula*; also liable to damage seedling trees and nursery stock. Widespread in Europe.

DESCRIPTION **Adult:** 10–13 mm long; head and thorax black, the abdomen reddish-brown; antennae black; legs brownish-black (Fig. **235**). **Larva:** 22–30 mm long; shiny yellowish-brown with a darker head; relatively broad and sub-cylindrical, with tip of body bifid (Fig. **236**).

LIFE HISTORY Adults occur from mid-May to July, and are more commonly encountered than those of many other elaterids, often flying during the daytime. Eggs are deposited in groups in moist soil, usually beneath the shelter of vegetation. They hatch in about a month, the larvae then attacking the underground parts of plants and also feeding on other vegetative matter in the soil. Development is slow, usually extending over four or five years, individuals causing most damage in spring and autumn. When fully grown, usually in mid- to late summer, larvae pupate in earthen cells. Adults are produced about a month later but they normally remain in their cells until the following spring.

DAMAGE Wireworms bite through the roots and bore into the base of plants, corms, bulbs, rhizomes, stolons and tubers, sometimes causing plants to wilt and die; occasionally, on chrysanthemum and other relatively fleshy-stemmed plants, they may also tunnel into the stems well above soil level. Damage is especially serious on young plants in spring and autumn.

CONTROL Thorough cultivation and good weed control will reduce the likelihood of attacks developing. If necessary, apply a soil insecticide 2–3 weeks before planting.

Agriotes lineatus (Linnaeus)
Common click beetle
This generally common, often important agricultural pest, also causes damage to horticultural crops, including various ornamentals. The larvae are usually most abundant in grassland areas, often causing damage to crops growing in recently broken-up pasture. They are distinguished from those of *Athous haemorrhoidalis* by the more cylindrical form and pointed hind end; they also take four or five years to reach maturity.

235 Garden click beetle, *Athous haemorrhoidalis*.

236 Larva of garden click beetle, *Athous haemorrhoidalis*.

3. Family NITIDULIDAE

A large family of mainly small beetles, the antennae usually terminating in a three-segmented club.

Meligethes spp.
Pollen beetles
Adult pollen beetles, mainly *Meligethes aeneus* (Fabricius), are increasingly reported invading garden flowers during the summer months. These small (*c.* 3 mm long), bronzy or greenish-black insects breed mainly on brassica seed crops and cruciferous weeds such as charlock (*Sinapis arvensis*), but often migrate in numbers to various other flowering plants in search of pollen. In gardens, they are often numerous on ornamentals such as *Hypericum*, rose (*Rosa*) and sweet pea (*Lathrus*

odoratus). Although causing little or no direct damage, the presence of the beetles on cut flowers can be a nuisance.

CONTROL On a small scale, the beetles can be dislodged from cut flowers by shaking. Alternatively, the flowers can be stood temporarily in water in a garage or shed during the daytime, the beetles being attracted to the brighter light coming from open doors or windows. Application of an insecticide is not recommended.

4. Family **BYTURIDAE**

A group of small, hairy beetles with clubbed antennae; the larvae develop in flowers and fruits of *Rubus*.

Byturus tomentosus (Degeer)
Raspberry beetle
In spring, following their emergence from hibernation, adults of this important raspberry pest often feed on the flowers of rosaceous trees and shrubs, including hawthorn (*Crataegus*) and ornamental cherry (*Prunus*); they also attack flowering shrubs such as lilac (*Syringa vulgaris*). Although sometimes numerous, the beetles do not cause significant damage and soon depart for *Rubus* hosts upon which they will eventually breed. Adults are 3.5–4.5 mm long, yellowish-brown in appearance, with elongate-oval bodies and short, clubbed antennae.

237 Cereal leaf beetle, *Oulema melanopa*.

238 Larva of cereal leaf beetle, *Oulema melanopa*.

5. Family **CHRYSOMELIDAE** (leaf beetles)

A large family of mainly small, leaf-feeding beetles; also includes flea beetles. Adults are often rounded, shiny and brightly coloured, and many are metallic-looking. The larvae commonly feed in exposed positions on the leaves of host plants.

Oulema melanopa (Linnaeus)
Cereal leaf beetle
This widely distributed but minor pest of cereals is associated occasionally with ornamental grasses, such as ribbon grass (*Phalaris arundinacea* var. *picta*), the adults and larvae grazing away longitudinal sections of leaf tissue. Such damage is mildly disfiguring but unimportant. The adult beetles (4.0–4.5 mm long) have a black head and antennae, a brownish-orange thorax and metallic, blue-green elytra (Fig. 237). They appear in May, the females depositing eggs on the leaves of suitable host plants during June and July. Eggs hatch in about ten days. Larvae, the stage most likely to be noticed on ornamental grasses, are up to 6 mm long, with a brownish-black head and a dirty-yellow body; they are usually coated in slimy black excrement (Fig. 238). They feed on the upper surface of the leaves for 3–4 weeks before entering the soil to pupate. Young adults emerge in the late summer and feed briefly before hibernating. There is just one generation annually.

Lilioceris lilii (Scopoli)
Lily beetle

A destructive pest of lily, especially madonna lily (*Lilium candidum*); also a pest of other ornamentals such as fritillary (*Fritillaria*), *Nomaocharis saluenensis* and Solomon's seal (*Polygonatum multiflorum*). Widely distributed across Eurasia, but local and usually uncommon.

DESCRIPTION **Adult:** 6–8 mm long; thorax and elytra red; head, antennae, legs and underside of body black; thorax relatively narrow, with a characteristic lateral constriction (Fig. **239**). **Egg:** 1 mm long; subspherical, red when newly laid but soon turning brown (Fig. **240**). **Larva:** 8–10 mm long; head black, body orange-red with small black plates on each abdominal segment, coated with black slime.

LIFE HISTORY Adults emerge from hibernation in the early spring and may be found in association with host plants from late March or April onwards. Eggs are laid shortly afterwards. They hatch in 7–10 days and the larvae, whose bodies often become coated with slimy black faeces, develop rapidly, feeding ravenously for about two weeks; they then enter the soil to pupate, each in a silken cocoon. Young adult beetles emerge about three weeks later. In favourable conditions, there are two generations annually.

DAMAGE Leaves, stems, flowers and seed pods are grazed extensively, the tissue of heavily infested plants becoming tattered and often totally destroyed.

CONTROL On a small scale, egg batches and young larvae can be destroyed by hand; where necessary, apply a contact or a stomach-acting insecticide.

Chrysolina polita (Linnaeus)

Although mainly an inhabitant of damp meadows, adults of this generally common species occur occasionally on ornamental lime (*Tilia*) and willow (*Salix*), causing minor damage to the leaves. Individuals are 6.5–8.5 mm long, broad and convex, with a golden-green, sometimes reddish-flushed head and thorax, and dull-metallic, brownish-red, irregularly punctured elytra (Fig. **241**) (cf. *Chrysomela populi*, p. 118).

239 Lily beetle, *Lilioceris lilii*.

240 Eggs of lily beetle, *Lilioceris lilii*.

241 Adult of *Chrysolina polita*.

117

Chrysomela populi Linnaeus

Red poplar leaf beetle

A locally common pest of poplar (*Populus*), including aspen (*Populus tremula*), and willow (*Salix*). Widespread throughout Europe.

DESCRIPTION **Adult:** 8–12 mm long; thorax bluish-black or greenish; elytra reddish with a black spot at the extreme tip (Fig. **242**) (cf. *Chrysomela tremula*). **Egg:** 1.0 × 0.56 mm; elongate-oval, pale yellow, yellow to brownish. **Larva:** 12–15 mm long; body creamish-white, with a mainly black prothoracic plate and prominent black verrucae; head black.

LIFE HISTORY Adults occur from May onwards, depositing eggs on the leaves of host plants. They show a particular liking for young trees and are sometimes common in poplar stool beds. Larvae graze the surface layers of the leaves but later form holes right through the lamina. If disturbed, larvae are capable of exuding globules of liquid from a lateral series of tube-like verrucae. Fully fed individuals pupate in the soil, adults of the next generation appearing in about July. A second generation of larvae feed in July and August, producing adults in September. These eventually hibernate and reappear in the following spring.

DAMAGE Adults and larvae skeletonize the leaves, infestations often causing extensive damage; young aspen trees are especially susceptible.

CONTROL If necessary, apply a contact or a stomach-acting insecticide, ideally in spring or early summer against the adults and young larvae. On a small scale, adults, eggs and larvae can be destroyed by hand.

Chrysomela aenea Linnaeus

Widely distributed on alder (*Alnus*) but also sometimes damaging to the foliage of other trees, including willow (*Salix*). There are two generations annually. Adults are 5–8 mm long and coppery-blue or metallic green (Fig. **243**); the larvae, which often skeletonize the leaves, are up to 10 mm long and blackish to whitish, with prominent black verrucae and a shiny black head and prothoracic plate.

242 Red poplar leaf beetle, *Chrysomela populi*.

243 Adult of *Chrysomela aenea*.

Chrysomela tremula Fabricius

syn. *longicollis* (Suffrian)

Adults of this locally distributed beetle are very similar to those of *Chrysomela populi* but are smaller (6–9 mm long) and narrower-bodied; they also lack the black mark at the tip of the elytra. They occur on aspen (*Populus tremula*) in various parts of Europe, sometimes causing damage to young cultivated trees.

Phyllodecta vitellinae (Linnaeus)
Brassy willow leaf beetle

Generally common on poplar (*Populus*) and willow (*Salix*); often a pest on ornamental trees and in stool beds. Widely distributed in Europe.

DESCRIPTION **Adult:** 3.5–5.0 mm long; thorax and elytra metallic bluish, brassy or coppery; punctation on elytra forming striae (cf. *Plagiodera versicolora*, p. 121); legs and antennae black. **Egg:** 1 mm long; elongate-oval, creamish-white (Fig. **244**). **Larva:** 5–6 mm long; whitish to dirty greyish-brown, with numerous black plates and verrucae; body rather flat and tapered (Fig. **245**). **Pupa:** 4–5 mm long; white (Fig. **246**).

LIFE HISTORY Adults hibernate under loose bark, or in similar situations, emerging in the following May or June. The beetles are then commonly found on host plants where, after mating, the females eventually deposit batches of eggs on the leaves. The eggs hatch in about two weeks, larvae then feeding in groups on the expanded leaves. When fully grown, usually in 2–3 weeks, the larvae enter the soil and pupate. A second generation of larvae occurs during the late summer or autumn.

DAMAGE Adults and larvae graze the tissue from the surface of leaves, the remaining tissue turning brown. Such damage is often very noticeable and is especially unsightly on ornamentals; attacks have little effect on established trees but damage can check the growth of seedlings and of young plants in stool beds.

CONTROL Contact and stomach-acting insecticides are effective against both adults and larvae, but treatment may not be practical except on nursery stock and in stool beds.

Phyllodecta laticollis Suffrian
Small poplar leaf beetle

Locally common on poplar, especially aspen (*Populus tremula*) and Lombardy poplar (*Populus nigra* 'Italica'), and sallow (*Salix*). Adults occur from early spring onwards, at first attacking unopened buds but later feeding on the expanded leaves. Eggs are laid in early June. Larvae (up to 6

244 Eggs of brassy willow leaf beetle, *Phyllodecta vitellinae*.

245 Larvae of brassy willow leaf beetle, *Phyllodecta vitellinae*.

246 Pupae of brassy willow leaf beetle, *Phyllodecta vitellinae*.

119

mm long) are mainly greyish, with numerous black plates and verrucae and a black head (Fig. **247**). They feed during the summer, at first grazing away the tissue on the underside of the leaves, the upper surface turning brown. Young larvae typically feed gregariously but towards the end of their development individuals often become solitary. Heavy infestations cause extensive skeletonization of the foliage. Fully fed larvae drop to the ground to pupate, adults emerging in the autumn. Adults are 3.5–5.0 mm long and mainly metallic bluish with black legs; the head has a characteristic longitudinal depression and the elytra are somewhat elongate. In most areas this species is single brooded but in especially favourable districts there may be two generations annually.

247 Larva of small poplar leaf beetle, *Phyllodecta laticollis*.

Phyllodecta viminalis (Linnaeus)

Minor infestations of this beetle sometimes occur on sallow (*Salix*), damage caused by the adults and larvae being similar to that described for related species. The adults are 5.0–7.5 mm long, mainly yellowish-red, with a black head, black markings on the thorax and 3–5 black spots on each elytron (Fig. **248**.)

Phyllodecta vulgatissima (Linnaeus)
Blue willow leaf beetle
A locally common pest of poplar and willow, especially aspen (*Populus tremula*), Lombardy poplar (*Populus nigra* 'Italica'), hybrid black poplar (*Populus* × *canadensis*) and osier (*Salix viminalis*); often of importance in nursery beds, where young plants can be destroyed, but apparently more restricted in its range than formerly. Adults occur from mid-April or May onwards, eggs being deposited in double rows between two major veins on the underside of the leaves. Eggs hatch in about 1–2 weeks, the larvae then feeding on the leaves for about four weeks before pupating in the soil. Young adults appear in July and August, producing a second generation of larvae which reach adulthood in the autumn. The adults (4–5 mm long) are distinguished from those of *Phyllodecta laticollis* (see above) by the slightly narrower body and the blue or green colour; punctation at the side of the elytra is irregular.

248 Adult of *Phyllodecta viminalis*.

Plagiodera versicolora (Laicharting)
This species occurs on poplar (*Populus*) and willow (*Salix*) in various parts of Europe. Adults occur mainly from April or May to September, depositing

249 Young larvae of *Plagiodera versicolora*.

250 Larvae of *Plagiodera versicolora* feeding on leaf of *Salix helvetica*.

251 Adult of *Plagiodera versicolora*.

eggs in groups of 10–30 on the underside of the leaves. The eggs hatch in about a week. The larvae (like those of *Phyllodecta*, pp. 119–120) feed in groups on the underside of the leaves (Fig. **249**), grazing away the lower tissue, the upper surface remaining intact but turning brown. Young larvae are mainly black and shiny; older individuals, which may also feed singly on either surface of the leaf, are lighter in colour but have the body darkened by shiny black verrucae and plates (Fig. **250**). There are two main generations annually, larvae occurring from May onwards, but the developmental stages tend to overlap. The ladybird-like adults are 3.0–4.5 mm long, somewhat flattened, and metallic blue or bluish-green, with irregularly punctured elytra and black legs (Fig. **251**). They also graze on the foliage of host plants but damage, although unsightly on ornamentals, is usually insignificant. Adults over-winter under loose bark, amongst dry leaves and in other shelter.

252 Brown willow leaf beetle, *Galerucella lineola*.

CONTROL As for *Phyllodecta vitellinae* (p. 119).

Galerucella lineola (Fabricius)
Brown willow leaf beetle
An often common pest of willow (*Salix*); especially important in stool beds. Widely distributed in Europe.

DESCRIPTION **Adult:** 5–6 mm long; mainly yellowish-brown, with distinct black marks on the pronotum and shoulders of the elytra; elytra of similar width throughout, with the pubescence shiny and relatively closely set (Fig. **252**). **Egg:** 1 mm across; rounded, with a roughened surface; pale brownish-yellow (Fig. **253**). **Larva:** 7–8 mm long; head black; body

253 Eggs of brown willow leaf beetle, *Galerucella lineola*.

yellowish to black, with numerous black verrucae and plates (Fig. 254). **Pupa:** 4.5–5.5 mm long; yellow (Fig. 255).

LIFE HISTORY Adults overwinter in the soil, under dead bark and in various other places, emerging in the following May. They favour open, sunny situations and are often numerous in willow beds and on young willow trees and bushes. Eggs are laid on the shoot tips or on the upper surface of the leaves, in groups of about 20–25. They hatch about a week later. Larvae appear from about mid-May onwards, feeding gregariously but later becoming solitary. When fully fed, usually after about a month, they enter the soil to pupate, and new adults appear about a week later. There are normally two generations annually, with young adults of the final generation appearing in the late summer, but the species is single-brooded in northern districts.

DAMAGE The larvae skeletonize the leaves and also destroy the buds and shoots, causing considerable damage on heavily infested hosts.

CONTROL If necessary, apply a contact or a stomach-acting insecticide.

Galerucella luteola (Müller)
Elm leaf beetle
An often important and destructive pest of elm (*Ulmus*). Also associated with Caucasian elm (*Zelkova carpinifolia*) and horse chestnut (*Aesculus hippocastanum*). Widely distributed in mainland Europe, Asia Minor and North Africa, but absent from the British Isles; an introduced pest in North America.

DESCRIPTION **Adult:** 5–7 mm long; yellowish-brown to olive-brown, marked with black on head, thorax and elytra (Fig. 256). **Egg:** 1.0 × 0.5 mm; lemon-shaped; pale yellow to brownish-white. **Larva:** 8–10 mm long; head black, body yellowish to blackish, marked with numerous black verrucae and plates (Fig. 257). **Pupa:** 5–7 mm long; yellowish-orange.

LIFE HISTORY Beetles emerge from hibernation in April and May. They then fly to host plants and begin feeding on the leaves, forming irregular holes between the major veins. Eggs are laid in groups of 10–30 from May onwards, typically in 2–3 rows close to a secondary vein on the underside of a young leaf (Fig. 258). Each female is capable of depositing several hundred eggs, and some may successfully lay over a thousand. Eggs hatch in about ten days. The young larvae feed gregariously

254 Larva of brown willow leaf beetle, *Galerucella lineola*.

255 Pupa of brown willow leaf beetle, *Galerucella lineola*.

256 Elm leaf beetle, *Galerucella luteola*.

257 Larva of elm leaf beetle, *Galerucella luteola*.

258 Hatched eggs of elm leaf beetle, *Galerucella luteola*.

from May onwards, grazing on the underside of the leaves, at first forming small pits but, later, large patches which eventually turn brown. At the early stages of development larvae are unable to feed on the toughened, older foliage, but at the later stages of their development, when they become less gregarious, they attack leaves of various ages. Larvae are fully fed in about three weeks. They then pupate amongst leaf litter or in the soil, new adults emerging from early June onwards. These beetles produce a second generation of larvae which complete their development later in the summer. These eventually produce further adults which usually feed briefly before overwintering.

DAMAGE Adults and larvae cause considerable disfigurement of the foliage, and leaves may be completely destroyed. Heavy infestations affect the growth and vigour of host plants.

CONTROL Adults and larvae can be controlled by contact or by stomach-acting insecticides, but treatment may be practical only on nursery stock and young plants.

Galerucella nymphaeae (Linnaeus)
Water-lily beetle

An often important pest of water-lilies, including yellow water-lily (*Nuphar lutea*) and, especially, white water-lily (*Nuphar alba*). Present throughout much of Europe, including England, Ireland and Wales; an introduced pest in North America.

DESCRIPTION **Adult:** 6–8 mm long; dark brown to yellowish-brown (Fig. 259). **Egg:** 0.75 mm across; sub-spherical, pale yellow to yellowish-orange, with a finely reticulated surface. **Larva:** 7–9 mm long; dark brown or black but yellow below (Fig. 260).

259 Water-lily beetle, *Galerucella nymphaeae*.

260 Larvae of water-lily beetle, *Galerucella nymphaeae*.

Pupa: 5–7 mm long; black and shiny (Fig. **261**).

LIFE HISTORY Adult beetles hibernate in the close vicinity of ponds and pools, sheltering amongst vegetation and in dead plant stems. They appear in the following May or June, to aggregate and feed on the leaves of water-lilies. Eggs are deposited on the upper surface of the leaves in groups of 12–18. They hatch in about a week. At first, the larvae feed in groups, attacking the leaves and grazing away the upper tissue; later, they feed singly and then bite completely through the laminae. Flowers may also be damaged. Pupation occurs on the upper surface of the leaves, young adults of the next generation emerging in July and August. There are normally two generations annually but there may be three in heated pools and in favourable southerly areas.

DAMAGE Most significant damage is caused by the larvae, attacked leaves becoming covered by irregular, elongate holes (Fig. **262**) and, if infestations are heavy, extensively shredded.

CONTROL Use of insecticides should be avoided because of the danger of killing fish and other aquatic life. If necessary, the pests can be removed by hosing infested plants with a strong jet of water.

Pyrrhalta viburni (Paykull)
Viburnum beetle
Generally common on wild guelder-rose (*Viburnum opulus*); also a pest of ornamental *Viburnum*. Widespread in central and northern Europe.

DESCRIPTION **Adult:** 4.5–6.5 mm long; yellowish-brown to light brown; elytra elongate, parallel-sided and regularly rounded posteriorly; body coated with a short, silky pubescence; antennae relatively long, especially in the male (Fig. **263**). **Egg:** 0.4 mm across; rounded, dark yellow to brownish. **Larva:** 6–9 mm long; shiny greenish-yellow to whitish, with numerous black pinacula and plates; body distinctly swollen, especially posteriorly (Fig. **2**

LIFE HISTORY Overwintering eggs hatch in May. The larvae then feed voraciously on the expanded leaves, becoming fully grown in 4–5 weeks. They then enter the soil to pupate in earthen cells some 30–50 mm below the surface. Adults appear about

261 Pupae of water-lily beetle, *Galerucella nymphaeae*.

262 Water-lily beetle damage to leaves of *Nuphar alba*.

263 Viburnum beetle, *Pyrrhalta viburni*.

264 Larva of viburnum beetle, *Pyrrhalta viburni*.

265 Viburnum beetle damage to foliage of *Viburnum*.

ten days later, usually in July. After mating, the females, which may survive until September or October, deposit eggs in the tips of one-year-old shoots. Each female is capable of depositing several hundred eggs, and infested bushes are commonly attacked by large numbers of larvae. There is just one generation annually.

DAMAGE Larvae cause considerable damage; they bite out irregular holes in the leaf laminae, between the major veins, much of the remaining tissue eventually turning brown. Such depredations are often extensive (Fig. **265**), affecting both the appearance and vigour of infested plants.

CONTROL If necessary, apply a contact or a stomach-acting insecticide.

266 Adult of *Lochmaea caprea*.

Lochmaea caprea (Linnaeus)

In mainland Europe a widely distributed and often common pest of sallow and willow (*Salix*); also occurs on birch (*Betula*), poplar (*Populus*) and, less frequently, alder (*Alnus*) and hornbeam (*Carpinus betulus*). Present in the British Isles but not regarded as a pest.

DESCRIPTION **Adult:** 4–6 mm long; elongate-oval, the elytra widened posteriorly; head matt black; thorax brownish-yellow, marked with black, and strongly pitted; elytra brownish-yellow, distinctly sculptured and widened posteriorly; legs and antennae black (Fig. **266**). **Egg:** 0.5 mm across; pale yellow to brownish-yellow. **Larva:** 5–6 mm long; head black; body pale greenish with numerous black plates and verrucae (Fig. **267**).

267 Larva of *Lochmaea caprea*.

LIFE HISTORY Adult beetles emerge from hibernation in early May. They then feed on the buds, leaves and young shoots. Mating takes place in June, eggs being laid in clusters of 10–15, usually just below the surface of the soil immediately beneath host plants. After hatching, larvae migrate onto the leaves to feed, usually grazing away the lower epidermis. Larvae are fully grown in 3–4 weeks. They then pupate in the soil, young beetles appearing from the end of August onwards. These adults feed briefly before overwintering. There is just one generation annually but, because of the extended oviposition period (old adults often surviving throughout the summer), there often appears to be a second brood. This species favours dry, sunny situations and, apparently, occurs in two races: one associated mainly with willow, the other with birch.

DAMAGE Affected leaves are severely disfigured (Fig. 268), such damage being especially important on young ornamentals and nursery stock.

CONTROL Larvae and adults can be controlled by application of a contact or a stomach-acting insecticide, but treatment may not be practical except on nursery stock or young plants.

Lochmaea crataegi (Förster)

This widely distributed but local species is associated with hawthorn (*Crataegus*), the larvae developing within the fruits, and eventually pupating in the soil. The adults are 4–5 mm long, mainly yellowish-brown to reddish-brown above and blackish below, with orange legs and black antennae (Fig. 269). They feed on the leaves from May onwards, and are sometimes found on nursery plants, but damage caused is not of significance.

Agelastica alni (Linnaeus)
Alder leaf beetle

An important pest of alder (*Alnus*) in mainland Europe, where infestations are often severe on wild and cultivated plants; also associated with beech (*Fagus sylvatica*), hazel (*Corylus avellana*), hornbeam (*Carpinus betulus*) and lime (*Tilia*). Probably now extinct in the British Isles.

268 *Lochmaea caprea* damage to leaf of *Salix*.

269 Adult of *Lochmaea crataegi*.

270 Alder leaf beetle, *Agelastica alni*.

DESCRIPTION **Adult:** 6–8mm long; rather bulbous, the elytra noticeably expanded towards the hind end; bluish to violet with black antennae, tibiae and tarsi (Fig. **270**). **Larva:** 10–11mm long; cylindrical and mainly black (Fig. **271**).

LIFE HISTORY Adults hibernate and emerge in the following spring, the females eventually laying eggs in large, scattered groups on the fully expanded leaves. Larvae feed gregariously (Fig. **272**) on both sides of the leaves, at first grazing the surface but later biting out holes between the major veins. They occur from June to July, individuals feeding for about three weeks. When fully grown, they drop to the ground to pupate, either on or just below the surface, adults appearing 1–2 weeks later. The adults graze on the foliage before overwintering. There is a single generation annually.

DAMAGE Attacked foliage becomes peppered with large, irregular holes (Fig. **273**), weakening and affecting the appearance of host plants. Damage is of particular significance on young trees.

CONTROL Application of a contact or a stomach-acting insecticide will give control but may not be practical except on nursery stock or young plants.

Phyllotreta cruciferae (Goeze)
Turnip flea beetle
A generally common pest of cruciferous plants, including ornamentals such as stock (*Mathiola*) and wallflower (*Cheiranthus cheiri*); also occurs on nasturtium (*Tropaeoleum*). Holarctic. Widespread in Europe.

DESCRIPTION **Adult:** 1.8–2.4mm long; metallic greenish-black and somewhat rounded in outline, with black legs; antennae 11-segmented, mainly black, with the second and third antennal segments yellowish-red. **Egg:** 0.3mm long; yellowish-white and oval. **Larva:** 5–6mm long; white, narrow-bodied, with blackish plates and pinacula on the thoracic and abdominal segments; head black; legs very short.

LIFE HISTORY Adults are active from early spring onwards, feeding on the leaves and cotyledons of various crucifers. Eggs are deposited in the soil close to host plants, usually in batches of 20–30; they hatch about two weeks later. The larvae attack the roots, feeding externally for about two weeks. Individuals then pupate and new adults emerge another two weeks later, usually in July or August. There is just one generation each year.

271 Larva of alder leaf beetle, *Agelastica alni*.

272 Larvae of alder leaf beetle, *Agelastica alni*.

273 Alder leaf beetle damage to leaves of *Alnus*.

274 Iris flea beetles damaging leaf of *Iris*.

275 Large blue flea beetle, *Altica lythri*.

DAMAGE Adults bite small holes into the cotyledons and leaves, these pit-like blemishes enlarging as the tissue grows. Attacks are usually most serious in May and June, especially on seedlings or recent transplants whose growth is retarded by lack of moisture. Larvae destroy the outer tissue of the roots, heavy attacks affecting plant vigour.

CONTROL If necessary, seedlings should be protected with an insecticide.

Aphthona nonstriata (Goeze)
syn. *coerulea* (Fourcroy)
Iris flea beetle
Locally common on yellow flag (*Iris pseudacorus*) and sometimes also associated with cultivated irises growing in parks and gardens. The beetles, which occur in the spring and summer, feed on the foliage, forming long whitish markings (Fig. **274**). Individuals are 2.5–3.0 mm long and mainly blue, with finely punctured elytra; the legs are pale brown, with distinctive black hind femora.

Altica lythri (Aubé)
Large blue flea beetle
An occasional pest of *Fuchsia* and certain other ornamentals, including large evening primrose (*Oenothera erythrosepala*) and *Potentilla fruticosa*, but associated mainly with wild willowherbs (Onagraceae). Widely distributed in Europe but local.

DESCRIPTION **Adult:** 4.5–5.5 mm long; bright blue and rather plump; elytra irregularly punctured and with slightly curved sides (Fig. **275**). **Egg:** 1.2 × 0.25 mm; whitish-orange (Fig. **276**). **Larva:** 10–12 mm long; head black; body yellowish with numerous darker plates (Fig. **277**).

276 Eggs of large blue flea beetle, *Altica lythri*.

277 Larva of large blue flea beetle, *Altica lythri*.

LIFE HISTORY Adults overwinter to reappear from May onwards. They commonly aggregate on host plants, especially wild hosts such as great willowherb (*Epilobium hirsutum*), marsh willowherb (*Epilobium palustre*) and rose-bay (*Chamaenerion angustifolium*). Eggs are laid singly or in small groups on the underside of the leaves. They hatch about 2–3 weeks later. The larvae feed on the leaves for several weeks before pupating in the ground, new adults appearing in the late summer.

DAMAGE Infested leaves become notched and covered in numerous holes, affecting the quality and marketability of nursery plants. Heavy infestations cause severe foliage damage which can lead to premature leaf-fall. Damage is also inflicted on the petals of host plants.

CONTROL Apply a contact or a stomach-acting insecticide as soon as adult feeding damage is seen. If feasible, destruction of willowherbs in the immediate vicinity of infested nursery beds can also be worth while.

Altica ericeti (Allard)
Heather flea beetle
A locally common species, associated with heather (*Erica*) and sometimes damaging to such plants in gardens and nurseries. Adults (4–5 mm long) are metallic bluish-green with distinctly punctured elytra.

Chalcoides aurata (Marsham)
Willow flea beetle
A widely distributed and locally common species on sallow and willow (*Salix*), affecting both mature trees and nursery stock. Infestations are also increasingly reported on poplar (*Populus*), damage occurring in nurseries, on windbreaks and on isolated trees. Adults, which overwinter in the ground, occur on host plants from May to September, grazing the foliage and producing small but noticeable holes in the expanded leaves; in spring, the beetles may also damage unopened buds. Individuals are 2.5–3.2 mm long and distinguished by the reddish thorax and greenish-black elytra, black or partly black antennae and mainly pale legs (Fig. **278**).

Chalcoides aurea (Fourcroy)
Poplar flea beetle
This locally common flea beetle is associated mainly with poplar (*Populus*), including aspen (*Populus tremula*), but will also occur on willow (*Salix*), causing minor damage to the foliage (Fig. **279**). Adults occur from May to early October. They are similar to those of the previous species but slightly larger (2.7–3.5 mm long), with mainly pale antennae, and the thorax and elytra metallic green with a reddish sheen (Fig. **280**).

278

278 Willow flea beetle, *Chalcoides aurata*.

279

279 Poplar flea beetle damage to leaf of *Populus alba*.

280

280 Poplar flea beetle, *Chalcoides aurea*.

6. Family ATTELABIDAE (weevils)

A small group of weevils whose antennae lack a long scape, each segment being of similar length.

Attelabus nitens (Scopoli)
syn. *curculionoides* Linnaeus
Oak leaf roller weevil

A locally common but minor pest of oak (*Quercus*); also found on alder (*Alnus*), hazel (*Corylus avellana*) and sweet chestnut (*Castanea sativa*). Widespread in mainland Europe, northwards to southern Scandinavia; in the British Isles most numerous in the southern half of England.

DESCRIPTION **Adult:** 5.0–6.5 mm long; bright red to dark red, with the head, antennae, legs, scutellum and underside of body black. **Egg:** 1.0–1.5 mm long; oval, yellow. **Larva:** head brownish-yellow; body yellow to orange-yellow, and strongly C-shaped (Fig. **281**).

LIFE HISTORY Adults appear in May, and are most commonly associated with young oak trees. In June, mated females deposit eggs singly or in small groups in the mid-rib of expanded leaves. At the same time, they initiate larval feeding shelters by severing the leaf lamina on both sides (cf. *Apoderus coryli*, p. 131) of the mid-rib at right-angles to the main axis, about a third of the distance from the base (the oviposition point being about 10 mm beyond the line of cut). The freed parts of the lamina on either side of the mid-rib then fold together, until their surfaces meet, and finally roll up from the tip to produce a characteristic pouch within which larval development takes place (Fig. **282**). Larvae feed within their shelters for about six weeks, the rolled leaf fragment eventually falling to the ground. Larvae then overwinter and pupate in the following spring.

DAMAGE The unusual larval shelters attract attention but damage to infested trees is not important.

281 Larva of oak leaf roller weevil, *Attelabus nitens*.

282 Larval habitation of oak leaf roller weevil on *Quercus*.

283 Hazel leaf roller weevil, *Apoderus coryli.*

284 Larva of hazel leaf roller weevil, *Apoderus coryli.*

Apoderus coryli (Linnaeus)
Hazel leaf roller weevil

A generally common but minor pest of hazel (*Corylus avellana*); at least in mainland Europe, attacks also occur on alder (*Alnus*), beech (*Fagus sylvatica*), birch (*Betula*), hornbeam (*Carpinus betulus*) and European hop-hornbeam (*Ostrya carpinifolia*). Present throughout Europe.

DESCRIPTION **Adult:** 6–8 mm long and mainly red, with a black, elongate, pear-shaped head, black legs and antennae (Fig. **283**). **Egg:** 1.0–1.5 mm long; oval, orange-coloured. **Larva:** 8–10 mm long and bright orange with a brown head (Fig. **284**). **Pupa:** 6–8 mm long; orange (Fig. **285**).

LIFE HISTORY Adults emerge in the spring to feed on the leaves of hazel. Eggs are laid in May and June, singly or in small groups, usually in the mid-rib towards the tip of an expanded leaf. The lamina is then severed near the base, the cut extending from one edge to or just beyond the mid-rib (cf. *Attelabus nitens*, p. 130); the cut tissue then curls laterally to remain suspended from the unsevered part of the lamina as a stumpy, cigar-like leaf roll (Fig. **286**). The larvae develop within these leaf rolls and then pupate, new adults appearing at the end of July or in early August. Larvae of a second generation complete their development in the autumn; they eventually overwinter on the ground, within their fallen habitations, and pupate in the spring.

DAMAGE Although infestations are sometimes established on ornamental plants and larval habitations may be very numerous, damage caused is unimportant.

285 Pupa of hazel leaf roller weevil, *Apoderus coryli.*

286 Larval habitation of hazel leaf roller weevil on *Ostrya carpinifolia.*

Rhynchites aequatus (Linnaeus)
Apple fruit rhynchites

A small (2.5–4.5 mm long), reddish-brown to purplish-bronze weevil (Fig. 287), associated primarily with hawthorn (*Crataegus*) but also attacking certain other rosaceous trees and shrubs, including crab apple (*Malus*), ornamental cherry (*Prunus*) and *Sorbus*. The adults feed in the spring on the buds, flowers and foliage, causing slight but minor damage. The females later attack the berries, depositing eggs in small pits bitten into the flesh. Larvae feed in the developing fruits for about three weeks, eventually dropping to the ground to pupate in the soil in earthen cells. There is just one generation annually. Infestations on ornamentals are not important.

287 Apple fruit rhynchites, *Rhynchites aequatus.*

Byctiscus betulae (Linnaeus)
syn. *betuleti* (Fabricius)
Poplar leaf roller weevil

A widespread but locally distributed weevil, associated mainly with hazel (*Corylus avellana*) and poplar (*Populus*) but also attacking other trees such as alder (*Alnus*), birch (*Betula*), elm (*Ulmus*) and sallow (*Salix*). Eggs are laid in leaf veins during late May or June, typically several per leaf; the female also partly severs the leaf petiole, the lamina then rolling into a pendulous, cigar-like tube within which larval development takes place. Fully grown larvae pupate in the ground, some adults emerging in the autumn but others not until the following spring. Infestations are sometimes noticed on cultivated plants but are usually of only minor importance. The adult weevils are 6–9 mm long, and dark blue, green or red with distinctly metallic legs; males possess a pair of characteristic, forwardly directed spines on the thorax.

288 Female birch leaf roller weevil, *Deporaus betulae.*

Byctiscus populi (Linnaeus)
Aspen leaf roller weevil

This smaller (5–6 mm long), green or reddish, black-legged weevil (the underside is characteristically bluish-black) is associated mainly with young aspen (*Populus tremula*) trees, causing similar damage to the previous species. Although infestations occur occasionally on ornamental plants, they are of little or no importance.

Deporaus betulae (Linnaeus)
Birch leaf roller weevil

An often common but minor pest of young birches (*Betula*); also associated with certain other trees, including alder (*Alnus*), beech (*Fagus sylvatica*), hazel (*Corylus avellana*) and hornbeam (*Carpinus betulus*). Present throughout Europe.

DESCRIPTION **Adult:** 3–5 mm long; black with rectangular, deeply pitted elytra (Fig. 288); hind femora of male greatly thickened. **Larva:** 5–7 mm long; whitish.

LIFE HISTORY Adults appear in the spring, to feed on the foliage of various host plants. Egg laying begins from early May onwards, when mated females initiate characteristic larval habitations. Firstly, one or two eggs are deposited in the upper surface of an expanded leaf; the selected leaf is then severed neatly across the lamina on both sides of the mid-rib, shortly beyond the base, the tissue rolling into an elongate, cone-like tube. The suspended material soon dries out, cones formed on beech turning rusty brown and becoming especially obvious (Fig. **289**). Larvae are fully grown in about three months. They then enter the soil where they pupate, adults emerging in the following spring. There is a single generation annually.

DAMAGE Infestations disfigure host plants and are often extensive, with all leaves on certain shoots affected, but damage is not of economic importance.

289 Larval habitation of birch leaf roller weevil on *Fagus*.

Deporaus tristis (Fabricius)
Maple leaf roller weevil

This relatively small (3.5–4.0mm long), bluish-black species is associated mainly with *Acer*, including sycamore (*Acer pseudoplatanus*); occasionally, beech (*Fagus sylvatica*) and oak (*Quercus*) are also hosts. The adults feed on the leaves, removing significant amounts of tissue; they also form characteristic, tightly wound leaf rolls (Fig. **290**) within which larval development takes place. Affected leaves are sometimes found on amenity trees in mountainous parts of Europe, from southern Germany southwards, but infestations are not of significance. The weevil is widely distributed in southern Europe.

290 Larval habitation of maple leaf roller weevil on *Acer*.

7. Family CURCULIONIDAE (weevils)

The main family of weevils, the antennae having a very long basal segment (scape). The larvae are apodous, with a distinct head; they often adopt a C-shaped posture.

Otiorhynchus clavipes (Bonsdorff)
syn. *fuscipes* (Olivier); *lugdunensis* Boheman
Red-legged weevil

Adults attack various woody hosts, including lilac (*Syringa vulgaris*), Japanese laurel (*Aucuba japonica*) and honeysuckle (*Lonicera*); the larvae are associated with the roots of soft-fruit crops, especially raspberry (*Rubus idaeus*), and ornamentals such as lilac, maple (*Acer*), *Viburnum* and *Weigelia*. Locally common in parts of Europe, including south-western England, Belgium, France, Germany, the Netherlands and Switzerland, especially on light soils.

DESCRIPTION **Adult:** 8–12 mm long; body blackish, oval and distinctly pointed posteriorly; sculpturing on thorax and elytra shallow; wingless; legs long and reddish (Fig. **291**). **Egg:** 0.5 × 0.6 mm; more-or-less spherical; whitish when laid but soon becoming black. **Larva:** 8–12 mm long; creamish-white with a brown head; body plump, wrinkled and strongly C-shaped.

LIFE HISTORY Adults appear from late spring to August, the period of appearance varying according to the timing of pupation. They are active at night, feeding on leaves and other aerial parts of various woody plants but drop to the ground immediately if disturbed. The weevils hide by day in grass tussocks, under stones and in other shelter. Eggs are laid in the soil, scattered at random close to the surface below host plants, each female depositing up to 300. Eggs hatch in about three weeks. The larvae then feed on plant roots, passing through five instars. They overwinter either as fully fed individuals or as young individuals which complete their development in the following spring; the former pupate in spring and the latter in the summer. Reproduction is either sexual or parthenogenetic.

DAMAGE Adults notch the leaves, and damage blossoms and unopened buds; they also check the growth of young shoots by gnawing at the bases. Larvae are more important; they browse on the roots, attacked plants wilting and sometimes dying.

CONTROL As for *Otiorhynchus sulcatus* (p. 136), but treatments against larvae are best delayed until September.

291 Red-legged weevil, *Otiorhynchus clavipes*.

292 Strawberry weevil, *Otiorhynchus ovatus*.

Otiorhynchus ovatus (Linnaeus)
Strawberry weevil

An often common pest of various herbaceous plants and young and seedling trees, especially Norway spruce (*Picea abies*) grown as Christmas trees. Eurasiatic. Present throughout Europe; introduced into Central and North America.

DESCRIPTION **Adult female:** 5.0–5.5 mm long; black to reddish-brown, with a slight yellow pubescence; disc of thorax distinctly furrowed (Fig. **292**). **Larva:** 6 mm long; creamish-white with a brown head.

LIFE HISTORY Weevils are active from April or May onwards, feeding on the foliage and young shoots of various hosts, especially seedling conifers and strawberry plants. Eggs are laid in the soil in the spring, the larvae attacking the fine roots, causing most damage in May and June, and usually completing their development in the summer. There is a single generation annually, but details of the life-cycle vary according to local conditions.

DAMAGE **Conifers:** adults destroy the young growth and needles, affecting the quality and vigour of host plants; the larvae destroy the smaller roots, weakening and often killing seedlings.

CONTROL As for *Otiorhynchus sulcatus* (p. 136), but treatments against larvae are necessary earlier in the year.

Otiorhynchus singularis (Linnaeus)
syn. *picipes* (Fabricius)
Clay-coloured weevil

An often troublesome pest of ornamentals such as buddleia (*Buddleja*), cherry (*Prunus*), *Clematis*, crab apple (*Malus*), *Primula*, *Rhododendron*, *Hydrangea*, rose (*Rosa*), *Wisteria* and yew (*Taxus baccata*); also damaging to conifers, especially western hemlock (*Tsuga heterophylla*). Widespread in Europe, especially in grassy lowland sites; an introduced pest in North America.

DESCRIPTION **Adult:** 6–7 mm long; shiny black and strongly sculptured, but covered with greyish-brown scales that give the body an overall dull, light and irregular pattern; also, the body is often encrusted with mud; wingless (Fig. **293**). **Larva:** 8 mm long; creamish-white with a brown head; body plump and wrinkled.

LIFE HISTORY Adults occur from April to October, hiding by day in cracks in the soil, under straw mulches and in other shelter beneath host plants. At night, they feed on the foliage, buds and bark, causing most harm from April to June. Eggs are laid in the soil during the summer. Larvae then feed on the roots of many plants, including various weeds, from late summer onwards, pupating in the early spring. Adults appear shortly afterwards. Some individuals survive for two or more seasons, hibernating during the winter months. Males are very rare and reproduction is mainly partheno-genetic.

DAMAGE Larvae attack the roots, sometimes seriously weakening herbaceous plants in rockeries. Generally, however, adults are more important pests, the precise type of injury caused varying from host to host. On young trees, large irregular areas of bark are removed, stems sometimes being ring-barked and plants killed; the weevils also notch the leaves and destroy the buds of young grafts. Leaf petioles on young shoots are sometimes gnawed, causing them to fold over, wilt and drop off; buds on young bushes are also attacked, producing misshapen plants with forked shoots or numerous unwanted laterals; new growth is often girdled and killed.

CONTROL Control of adult damage is often difficult but drenching the base and trunks of susceptible trees and shrubs in April, May or early June is recommended, using an insecticide effective against chewing insects. For control of larvae, see under *Otiorhynchus sulcatus* (p. 136).

293 Clay-coloured weevil, *Otiorhynchus singularis*.

294 Vine weevil, *Otiorhynchus sulcatus*.

Otiorhynchus sulcatus (Fabricius)
Vine weevil

A major pest of glasshouse and outdoor ornamentals, including alpines (especially *Saxifraga*, *Sedum* and *Sempervivum*), balsam (*Impatiens*), *Begonia*, *Calceolaria*, *Camellia*, *Cotoneaster*, *Cyclamen*, *Elaeagnus*, ferns (e.g. *Adiantum*), heather (*Erica*), lily of the valley (*Convallaria majalis*), Michaelmas daisy (*Aster*), peony (*Paeonia*), *Phlox*, *Polyanthus*, *Primula*, *Rhododendron* and various tree seedlings (coniferous and deciduous species). Widely distributed in Europe; an introduced pest in Australasia and North America.

DESCRIPTION **Adult:** 7–10 mm long; black and shiny, the elytra parallel-sided and coated with patches of yellowish hairs; body sculpturing deep; wingless (Fig. **294**). **Egg:** 0.7 mm across; more-or-

135

less spherical; white at first, soon turning brownish (Fig. **295**). **Larva:** 8–10mm long; creamish to brownish-white, with a reddish-brown head (Fig. **296**). **Pupa:** 7–10mm long; white (Fig. **297**).

LIFE HISTORY In outdoor situations, adult females of this parthenogenetic weevil emerge in May and June. They are active at night, feeding on the foliage of various plants. During the daytime, they tend to hide away beneath debris, in crevices in walls or in other shelter. The weevils mature a few weeks after commencing feeding, each being capable of depositing several hundred eggs. The eggs are laid in soil near host plants from late July onwards. They hatch in 2–3 weeks and the larvae then attack plant roots, corms or rhizomes. Larval feeding extends over many months, development being completed in the following spring and fully grown individuals pupating in oval subterranean cells formed in the soil; transformation to the adult stage occurs from mid-April to June. Most weevils die before the onset of winter but some may survive for two seasons or longer, overwintering in sheltered situations and reappearing in the spring. Young adults can be distinguished from older individuals by the presence of a thick spine on each mandible, which usually breaks off soon after emergence. In heated glasshouses, adults often emerge in large numbers in the autumn; also, eggs are laid over a more extended period, and all stages of the pest may occur at one and the same time.

DAMAGE Adults notch the leaves of host plants, often causing extensive damage to ornamentals such as camellia, lily of the valley and rhododendron. They may also ring-bark young plants. Most important damage, however, is caused by the larvae which destroy the finer roots, or burrow into the fleshy parts of corms and rhizomes; affected plants lack vigour and may suddenly wilt, or collapse and die. Damage is especially important on outdoor containerized plants growing in peat compost and on glasshouse pot plants, but may also be extensive in nursery beds and rock gardens.

CONTROL Control of adults is difficult because of the extended period of activity. However, to reduce the incidence of leaf notching and ring-barking, spray with an insecticide at the first signs of damage; such treatments are most effective if applied in the evening. Some protection of vulnerable plants can be achieved by physical barriers but these are not always a practical method of control. Best chemical control of larvae is given (where appropriate) by compost-incorporated treatments, although topical

295 Eggs of vine weevil, *Otiorhynchus sulcatus.*

296 Larva of vine weevil, *Otiorhynchus sulcatus.*

297 Pupa of vine weevil, *Otiorhynchus sulcatus.*

application of insecticide drenches or granules can also be effective; the latter treatments should be applied against the young larvae (usually in August but perhaps earlier in the year on glasshouse plants). If necessary, on infested sites, a soil insecticide should be worked into the soil before replanting. Biological control agents, such as entomopathogenic fungi or entomophilic nematodes, are also effective against vine weevil; the commercial use of these products has greatest potential on container-grown plants.

Otiorhynchus crataegi Germar
Mediterranean hawthorn weevil

This small species is a pest of various ornamentals in eastern Europe, the weevils grazing on foliage and the larvae feeding on plant roots. In parts of France, lilac (*Syringa vulgaris*) and privet (*Ligustrum vulgare*) are also attacked; in 1985, following an accidental introduction, extensive but localized damage was reported on privet hedges, ornamental shrubs and certain other plants, in south-central England. Adults are 5 mm long, and mainly dark brown, marked irregularly with golden-brown; the narrow, rounded thorax, bulbous abdomen and relatively narrow waist (Fig. **298**) are characteristic.

Otiorhynchus niger (Fabricius)

A common forestry pest in mainland Europe, especially in mountainous areas; sometimes found on amenity trees but not an important pest of ornamentals. Adults are 9–11 mm long and mainly black, with dark orange legs and deeply sculptured elytra (Fig. **299**).

Otiorhynchus rugosostriatus (Goeze)
Lesser strawberry weevil

Although most frequently reported as a pest of strawberry, this widely distributed and locally common weevil also attacks pot plants and ornamental shrubs, the adults notching the edges of expanded leaves and the larvae feeding on the roots. Damage is most commonly reported on *Begonia*, *Cyclamen*, lilac (*Syringa vulgaris*), *Primula* and privet (*Ligustrum vulgare*). Adults (6–7 mm long) are blackish to reddish-brown, and strongly sculptured (Fig. **300**); they are most numerous from June to September. The larvae, which feed throughout the autumn and winter, are similar to those of *Otiorhynchus sulcatus* (p. 136) but smaller,

298 Mediterranean hawthorn weevil, *Otiorhynchus crataegi*.

299 Female of *Otiorhynchus niger*.

300 Lesser strawberry weevil, *Otiorhynchus rugosostriatus*.

reaching a maximum length of about 8 mm (Fig. **301**).

Phyllobius argentatus (Linnaeus)
Silver-green leaf weevil
A common but minor pest of various trees and shrubs, including alder (*Alnus*), beech (*Fagus sylvatica*), birch (*Betula*), cherry (*Prunus*), hawthorn (*Crataegus*) and *Sorbus*, often troublesome on young ornamentals and nursery stock; sometimes also found on conifers. Widespread in Europe.

DESCRIPTION **Adult:** 4–6 mm long; body black but appearing bright green due to a covering of shiny, golden-green, rounded, disc-like scales; legs and antennae light brown or yellowish and partially clothed in golden-green scales; femora toothed (Fig. **302**) (cf. *Polydrusus sericeus* p. 140).

LIFE HISTORY Adults feed in the spring and early summer on the foliage and flowers of various trees and shrubs. In sunny weather, the weevils often bask openly on the leaves; in dull conditions they tend to hide amongst folded or crinkled foliage, often sheltering in vacated habitations of tortricid moths and other leaf-deforming pests, but they are readily detected if a branch is tapped over a cloth or tray. Later in the season, the weevils migrate to other hosts, including herbaceous plants. Eggs are laid in the soil during the early summer, the larvae feeding on the roots of various herbaceous weeds and grasses. They pupate in the spring within earthen cells, adults emerging a few weeks later.

DAMAGE Adults bite holes into the leaves and flower petals, sometimes causing extensive damage to young trees and shrubs; most damage is caused from April to July, but the extent of injury varies considerably from year to year and is not apparently related to the size of populations invading the trees.

CONTROL Treatment of established trees and shrubs is rarely justified but, if necessary, apply an insecticide as soon as damage occurs. Do not spray at the flowering stage because of the risk to bees and other beneficial insects.

Phyllobius maculicornis Germar
Often common on young deciduous trees, especially beech (*Fagus sylvatica*), birch (*Betula*) and hawthorn (*Crataegus*). The weevils are usually less numerous than *Phyllobius argentatus* (p.000), but are capable of causing extensive damage to the foliage of suitable hosts. Both species are of similar appearance and size, but *Phyllobius maculicornis* is distinguished

301 Larva of lesser strawberry weevil, *Otiorhynchus rugosostriatus*.

302 Silver-green leaf weevil, *Phyllobius argentatus.*

by the black femora and stouter, partly blackish antennae (club and apex of the scape) and more prominent eyes (Fig. **303**).

Phyllobius oblongus (Linnaeus)
Brown leaf weevil
An often abundant species occurring throughout Europe on various shrubs and trees, including elm (*Ulmus*), lime (*Tilia*), maple (*Acer*), poplar (*Populus*) and willow (*Salix*), but most often noted on rosaceous species such as crab apple (*Malus*), hawthorn (*Crataegus*) and ornamental cherry (*Prunus*); frequently a problem on nursery trees. Adults (3.5–6.0 mm long) are black with reddish-brown, slightly pubescent elytra, and brownish legs and antennae (Fig. **304**); unlike other members of the genus, the elytra are devoid of scales.

303 Adults of *Phyllobius maculicornis.*

304 Brown leaf weevil, *Phyllobius oblongus.*

Phyllobius pyri (Linnaeus)
Common leaf weevil

Heavy infestations of this species often occur on trees and shrubs, including alder (*Alnus*), ash (*Fraxinus excelsior*), beech (*Fagus sylvatica*), birch (*Betula*), elm (*Ulmus*), hawthorn (*Crataegus*), horse chestnut (*Aesculus hippocastanum*), ornamental cherry (*Prunus*) and *Sorbus*. The weevils often occur in association with *Phyllobius argentatus* (p. 138), sometimes causing extensive damage to the foliage and blossom of ornamentals and nursery stock. In early summer, they are especially partial to nettles. Adults are 5–7mm long, and black but more-or-less covered with elongate, coppery, golden or greenish-bronze scales (Fig. **305**); the femora are distinctly toothed.

305 Common leaf weevil, *Phyllobius pyri.*

Phyllobius roboretanus Gredler
syn. *parvulus* (Olivier)

This relatively small species is common on young deciduous trees, especially oak (*Quercus*), upon which it may be found from May onwards; also associated with nettle (*Urtica*). Adults are 3–4mm long, with untoothed femora and the elytra coated in green scales (Fig. **306**); they are readily distinguished by the black, unscaled abdomen which bears just a few green hairs (cf. *Phyllobius viridiaeris*, pp. 139–140).

Phyllobius viridiaeris (Laicharting)
syn. *pomonae* (Olivier)

A generally common weevil, sometimes causing minor damage to the foliage of young ornamental trees and shrubs. Adults (3.0–4.5mm long) are black, covered with green or yellowish-green scales both above and below; the antennae and legs are

306 Adult of *Phyllobius roboretanus.*

mainly reddish and the femora untoothed (Fig. 307).

Polydrusus sericeus (Schaller)
Green leaf weevil
A locally common but minor pest of various trees and shrubs, including alder (*Alnus*), birch (*Betula*), elm (*Ulmus*), hazel (*Corylus avellana*), oak (*Quercus*), poplar (*Populus*) and willow (*Salix*); also occurs on conifers. Widespread in Europe; introduced into North America.

DESCRIPTION **Adult:** 5–8 mm long; body black but covered with shiny, green scales; legs brown; femora untoothed. Furrow (strobe) on each side of the rostrum, distinct and curved downwards posteriorly, below the eye — typical of *Polydrusus* (in *Phyllobius*, the strobes are straight and often filled with scales).

LIFE HISTORY Adults occur during the spring and summer, feeding on the buds, flowers and foliage of trees and shrubs. Eggs are laid in the soil, the larvae feeding on plant roots from summer onwards; they hibernate during the winter and pupate in the early spring.

DAMAGE Although adults are sometimes present in large numbers, damage to ornamental trees and shrubs is usually slight.

CONTROL As for *Phyllobius argentatus* (p. 138).

Polydrusus pterygomalis (Boheman)
This local but widely distributed weevil occurs on young trees, including hawthorn (*Crataegus*), oak (*Quercus*), ornamental cherry (*Prunus*) and willow (*Salix*), contributing to leaf damage caused by other more numerous leaf weevils. The adults (3–5 mm long) are black-bodied, coated in shiny green scales, with yellow legs and antennae; the femora are untoothed and the head has distinctly swollen temples.

Barypeithes araneiformis (Schrank)
Smooth broad-nosed weevil
Adults of this widely distributed but local weevil are sometimes damaging during the spring and summer in conifer nursery beds, when they destroy the initial shoots and also ring-bark the young seedlings. They may also attack young deciduous trees, such as birch (*Betula*), horse chestnut (*Aesculus hippocastanum*) and oak (*Quercus*), damaging the buds and severing or forming holes in the

307 Adult of *Phyllobius viridiaeris*.

308 Smooth broad-nosed weevil damage to leaf of *Betula*.

expanded leaves (Fig. **308**). The larvae feed on the roots of herbaceous plants, including various weeds. Adults are 3–4 mm long, shiny brownish-yellow to black, with a pointed abdomen.

CONTROL Spray with a contact or a stomach-acting insecticide at the first signs of damage.

Barypeithes pellucidus (Boheman)
Hairy broad-nosed weevil
This locally distributed weevil is also sometimes damaging to young or seedling trees. Adults (Fig. **309**) are distinguished from those of *Barypeithes araneiformis* by the longer, denser and more distinct pubescence.

Strophosomus melanogrammus (Förster)
syn. *coryli* (Fabricius)
Nut leaf weevil
An often common pest of various trees and shrubs, including beech (*Fagus sylvatica*), birch (*Betula*), *Rhododendron* and various conifers, but especially abundant on hazel (*Corylus avellana*); sometimes an important pest of young trees and nursery stock, especially in wooded areas. Widely distributed in Europe.

DESCRIPTION **Adult female:** 4–6mm long; body bulbous and mainly black or brown, coated in greyish-brown scales and with a distinctive, black, longitudinal line down the thorax and between the elytra; head broad, with protruding black eyes (Fig. **310**).

LIFE HISTORY Adult females of this mainly parthenogenetic species emerge from hibernation in the spring. They browse on the leaves and attack the shoots of various trees and shrubs, feeding throughout the spring and summer months. Eggs are laid in the soil, the larvae developing on the roots of herbaceous weeds and usually pupating in the late summer; young adults emerge shortly afterwards.

DAMAGE Leaf damage is of no importance but shoots are often ring-barked so that they wither and die, infestations being of particular significance on larch (*Larix*) seedlings. Infestations are also often of importance on western hemlock (*Tsuga heterophylla*).

CONTROL Spray with an insecticide in April or early May, when damage is first seen.

309 Hairy broad-nosed weevil, *Barypeithes pellucidus*.

310 Nut leaf weevil, *Strophosomus melanogrammus*.

Barynotus obscurus (Fabricius)
Adults of this widely distributed and locally common weevil feed on the flowers of various plants, including ornamentals such as *Primula*, rose (*Rosa*) and *Viola*. Damage occurs from spring to mid-summer, most serious attacks having been reported on rose in England. The weevils are 8–10mm long and black or brownish, with a coating of round, close-set, yellow (sometimes greenish or coppery) scales (Fig. **311**); they are superficially similar in appearance to *Otiorhynchus sulcatus* (p. 135) but the thoracic region is relatively broad, with only a slight constriction between thorax and abdomen, and the femora are untoothed.

311 Female of *Barynotus obscurus*.

Hypera arator (Linnaeus)
syn. *polygoni* (Linnaeus)

This widely distributed species is associated with various members of the Caryophyllaceae and is sometimes a minor pest of carnation and pink (*Dianthus*). The yellowish-grey to greenish larvae (up to 7 mm long) feed within the flowers, damage on cultivated plants sometimes being mistaken for that caused by larvae of the carnation tortrix moth (*Cacoecimorpha pronubana*, p. 239). Pupation occurs in oval, yellowish-white cocoons formed on the foodplant, young weevils appearing about two weeks later. The adults (5–7 mm long) are yellowish-brown with pale, dark-edged longitudinal bands on the thorax and elytra and a black line along the elytral suture. They may be found from early May to the end of September.

312 Figwort weevil, *Cionus scrophulariae.*

Cionus scrophulariae (Linnaeus)
Figwort weevil

A generally common but minor pest of buddleia (*Buddleja globosa*), mullein (*Verbascum*) and *Phygelius*; most abundant on wild figwort (*Scrophularia*). Widely distributed in Europe.

DESCRIPTION **Adult:** 4.0–4.5 mm long; thorax small, elytra square; body black but largely coated with a purplish-grey pubescence; elytra with a regular pattern of black markings, including two large circular patches; antennae dark red (Fig. **312**). **Larva:** 5–6 mm long; yellowish with a black head and prothoracic plate; body coated with greenish-yellow slime (Fig. **313**).

313 Larva of figwort weevil, *Cionus scrophulariae.*

LIFE HISTORY Adults occur on host plants from May or June onwards, feeding on the leaves, petioles and flowers. Larvae feed during the summer on the leaves, grazing just one surface, the other remaining intact. When fully grown, they construct more-or-less spherical, brownish to yellowish-brown, parchment-like cocoons on the stems and leaves (Fig. **314**), and then pupate. Adults emerge 2–3 weeks later and eventually produce a further generation of larvae. Adults of the final generation enter hibernation to reappear in the following spring.

DAMAGE Attacked plants are disfigured, and leaf tissue infested by larvae may eventually split; flower buds may also be destroyed.

CONTROL If necessary, spray with an insecticide as soon as damage is seen.

314 Pupal cocoon of figwort weevil, *Cionus scrophulariae.*

315 Mullein weevil, *Cionus hortulanus*.

316 Ash leaf weevil, *Stereonychus fraxini*.

Cionus hortulanus (Fourcroy)
Mullein weevil

This locally common weevil also infests cultivated mullein (*Verbascum*) but has a more restricted host range than the previous species, being found most often on dark mullein (*Verbascum nigrum*). Adults of both species of *Cionus* are similar in appearance but *Cionus hortulanus* is slightly smaller (3.5–4.0 mm long) and paler in colour, the elytra appearing mainly yellowish-grey or greenish-grey, interrupted by two black patches and a distinct pattern of small black markings (Fig. **315**).

Stereonychus fraxini (Degeer)
syn. *rectangulus* (Herbst)
Ash leaf weevil

A southerly-distributed pest of ash (*Fraxinus excelsior*), including ornamental trees; also associated with certain other plants, including lilac (*Syringa vulgaris*) and *Phillyrea*. Widespread in south-central and southern Europe, including parts of France and southern Germany; also found in Asia Minor and North Africa. Not present in the British Isles.

DESCRIPTION **Adult:** 3 mm long; greyish-brown to reddish-brown, with a blackish thoracic disc and an elongate, cylindrical rostrum (Fig. **316**). **Larva:** 4 mm long; greenish-yellow with a black head.

LIFE HISTORY Adults overwinter in debris on the ground, and in other shelter, emerging in the spring. They then feed on the buds and, later, on the leaves and petioles. Eggs are deposited close to the veins on the underside of the expanded leaves. The larvae graze on the surface of the leaves, forming a series of window-like patches (Fig. **317**). When fully fed they pupate, each in an oval, transparent, yellowish-brown to brownish, parchment-like cocoon formed on the leaf surface (Fig. **318**).

317 Ash leaf weevil damage to leaf of *Fraxinus*.

318 Pupal cocoon of ash leaf weevil.

Young adults appear about ten days later; they feed on the foliage before entering hibernation. There is one generation annually.

DAMAGE Weevils feeding on unopened buds in the early spring can delay the appearance of the new growth. Adult and larval damage to the expanded foliage is disfiguring and, if attacks are heavy, will reduce plant vigour.

CONTROL On young plants, apply an insecticide at the first signs of leaf damage. Control measures on larger trees are not worth while.

Cryptorhynchus lapathi (Linnaeus)
Osier weevil

A locally important pest of poplar, especially the hybrid *Populus serotina*, and willow (*Salix*); also associated with alder (*Alnus*) and, less frequently, birch (*Betula*). Often damaging in nursery beds, including basket-willow beds. Widely distributed in Europe, including Czechoslovakia, England, France, Germany, Hungary, Italy, the Netherlands, Poland, Spain, Wales and Yugoslavia; also introduced into North America.

DESCRIPTION **Adult:** 8–9 mm long; robust; mainly black, interspersed with white or yellowish-white scales, especially towards the base and at the apex of the elytra (Fig. **319**). **Egg:** 1 mm long; creamish-white. **Larva:** creamish-white with a brown head.

LIFE HISTORY In subtropical areas and southern Europe, this insect has a one-year life-cycle. In most parts of Europe, however, development usually extends over two years. Adults emerge in the spring, from May onwards. They usually remain hidden beneath host plants during the day and become active in the evening, crawling up the stems to feed on the young shoots. They are very sluggish, dropping to the ground if disturbed; although fully winged, they rarely, if ever, fly. Adults continue to feed throughout the summer and autumn, mated females eventually laying eggs singly in the rods and stools of host plants. Surface wounds, leaf scars and lenticels are often chosen as oviposition sites, the females boring out a cavity before depositing a single egg a few millimetres beneath the surface. Eggs hatch in about three weeks. Irrespective of when eggs are laid, the larvae feed only briefly during the first season; they then enter diapause, still in their first instar, to resume activity in the following spring. The larvae feed within the host tissue for several months, ejecting whitish frass through a hole bored through the bark. They often penetrate to the pith of the branches and trunks of host plants, also burrowing

319 Osier weevil, *Cryptorhynchus lapathi.*

320 Osier weevil damage to shoot of *Populus.*

extensively within the rods and stools, fully fed larvae eventually pupating at the end of their galleries from mid-July onwards. Adults emerge from the pupae in late July or August but do not vacate the pupal cells until the following spring.

DAMAGE Young terminal shoots are partly severed by the adult weevils, the tips keeling over and eventually turning black (Fig. **320**). Shoots may also become riddled with holes; such damage is of particular importance in willows intended for basket making, affected rods having to be discarded. The larvae reduce the production of new shoots from infested stools; they are sometimes of importance in nurseries, heavy attacks causing death of plants.

CONTROL Chemical control of adults is difficult and often impractical, requiring several applications to coincide with the main periods of activity; control of larvae is also difficult, but likely to be most successful if an insecticide is applied to stools or trunks of young trees in April at the commencement of shoot growth and before larvae have attained the second instar.

Dorytomus taeniatus (Fabricius)

Generally common on sallow and willow (*Salix*) from May onwards. The adults graze on the leaves, removing patches of tissue to expose a network of fine veins; they often feed in groups but damage caused, although noticeable and sometimes extensive, is insignificant. Eggs, which are laid during the autumn in the axils of the catkin buds, hatch early in the following spring. The larvae then mine within the catkins and eventually pupate in the soil, young weevils appearing in May and June. The weevils often hide during the day amongst curled leaves, dropping to the ground if disturbed. Adults are 4–5 mm long, black to brownish-black with elongate, parallel-sided elytra, an elongate rostrum and toothed femora; the elytra are also patterned with small patches of lighter hairs (Fig. **321**). At least in mainland Europe, the weevils will also attack alder (*Alnus*), birch (*Betula*) and poplar (*Populus*).

Anthonomus brunnipennis (Curtis)

syn. *comari* Crotch

A generally common and widely distributed species, associated with wild and cultivated cinquefoil (*Potentilla*). The weevils feed on the foliage and flower petals, producing numerous small holes. Individuals are about 2 mm long and mainly black, with the thorax and elytra covered extensively in fine punctures. Infestations have no significant effect on plant growth but damage, especially to petals of the opened blossoms (Fig. **322**) is unsightly.

Rhynchaenus alni (Linnaeus)

syn. *ferrugineus* (Marsham); *saltator* (Fourcoy)

Elm leaf-mining weevil

A widely distributed pest of elm (*Ulmus*); at least in parts of mainland Europe, also associated with alder (*Alnus*) and hazel (*Corylus avellana*). Present throughout much of mainland Europe, from Denmark southwards; in the British Isles most numerous in the southern half of England.

DESCRIPTION **Adult:** 2.5–3.0 mm long, with yellowish-red or red, black-marked elytra and black legs and head (Fig. **323**); a yellowish variety also

321 Adult of *Dorytomus taeniatus*.

322 *Anthonomus brunnipennis* damage to flower of *Potentilla*.

323 Elm leaf-mining weevil, *Rhynchaenus alni*.

324 Larva of elm leaf-mining weevil, *Rhynchaenus alni*.

325 Mine of elm leaf-mining weevil on *Ulmus*.

occurs. **Larva:** 5–6mm long; whitish to dirty yellowish-white, with a black head (Fig. **324**).

LIFE HISTORY Adults appear in the spring, to feed on the young leaves. Eggs are laid shortly after mating, each in a main vein on the underside of a leaf. The larvae mine within the lamina, each gallery commencing as a narrow channel that widens into a prominent brown blotch and usually terminates at the apex of the leaf (Fig. **325**). The larvae feed for about two weeks. They then pupate, each in a rounded cocoon spun within the mine. New adults emerge in the summer. They feed on the leaves before eventually overwintering.

DAMAGE Adult feeding is usually of minor importance; the larval mines are disfiguring and cause noticeable distortion of affected leaves.

326 Beech leaf-mining weevil, *Rhynchaenus fagi*.

CONTROL Some control of adults can be achieved by spraying in the spring with a contact or a stomach-acting insecticide; young larvae are susceptible to certain systemic insecticides. On ornamentals, chemical control is practical only in nurseries and on small trees.

Rhynchaenus fagi (Linnaeus)
Beech leaf-mining weevil
larva = beech leaf miner
A common and widely distributed pest of beech (*Fagus sylvatica*). Present throughout Europe.

DESCRIPTION **Adult:** 2.2–2.8mm long; body black, covered with a greyish-brown pubescence; antennae brown; elytra distinctly broader than pronotum; eyes close-set; hind tibiae robust (Fig. **326**). **Larva:** 5mm long; body white, shiny and distinctly tapered posteriorly; head blackish (Fig. **327**).

327 Beech leaf miner, *Rhynchaenus fagi*.

LIFE HISTORY Adults overwinter under bark, in the ground and in other shelter, emerging from hibernation in late April or early May. They then frequent beech trees, feeding on the foliage. Eggs, usually 30–35 per female, are laid singly on the underside of the mid-rib of expanded leaves; less often, they are placed beneath a lateral vein. The larvae mine within the leaves from May to June, forming brownish-black galleries which commence as a narrow channel running from the mid-rib to the leaf margin, where it expands into a small brown blotch (Fig. **328**). There are three larval instars, fully fed larvae pupating within the feeding gallery, each in a white, spherical cocoon (Fig. **329**). The overwintered adults survive until the end of May or early June, the new generation of adults appearing in mid-June. These weevils continue feeding on the foliage for several weeks before entering their overwintering sites, often causing considerable leaf damage during June and July. Most weevils enter hibernation in late July or August, but some delay their departure until mid-September. There is just one generation annually.

DAMAGE Adults pepper the foliage with small holes and, if the mid-rib is damaged, leaves may wilt. Larval galleries cause distortion, the leaf tips also turning brown; such damage is often extensive on woodland trees and is also of considerable significance on ornamental trees and hedges.

CONTROL Many pesticides have proved ineffective against this pest but some control of adults can be achieved with contact and stomach-acting insecticides; young larvae are susceptible to certain systemic materials.

328 Mines of beech leaf miner on *Fagus*.

329 Pupal cocoon of beech leaf-mining weevil, *Rhynchaenus fagi*.

Rhynchaenus populi (Fabricius)
Poplar leaf-mining weevil
An often common species, the larvae mining within the leaves of poplar (*Populus*) and willow (*Salix*) to form large, disfiguring, brownish-black blotches (Fig. **330**). Adults (2.0–2.5 mm long) are mainly black with a whitish scutellum and yellowish-red antennae and legs. There is one generation a year, larvae occurring during the summer and new adults, which eventually overwinter, appearing in September.

330 Larval mines of poplar leaf-mining weevil on *Salix*.

331 Larval mine of oak leaf-mining weevil on *Quercus*.

332 Larva of oak leaf-mining weevil, *Rhynchaenus quercus*.

Rhynchaenus quercus (Linnaeus)
Oak leaf-mining weevil

This common species occurs on oak (*Quercus*) and is noted occasionally on young ornamental trees. The weevils occur from April or May onwards, grazing the surface of the leaves. Eggs are laid singly, usually in the underside of the mid-rib. On hatching, the larva mines towards the leaf apex to form an elongate gallery which passes along the mid-rib and eventually moves into the lamina and transfers from the lower to the upper leaf surface; the gallery terminates in an irregular brownish blotch, clearly visible from above (Fig. **331**). The larvae are transparent, greenish-white, with a brown head (Fig. **332**). When fully fed, usually in June or July, they pupate in rounded transparent cocoons formed within the mine. Young weevils emerge 1–2 weeks later. They feed on the leaves and then hibernate, reappearing in the following spring. Adults are 2.5–3.5 mm long, and mainly reddish to brownish-black, with a yellow pubescence.

Rhynchaenus rusci (Hübner)
Birch leaf-mining weevil

Associated mainly with birch (*Betula*) and oak (*Quercus*), adults occurring from March onwards. Unlike the previous four species, the larval gallery follows the leaf margin and then turns inwards to terminate in a more-or-less circular blotch (Fig. **333**). When fully grown, the larva cuts out a circular section from above and below the mine and forms its cocoon on the ground between these two epidermal discs. Leaves formerly containing larvae are recognized by the presence of a circular hole at the end of the deserted mine (Fig. **334**). The

333 Larval mine of birch leaf-mining weevil on *Quercus*.

334 Remains of larval mine of birch leaf-mining weevil on *Betula*.

adult weevils are 2.0–2.5 mm long, and rather oval-bodied, with a black pronotum and elytra, the latter with two bands of white or yellow scale-like hairs; the antennae are red and the legs black with brownish tarsi.

Rhynchaenus salicis (Linnaeus)
Sallow leaf-mining weevil
A generally common species, the bright yellow larvae (up to 3 mm long) (Fig. 335) forming brown blotches in the leaves of broad-leaved willows, such as common sallow (*Salix atrocinerea*), and sometimes causing noticeable distortion of the foliage of nursery plants (Fig. 336). Adults (2.0–2.5 mm long) are mainly black with a prominent white scutellum; the elytra are patterned with white and may also be marked with yellow anteriorly (Fig. 337).

335 Larva of sallow leaf-mining weevil, *Rhynchaenus salicis*.

336 Sallow leaf-mining weevil damage to leaf of *Salix*.

8. Family SCOLYTIDAE (bark beetles)

Small, cylindrical, wood-boring beetles (some known as ambrosia beetles and others as shot-hole borers) with relatively short, distinctly clubbed antennae. The larvae have an enlarged prothoracic region, giving them a distinct, hunch-backed appearance. Development takes place entirely within the host, some species being dependent upon the presence in their galleries of hyphae of the ambrosia fungus upon which the larvae feed.

Scolytus scolytus (Fabricius)
Large elm bark beetle
An important pest of elm (*Ulmus*), the beetles acting as vectors of Dutch elm disease; dying or at least weakened ash (*Fraxinus excelsior*), oak (*Quercus*), poplar (*Populus*) and various other dying trees are also attacked. Widely distributed in Europe.

DESCRIPTION **Adult:** 5–6 mm long; mainly black with reddish-brown elytra; front tibiae toothed; a median peg-like tubercule on the hind margin of the third and fourth sternites. **Larva:** white, C-shaped and legless; head dark brown.

337 Sallow leaf-mining weevil, *Rhynchaenus salicis*.

LIFE HISTORY Young adult beetles appear from mid-May to October, congregating at the tops of elm trees to feed on the small twigs. Beetle activity is greatest during warm weather and, if carrying Dutch elm disease spores on their mouthparts or bodies, the insects will readily introduce the fungus into healthy tissue. Infections in May or June are most important, the spores usually gaining entry through the feeding channels formed by the beetles in the crotches between the twigs. Later in the summer, the beetles burrow into the trunks and branches of weak or dying hosts to form short (25mm long) perpendicular breeding chambers immediately beneath the bark. After mating, each female lays about 50 eggs along the length of her chamber, the eggs hatching about ten days later. Larvae then burrow away from the maternal chamber, between the bark and sap wood, to produce a series of feeding galleries. These galleries form an irregular, fan-like pattern, on either side of the main chamber, which is characteristic for the species and clearly visible when the covering of dead bark is peeled away. The larvae are fully grown by the winter or following spring, their rate of development varying considerably from site to site. They then pupate, each in a slight bulb at the end of its burrow, adults eventually emerging via small, rounded flight holes which they excavate through the bark. Host trees remain suitable as breeding sites for bark beetles and, hence, as potential reservoirs of Dutch elm disease for about two years after death.

DAMAGE **Elm:** infestations cause the die-back of branches and hasten the decline of host trees. If Dutch elm disease takes a hold, the leaves of infected branches suddenly turn yellow, the fungus usually spreading and eventually killing the whole tree. **Other hosts:** attacks, which lead to the eventual death of branches or complete trees, are usually initiated only in unhealthy hosts or in those under severe root stress.

CONTROL Planting of susceptible trees in poor, waterlogged sites should be avoided, and any wounds found on trees treated promptly with a proprietary wound paint. Limited reduction in bark beetle infestations can be achieved by pruning and burning affected branches, but heavily infested trees should be grubbed completely. Protective injections against Dutch elm disease are feasible on important amenity or specimen trees, at least as part of an overall control strategy, but are not generally advocated.

Scolytus mali (Bechstein)
Large fruit bark beetle
This widespread species is found occasionally on ornamentals such as *Chaenomeles*, cherry (*Prunus*), *Cotoneaster*, hawthorn (*Crataegus*) and *Sorbus*, but is more commonly associated with fruit trees. Breeding colonies usually occur in the trunks and larger branches of already weak or dying host plants. Each colony consists of a long maternal chamber and a series of about 50–60 more-or-less perpendicular larval galleries. These spread upwards and downwards, immediately beneath the bark, each terminating in a slight bulb in which pupation occurs.

Scolytus multistriatus (Marsham)
Small elm bark beetle
A relatively small species (2.0–3.5mm long), associated mainly with elm (*Ulmus*) but also capable of damaging other trees such as oak (*Quercus*) and poplar (*Populus*). The adults are structurally similar to those of *Scolytus scolytus* (p.000), the other elm-feeding species, but have a pair of lateral teeth on the hind margin of the second, third and fourth abdominal sternites. The biology of both species is similar, and both act as vectors of Dutch elm disease, but the larval galleries of *Scolytus multistriatus* are slightly smaller and characterized by their more regular appearance.

Scolytus rugulosus (Müller)
Fruit bark beetle
Although most commonly associated with fruit trees, especially cherry and plum, infestations of this widespread species may also occur on ornamental *Prunus* and other ornamentals. Attacks typically occur in the trunks of small trees and in branches up to 6cm in diameter. The larval galleries, which lead away from the maternal chamber between the bark and sap wood, commonly overlap, especially towards their extremities.

Order **DIPTERA** (true flies)

1. Family **TIPULIDAE** (crane flies)

Slow-flying insects with elongate bodies, wings and legs; often called 'daddy-longlegs'. The larvae, commonly known as 'leatherjackets', are soil-inhabiting, soft-bodied but tough-skinned.

Tipula paludosa Meigen
Common crane fly
An important and often common pest of protected ornamentals, hardy nursery stock and herbaceous garden plants; also damaging to lawns and sports turf. Holarctic. Present throughout Europe.

DESCRIPTION **Adult:** 17–25 mm long; body grey with a brownish or yellowish-red tinge; thorax with faint longitudinal stripes; wings (13–23 mm long) shorter than body; legs fragile, brown and very long; antennae 14-segmented (Fig. **338**). **Egg:** 1.0×0.4 mm; oval, black and shiny. **Larva:** 35–45 mm long; brownish-grey with (unless wet) a dull, dusty appearance; body fat and slightly tapered anteriorly, with a soft but tough, leathery skin; head black, small and indistinct (cf. Bibionidae, p. 153); anal segment with a single pair of elongate anal papillae (Fig. **339**). **Pupa:** 20–30 mm long; brown and elongate, with paired respiratory horns on the head.

LIFE HISTORY Adults emerge from June onwards but are most abundant in late summer or early autumn. Most eggs are laid just below ground level from mid-August to the end of September, each female depositing about 300 in batches of five or six. Eggs hatch about 14 days later, the larvae then attacking plant roots near or at the soil surface. In mild, muggy nights they also appear on the soil surface to graze on the base of plant stems. Feeding usually ceases during the winter to be resumed in the following spring, the larvae becoming fully grown by about June and then pupating. If adults invade and lay eggs in glasshouses in the late summer and autumn, the resulting larvae often complete their development within a few months and new adults may emerge from late March onwards.

338 Female common crane fly, *Tipula paludosa*.

339 Larva of common crane fly, *Tipula paludosa*.

DAMAGE Leatherjackets graze the roots, corms, rhizomes and basal parts of stems, often causing plants to wilt; small plants may be killed, and seedlings completely destroyed. Large holes can also be produced in foliage lying close to the soil or compost (Fig. 340). On outdoor plants, leatherjacket damage is usually most extensive in the spring (but see notes on the other species) and in wet conditions (cf. cutworm damage p. 309); plants growing in recently ploughed or broken grassland or pasture are especially liable to be attacked. Damage can also occur on plants growing in containers, pots and seedboxes if any leatherjackets are introduced in the compost.

CONTROL **Open beds:** ground should be kept weed-free and regularly cultivated to reduce its attractiveness to egg-laying females. If necessary, an insecticide should be applied and incorporated into infested land prior to planting. Application of an insecticide or use of leatherjacket baits (e.g. mini pellets) will give some control if attacks develop on established plants; such treatment is most effective if applied in mild, humid weather in the autumn or spring. **Containers:** apply an insecticide to the surface or sheeting upon which containers will be stood; alternatively, scatter mini-pellets between the pots or seedboxes. **Lawns:** apply an insecticide in the autumn, whilst larvae are still young. Alternatively, small areas may be watered liberally in the evening and then covered overnight with a ground sheet or tarpaulin; any leatherjackets accumulating on the surface should be collected up and killed.

Tipula lateralis Meigen

Infestations of this generally common crane fly also occur in association with ornamentals, the larvae causing most damage to the rooting systems of containerized plants. Adults (wings 15–20 mm long) are mainly yellowish to greyish; the wing membrane is partly yellow and has a characteristic whitish patch in the stigmatic area. The larvae are distinguished from those of related species by the exceptionally long anal papillae.

Tipula oleracea Linnaeus
Large common crane fly

In some regions this widespread and often common species is the most important crane fly associated with ornamental plants. It appears earlier in the year than *Tipula paludosa* (p. 151), adults flying from May to August. The larvae, which possess two pairs of elongate anal papillae (cf. *Tipula paludosa*), often complete their development in the

340 Common crane fly damage to leaves of *Primula*.

341 Female spotted crane fly, *Nephrotoma appendiculata*.

autumn and pupate before the winter, thereby causing most damage in the same year as eggs were laid. Adults (15–23 mm long) are distinguished from those of *Tipula paludosa* by the 13-segmented antennae and thinner legs; also, the wings (18–28 mm long) are noticeably longer than the body.

Nephrotoma appendiculata (Pierre)
syn. *maculata* (Meigen)
Spotted crane fly

A widely distributed crane fly, particularly abundant in gardens, the larvae commonly causing damage to herbaceous plants and seedlings. Adults (12–20 mm long) are yellow and black, with greyish wings (Fig. 341); they occur from May to August, depositing their eggs at random on the soil surface. The greyish-brown larvae (up to 30mm long) feed mainly in the summer and autumn, often completing their development before the winter; they are distinguished from those of *Tipula paludosa*, *T. lateralis* and *T. oleracea* by the short, rounded anal papillae.

2. Family **BIBIONIDAE** (St. Mark's flies)

Mainly black, hairy, robust-bodied, medium-sized to large flies, the males often hovering rather sluggishly in the air in conspicuous groups. The larvae are cylindrical, with a prominent head, well-developed mouthparts and fleshy processes on each body segment.

342 St. Mark's fly, *Bibio marci*.

Bibio spp.

Several species of bibionid fly, including *Bibio johannis* (Linnaeus), *B. marci* (Linnaeus) and *B. nigriventris* (Haliday), are reported as pests of seedling broad-leaved and coniferous trees. Adults are slow-flying, robust, black-bodied flies (Fig. **342**) that often appear in vast swarms during sunny days in the spring (e.g. *Bibio marci*) or early summer (e.g. *Bibio johannis*). Eggs are deposited in groups in the soil, especially where there are accumulations of organic material, the egg-laying females frequently being attracted by decaying manure. The larvae often occur in considerable numbers in well-manured soil, feeding indiscriminately on vegetative matter and often damage the roots of plants. They occur throughout the summer, those of some species (e.g. *Bibio johannis*) overwintering and completing their development in the following spring. Bibionid larvae are superficially similar in appearance to leatherjackets but have a distinct black head and do not exceed 20mm in length; the body also bears several fleshy papillae (Fig. **343**); in the genus *Bibio*, the anal spiracles each have two pores.

343 Larva of *Bibio johannis*.

Dilophus febrilis (Linnaeus)
Fever fly

Larvae of this generally abundant bibionid also attack the roots of young shrubs and trees, occurring in vast masses both in heavily manured soil and amongst decaying vegetable matter. There are two, perhaps three, generations annually. The larvae are relatively small, up to 10mm long; they are distinguished from those of *Bibio* spp. by the presence of three pores per anal spiracle.

3. Family **CHIRONOMIDAE** (non-biting midges)

Delicate, gnat-like flies with poorly developed mouthparts, reduced wing venation and no ocelli; male antennae are plumose.

Bryophaenocladius furcatus (Kieffer)

A widely distributed, parthenogenetic midge, the larvae sometimes damaging the roots of glasshouse plants, including ornamentals such as *Primula* and *Rhododendron*. Infestations are most likely to develop in damp conditions. Fully grown larvae are 5mm long, yellowish-green with a dark brown head; the anal segment bears prominent ventral tubercules (cf. Sciaridae).

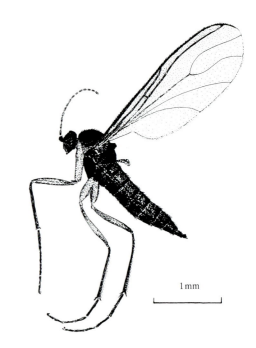

1 mm

344 Adult of a sciarid fly.

4. Family **SCIARIDAE** (fungus gnats or sciarid flies)

Small, delicate, gnat-like flies with a characteristic wing venation, 16-segmented antennae, large compound eyes (which usually meet above the antennae to form an 'eye bridge') and a somewhat humped thorax (Fig. **344**). The larvae are elongate and translucent-whitish, with a conspicuous black head (Fig. **345**).

345 Larva of a sciarid fly.

Bradysia spp.

Various species of sciarid fly, including *Bradysia amoena* (Winnertz), *B. aprica* (Winnertz), *B. paupera* (Tuomikoski) and *B. tritici* (Coquillett), are associated with ornamental plants. Larvae of these species may sometimes damage cuttings, seedlings and young pot plants by tunnelling into the roots, corms or main stems. However, such attacks are usually limited to plant material already invaded by moulds or other micro-organisms. Attacked cuttings and seedlings are checked; they may also collapse and die. Adult sciarids are very active; they are often seen flitting or scurrying about the bases of cuttings, seedlings or older plants. The egg-laying females are much attracted by dried blood fertilizer and to steam-sterilized soil, each depositing 100 or more eggs in the soil close to host plants. The eggs are small (c. 0.2×0.1mm), oval and translucent-whitish; they hatch several days later, the incubation time varying considerably according to temperature. The larvae feed for 3–4 weeks, usually attacking the root hairs; they then construct silken cocoons within which to pupate, adults emerging about a week later. More rarely, eggs may be placed directly onto plant tissue; the larvae then feed on the leaf tissue and may eventually pupate on the foliage. Breeding is continuous under suitable conditions.

CONTROL Vulnerable sites, including beds, capillary matting and sheeting, should be kept clean and free from decaying plant material; structures should also be kept well ventilated. Steam-sterilized soil or compost should be kept covered or in sealed containers. Avoid over-watering and do not incorporate dried blood into the compost. Glasshouses should be monitored regularly for sciarid fly activity (if necessary using yellow sticky traps). N.B. Chemical treatments against larvae (including incorporating of an insecticide into the compost) give better overall control than sprays or fumigants against adults.

346 Larvae of hawthorn button-top midge, *Dasineura crataegi*.

5. Family CECIDOMYIIDAE (gall midges)

Minute to small, delicate flies with bead-like flagellar antennal segments, broad and often hairy wings (the venation much reduced), long, thin legs and prominent genitalia. The larvae are short, narrowed at both ends, and have a small, inconspicuous head; they usually possess a sternal spatula ('anchor-process' or 'breast-bone'), which is often of characteristic shape for the genus or species.

347 Gall of hawthorn button-top midge on *Crataegus*.

Dasineura crataegi (Winnertz)

Hawthorn button-top midge
A generally common pest of hawthorn (*Crataegus*), infestations often occurring in abundance on hedges and in nurseries. Eurasiatic. Widely distributed in Europe.

DESCRIPTION **Adult:** 2.0–2.5 mm long; brownish with darker crossbands on the abdomen. **Larva:** 2–3 mm long; reddish, but whitish when young (Fig. 346).

LIFE HISTORY Adult midges appear in April or May, females depositing eggs in the tips of young shoots of hawthorn. Attacked shoots fail to elongate and, instead, develop into compact, rosette-like galls (Fig. 347); these galls eventually turn black, remaining on the host plants long after the causal larvae have departed. The galls are often noted on nursery stock, and may be especially numerous on regularly trimmed hedges which provide an abundance of suitable egg-laying sites. Several larvae develop within each gall, becoming fully fed in 2–3 weeks. They then drop to the ground and pupate inside silken cocoons, adults emerging shortly afterwards. There are further generations during the summer, autumn larvae overwintering in their cocoons and pupating in the following spring.

DAMAGE The galls disfigure host plants; also, by interfering with normal shoot development, they also affect the shape and, hence, quality of nursery stock. Leaves immediately below the main galls often unfurl but remain disfigured by numerous red pimples (Fig. **348**). Attacks on established plants are rarely troublesome, even when infestations are heavy but damage to nursery stock may be unacceptable.

CONTROL If necessary, galls on nursery stock should be cut off and burnt; treatment with an insecticide is of limited value.

348 Hawthorn button-top midge damage to leaf of *Crataegus*.

Dasineura abietiperda (Henschel)
This widely distributed midge attacks the young growth of various spruces, including Norway spruce (*Picea abies*). Adults appear in April and May, the females depositing eggs near the growing points of the new shoots. The reddish larvae (2–3 mm long) feed singly, each within a cavity formed in the bark or new wood; they may also burrow into the buds at the shoot tips. The larvae usually complete their development in the autumn. They then overwinter, pupating in the spring shortly before the emergence of the adults. Several galls usually occur close together, affected shoots remaining shorter than normal, bearing fewer needles and becoming slightly bent. Heavy infestations may cause considerable damage to young trees; losses are of particular importance on spruces grown as Christmas trees.

CONTROL Cutting off and destruction of infested shoots is recommended only in areas where natural parasitism of larvae is not exerting adequate control. Application of an insecticide in the spring, to kill larvae before they enter the host, can be effective.

Dasineura affinis (Kieffer)
Violet leaf midge
A widely distributed, important and often persistent pest of wild and cultivated violet (*Viola*). Infested plants become stunted and malformed, with flower production seriously affected if not entirely prevented; in severe cases, plants may eventually die. Adult midges occur from May onwards, depositing

eggs in the rolled margins of young leaves and, occasionally, in the developing flowers. Infested tissue fails to unroll and often becomes swollen into conspicuous galls, within which several whitish to pale orange larvae (each up to 2 mm long) develop. Larvae are fully fed in about six weeks; they then pupate in silken cocoons, still within the gall, adult midges emerging about 10–12 days later. There are usually four overlapping generations annually on outdoor plants, larvae of the final brood over-wintering and completing their development in the following spring. In glasshouses, there are commonly additional generations and, if conditions are suitable, breeding may continue throughout the year.

CONTROL On a small scale, hand-picking and destruction of galled leaves or roguing of heavily infested plants is recommended. Application of a systemic insecticide at the commencement of galling may have some effect but chemical treatment is not always reliable.

Dasineura alpestris (Kieffer)
syn. *arabis* Barnes
Arabis midge
A potentially important pest of certain kinds of arabis, including rock cress (*Arabis alpina*),

damaging infestations sometimes occurring in gardens and nurseries. Adults appear from May onwards, the females depositing reddish eggs between the leaf tissue in the developing buds. These eggs hatch within a few days, and pinkish or reddish larvae then feed in compact but open galls which develop as the leaf bases become enlarged and swollen. There are commonly 20–30 larvae in each gall, sometimes considerably more. Fully grown larvae pupate in white cocoons spun either within the gall or in the soil, midges of the next generation emerging shortly afterwards. There are three or four generations each year, larvae of the final generation overwintering in their cocoons and pupating in the spring. Heavily infested plants produce few flowers, their central crowns being destroyed and any new growth limited to weak, lateral shoots.

CONTROL As for *Dasineura affinis* (p. 156).

Dasineura fraxini (Bremi)
syn. *fraxini* (Kieffer)
Ash mid-rib pouch-gall midge
This widely distributed and often common species causes the underside of leaves of ash (*Fraxinus excelsior*) to swell into conspicuous galls (25–30 mm long) along either side of the mid-rib (Fig. **349**). Development commences in May or June, the galls usually containing from four to eight orange-coloured larvae, each located in its own cell. Maturity is reached in September. The upper surface of the gall then splits open longitudinally, the fully fed (2–3 mm long) larvae dropping to the ground and overwintering in the soil. Pupation occurs in the following spring, adult midges emerging shortly after. Infestations occur on both young and mature trees; although affected leaves may eventually turn brown and die prematurely (Fig. **350**), damage does not affect shoot growth.

Dasineura gleditchiae (Osten-Sacken)
Honeylocust gall midge
Formerly of North American origin but now well established on honeylocust (*Gleditsia*) in the Netherlands and in various other parts of continental Europe; infested plants have also been found in southern England. Adults emerge from late May onwards. Larvae feed gregariously on the leaves, causing them to develop into yellowish-green to purplish-red, pod-like galls (Fig. **351**) that eventually

349 Galls of ash mid-rib pouch-gall midge on *Fraxinus.*

350 Ash mid-rib pouch-gall midge damage to leaves of *Fraxinus.*

351 Young galls of honeylocust gall midge on leaves of *Gleditsia.*

turn brown. Infested shoots are also distorted (Fig. 352), the galls often causing severe disfigurement. Pupation occurs within the galls, adults appearing shortly afterwards. There are several overlapping generations throughout the summer, development from egg to adult being completed in 3–4 weeks. Fully grown larvae are whitish and 3.0–3.5 mm long; those reared in the autumn overwinter in subterranean cocoons, pupating in the following spring. The pest can also persist in the soil for more than a year before producing adults.

CONTROL Apply a soil insecticide in the spring before the commencement of adult emergence. Well-established infestations are difficult, if not impossible, to eliminate.

352 Galls of honeylocust gall midge on leaves of *Gleditsia*.

Dasineura rosarum (Hardy)
Rose leaf midge
A widely distributed midge, attacking both wild and cultivated rose (*Rosa*); the larvae develop within reddish-tinged, pod-like galls, each formed from an expanded but unfolded leaflet (Fig. **353**). Adult midges appear in the late spring, depositing eggs in unfurled leaves. Up to 50 orange-coloured larvae (Fig. **354**) occur in each infested leaflet, causing the galled tissue to change from green through reddish to dark brown. Individuals are fully grown by the late summer; they then drop to the ground where they overwinter and eventually pupate; there is just one generation annually. Galled leaves often occur in groups but any effect on plant growth is slight.

353 Galls of rose leaf midge on *Rosa*.

Dasineura thomasiana (Kieffer)
Minor infestations of this midge are sometimes noticed on lime (*Tilia*), the larvae feeding within distorted leaves at the tips of young shoots. Affected leaves remain folded, the veins becoming thickened and deformed (Fig. **355**); buds are also affected. Each gall contains several (2–3 mm long) whitish to orange or reddish larvae. There may be two or more generations annually.

354 Larvae of rose leaf midge, *Dasineura rosarum*.

355 Gall of *Dasineura thomasiana* on young leaf of *Tilia*.

356 Galls of lime leaf-roll gall midge on *Tilia*.

Dasineura tiliamvolvens (Rübsaamen)
Lime leaf-roll gall midge

The striking galls of this midge occur on lime (*Tilia*), developing on the young leaves from April or May onwards. Each gall appears as an upward rolling of the leaf edge, affected tissue turning dark red; galling may extend around most if not all of the leaf margin, producing considerable distortion; the leaf lamina may also become mottled with red (Fig. **356**) (cf. galls formed by the mite *Phytoptus tetratrichus*, p. 383). The galls, which eventually turn black (Fig. **357**), enclose several whitish to orange (2–3 mm long) larvae. These larvae complete their development in June; they then enter the soil where they eventually overwinter. Adult midges appear in late April and May, there being just one generation annually.

357 Mature galls of lime leaf-roll gall midge on *Tilia*.

Arnoldiola quercus (Binnie)
Oak terminal-shoot gall midge

Associated with oak (*Quercus*), infestations causing the death of terminal shoots. The larvae are whitish and up to 3 mm long. They feed gregariously amongst the unfurling leaves, infested tissue failing to open and eventually turning black (Fig. **358**). There are usually two generations annually, infestations coinciding with the development of new leaf growth in the spring and summer. Attacks are of little or no importance on mature trees, but they can affect the growth of shoots on young trees and may be of significance on nursery stock.

CONTROL Rarely worth while; if necessary on nursery trees known to be at risk, apply an insecticide in spring and again in summer to protect the young shoots.

358 Oak terminal-shoot gall midge damage to young shoot of *Quercus*.

Didymomyia tiliacea (Bremi)
syn. *reaumuriana* (Loew)
Lime leaf gall midge

A local species, associated with lime (*Tilia*), forming conspicuous, greenish-white, pustule-like galls on the leaves, visible from above (Fig. **359**) and from below. The galls, which occur from May onwards, are about 4 mm in diameter. The larvae, which are 2–3 mm long and white to pale yellow, develop singly within the galls, to become fully grown by the late summer. At maturity, a characteristic capsule containing the fully fed larva is ejected through the underside of the gall (Fig. **360**). Individuals overwinter on the ground within these protective structures, adult midges appearing in the spring from late April onwards. There are commonly up to 40 galls on an infested leaf but, apart from possible distortion of the lamina, growth is not affected.

359 Young galls of lime leaf gall midge on *Tilia*.

360 Pupal capsule of lime leaf gall midge, *Didymomyia tiliacea*.

Hartigola annulipes (Hartig)
syn. *tornatella* (Bremi); *piligera* (Loew)
Beech hairy-pouch-gall midge

Beech (*Fagus sylvatica*) is commonly attacked by this widespread gall midge. The adults, which occur from mid-May to early June, insert eggs through the upper epidermis of expanded leaves, generally close to the mid-rib. In July, inner tissue of the leaf bursts through the upper surface at each oviposition point to form yellowish, hairy, cylindrical galls (*c.* 5 × 3 mm) (Fig. **361**), each enclosing a small (2–3 mm long) white larva. The galls eventually turn reddish-brown and in the late summer, when mature, they break away from the leaf and drop to the ground. Larvae overwinter within the galls, pupating in the spring. An infested leaf may bear several galls, sometimes as many as 50, but plant growth is not affected.

361 Gall of beech hairy-pouch-gall midge, *Hartigiola annulipes*.

362 Gall of beech smooth-pouch-gall midge, *Mikiola fagi*.

363 Larva of beech smooth-pouch-gall midge, *Mikiola fagi*.

Mikiola fagi (Hartig)
Beech smooth-pouch-gall midge

This midge also occurs on beech (*Fagus sylvatica*), the larvae developing in bulbous leaf galls, each 4–10mm long. The galls, which have a smooth, often plum-like surface (Fig. **362**), occur on the leaves from May to late September or October and, if numerous, can cause the death of young plants. The larvae are whitish and plump (*c.* 4mm long when fully grown) (Fig. **363**), and feed within the galls throughout the summer. They then overwinter within the fallen galls, eventually pupating and producing adults in the following March or April. Although present in England, this pest appears more common and of greater significance in mainland Europe.

CONTROL If practical, galls or infested leaves on nursery trees should be cut off and burnt.

364 Galls of *Iteomyia capreae* on leaf of *Salix*, viewed from above.

Iteomyia capreae (Winnertz)
Larvae of this widely distributed and often common species occur in galls formed on the leaves of eared sallow (*Salix aurita*), pussy willow (*Salix caprea*) and certain other species of *Salix*. The galls occur from May or June onwards. From above, they appear as polished, yellowish-green, red-flushed swellings, each 2–3mm in diameter (Fig. **364**); below, they are pale green and volcano-like, each with a purplish-red rim surrounding a prominent central entrance hole (Fig. **365**). Each gall contains a small (1–2mm long) translucent-whitish, and finally red, larva which feeds throughout the summer, eventually dropping to the ground to pupate. There is just one generation annually. The galls are sometimes invaded by mites, including the willow leaf gall mite (*Aculops tetanothrix*) (p. 398).

365 Galls of *Iteomyia capreae* on leaf of *Salix*, viewed from below.

366 Galls of *Janetiella lemei* on leaf of *Ulmus*.

367 Galls of *Janetiella lemei* on shoot of *Ulmus*.

Janetiella lemei (Kieffer)

A widely distributed but local species, associated with elm (*Ulmus*). The yellow-coloured larvae develop during the summer in pale, flask-like galls formed as swellings of the veins on the underside of leaves (Fig. 366). The galls also occur on the petioles and young shoots (Fig. 367). When fully grown, the larvae drop to the ground and eventually pupate in the soil, adult midges appearing in the spring from April or early May onwards.

Rhabdophaga rosaria (Loew)

syn. *cinerearum* (Hardy)

Willow rosette-gall midge

A generally common pest of willow, especially crack willow (*Salix fragilis*), pussy willow (*Salix caprea*) and white willow (*Salix alba*). Present throughout Europe.

DESCRIPTION **Adult female:** 2.2 mm long; head black; thorax blackish to reddish; abdomen red; wings clear, 2.8 mm long. **Larva:** 4 mm long; whitish to pinkish-orange (Fig. 368).

LIFE HISTORY Adults occur in late April or early May, depositing eggs in association with vegetative buds. Infested buds develop during the summer into leafy, cabbage-like rosette galls (Fig. 369), each containing a single larva (Fig. 370) (cf. *Rhabdophaga heterobia*, p. 163). The galls eventually turn brown and remain on the shoots throughout the winter. They are especially obvious in the early spring (Fig. 371) just before the emerging new green growth forces them to drop off; at this stage they are sometimes referred to as 'camellia galls'. Larvae overwinter within the galls and pupate in the spring, shortly before adults emerge.

368 Larva of willow rosette-gall midge, *Rhabdophaga rosaria*.

369 Gall of willow rosette-gall midge, *Rhabdophaga rosaria*.

370 Section through gall of willow rosette-gall midge, *Rhabdophaga rosaria*.

371 Old gall of willow rosette-gall midge, *Rhabdophaga rosaria*.

DAMAGE The galls disfigure host plants and cause the death of leading shoots, but tend to be of significance only on small bushes.

Rhabdophaga clausilia (Bremi)
syn. *inchbaldiana* (Mik)
Willow leaf-folding midge
This widely distributed midge causes a discontinuous marginal rolling of willow (*Salix*) leaves. Each roll-gall, unlike those of *Rhabdophaga marginem-torquens* (p. 164), consists of a short length of folded tissue which encloses a single yellowish-orange larva. The galls, which mature in the autumn, occur mainly on white willow (*Salix alba*). They are noted occasionally on cultivated plants but damage is not important.

Rhabdophaga heterobia (Löw)
syn. *saligna* (Hardy)
Willow button-top midge
A widespread species, breeding in galls formed from the terminal buds, lateral buds and male catkins of certain willows, especially almond willow (*Salix triandra*). Adults occur in two main generations, from late April to September, depositing eggs on or close to the growing points. Infested terminal buds develop into loose rosette galls (Fig. **372**), each enclosing up to 40 small (*c.* 2 mm long), whitish to orange-red larvae (Fig. **373**) (cf. *Rhabdophaga rosaria*, p. 162). When fully fed, the larvae pupate in white cocoons formed within the gall, adults emerging shortly afterwards. Larvae of the second generation induce further terminal and lateral galls; they eventually overwinter, and pupate in the following spring, usually still within the protective remnants of the dead rosette leaves which remain attached to the host plant long after leaf-fall.

372 Gall of willow button-top midge, *Rhabdophaga heterobia*.

373 Section through gall of willow button-top midge, *Rhabdophaga heterobia*.

Rhabdophaga marginemtorquens (Bremi)

syn. *marginemtorquens* (Winnertz)
Osier leaf-folding midge

Leaves of osier (*Salix viminalis*) are frequently infested by larvae of this common and widespread species. Adults occur in the spring, depositing eggs on young leaves. Attacked leaves develop long, continuous marginal folds which become yellowish-green and tinged with red; the galls enclose several orange (2 mm long) larvae (cf. *Rhabdophaga clausilia*, p. 163), the position of each being disclosed by a local swelling (Fig. **374**). Fully fed larvae vacate the galls in August or September; they then overwinter in the soil and pupate in the following spring. The galls do not affect plant growth, although severe infestations in nursery beds can give an impression of poor plant quality.

374 Galls of osier leaf-folding midge, *Rhabdophaga marginemtorquens*.

Taxomyia taxi (Inchbald)

Yew gall midge

A common pest of yew (*Taxus*); also associated with cowtail pine (*Cephalotaxus*). Present throughout Europe.

DESCRIPTION **Adult:** 3–4 mm long; yellowish-orange. **Larva:** 2–3 mm long; yellowish-orange, lacking a sternal spatula (Fig. **375**).

LIFE HISTORY Adult midges occur in May and June, depositing eggs singly in the terminal buds of yew trees. The eggs hatch about ten days later, the young larvae then burrowing into the immature buds to begin feeding. Each infested bud develops into a characteristic 'artichoke' gall, formed from a tight cluster of terminal leaves within which a single larva develops; galled buds remain relatively small in the first year but in the next they become larger and less compact, and may then exceed 15–20 mm in diameter (Fig. **376**). Some larvae develop quickly, pupating within the galls to produce adults in the following spring, but most develop slowly and do not mature for another year. Occupied galls occur throughout the year on both young and mature hosts; vacated galls, which eventually turn brown (Fig. **377**), often persist *in situ* for two or more seasons.

DAMAGE The galls abort shoot growth and are disfiguring; host plants are stunted by persistently heavy infestations. Attacks are especially damaging in nurseries and on plants already infested by the gall mite *Cecidophyopsis psilapsis* (p. 401).

CONTROL Galls on young trees should be removed during the winter and destroyed.

375 Larva of yew gall midge, *Taxomyia taxi*.

376 Gall of yew gall midge, *Taxomyia taxi*.

377 Old gall of yew gall midge, *Taxomyia taxi.*

378 Galls of hornbeam leaf gall midge, *Zygiobia carpini.*

Zygiobia carpini (Löw)
Hornbeam leaf gall midge

Galls of this widely distributed midge develop on the leaves of hornbeam (*Carpinus betulus*), either occurring singly or as a series of swellings along the length of the mid-rib (Fig. **378**). Infestations are sometimes noticed on cultivated plants but, although disfiguring, they have no apparent effect on plant growth. The whitish to yellowish-white larvae (2–3 mm long), one per gall, develop throughout the summer; they vacate the galls in September, eventually pupating in the soil. Infestations also occur on European hop-hornbeam (*Ostrya carpinifolia*).

Anisostephus betulinum (Kieffer)
Birch leaf gall midge

A widespread species associated with birch, including downy birch (*Betula pubescens*) and silver birch (*Betula pendula*); the larvae feed in distinctive galls formed on the leaves. The galls develop as yellowish-red to maroon pustules (Fig. **379**), visible from above and from below; infestations are sometimes found on the foliage of young trees in nurseries or amenity areas. The larvae (2–3 mm long) are whitish at first but later become bright yellow (Fig. **380**); they develop singly, from June to early August, eventually escaping through a small hole in the side of the gall and dropping to the ground to pupate in the soil. There is one generation annually.

379 Galls of birch leaf gall midge, *Anisostephus betulinum.*

380 Larva of birch leaf gall midge, *Anisostephus betulinum.*

Contarinia acerplicans (Kieffer)
Sycamore leaf-roll gall midge

This widely distributed European midge infests maple (*Acer*) and sycamore (*Acer pseudoplatanus*), the larvae developing gregariously within reddish marginal leaf-roll galls (Fig. 381). Galls may also develop on the leaves as elongate folds, typically glabrous and shiny above and hairy below, affected leaves often being considerably distorted (Fig. 382). The galls, which eventually turn black, first appear on the young leaves in May. Fully fed larvae, which are whitish and about 2 mm long, eventually vacate the galls and drop to the ground to pupate in the soil. There are two generations annually. Although infestations sometimes occur on young ornamental or specimen trees, they do not cause significant damage.

381 Young galls of sycamore leaf-roll gall midge on *Acer pseudoplatanus*.

Contarinia petioli (Kieffer)
Poplar gall midge

Widely distributed in association with aspen (*Populus tremula*) and, less frequently, white poplar (*Populus alba*), the orange-coloured larvae (each 3–4 mm long) developing in galls on the young twigs or leaf petioles. Galls on the petioles are globular and there are often several fused together; those on the twigs develop singly, each forming a localized lateral swelling. At maturity, each gall develops a lateral aperture through which the causal organism escapes to pupate in the soil. Although the galls attract attention, particularly when present on young trees, they are not considered harmful.

382 Sycamore leaf-roll gall midge damage to leaf of *Acer pseudoplatanus*.

Contarinia quinquenotata (Löw)

A pest of day lily (*Hemerocallis fulva*); widely distributed in mainland Europe, and recently discovered causing extensive damage in southern England. The whitish larvae feed gregariously within the flower buds during the early summer. Fully grown larvae enter the soil and overwinter, adults emerging in the following year; there is one generation annually. Infested buds become greatly swollen and fail to open, those attacked at a very early stage in their development sometimes withering and turning brown. Attacks are especially common on yellow-flowered varieties, infestations reaching their peak by late June; buds produced after mid-summer are unaffected.

CONTROL Hand-picking and destruction of infested buds will help to reduce pest numbers.

383 Mature gall of lime leaf-petiole gall midge, *Contarinia tiliarum*.

384 Section through gall of lime leaf-petiole gall midge, *Contarinia tiliarum*.

Contarinia tiliarum (Kieffer)
Lime leaf-petiole gall midge

This moderately common midge forms globular galls (about 10 mm across) on the petioles of lime (*Tilia*), usually just before the leaf blade; galling can also occur on the young stems. The galls are at first pale but later turn red (Fig. **383**) and, finally, black. They contain several (often up to 20) yellowish-orange larvae (2–3 mm long), each individual developing in a separate chamber (Fig. **384**). The larvae commence their development in May or June and are usually fully fed by the end of July; they then enter the soil and pupate, adult midges appearing in the following spring, usually in May. There is one generation annually. Leaves on galled petioles fail to develop properly, often becoming distorted and noticeably hairy; the underside of the mid-rib may also be swollen. Heavy attacks of this pest can prevent trees flowering. However, the galls are usually most common on sucker growth around the base of older trees, and this reduces the importance of infestations.

385 Galls of oak fold-gall midge, *Macrodiplosis dryobia*.

386 Larva of oak fold-gall midge, *Macrodiplosis dryobia*.

Macrodiplosis dryobia (Loew)
An oak fold-gall midge

Adults of this widely distributed and often common midge occur in the late spring, depositing eggs close to the tips of lateral veins of the leaves of oak (*Quercus*). Subsequently, the leaf tissue folds over to meet the underside of the lamina (Fig. **385**); up to four 2–3 mm long, yellowish-white, translucent larvae (Fig. **386**) feed within this protective envelope. The flap covering the larvae soon becomes mottled and, eventually, the upper surface of the lamina immediately above the gall also becomes discoloured (Fig. **387**). Fully grown larvae escape from the galls from July onwards, individuals eventually pupating in the soil. There is just one generation annually. Although often abundant on young trees, the galls do not affect plant growth.

387 Oak fold-gall midge damage to leaf of *Quercus*.

Macrodiplosis volvens Kieffer
An oak fold-gall midge
This species is essentially similar to *Macrodiplosis dryobia* but larval development takes place in somewhat broader and shallower folds which arise between the lateral veins and overlap the *upper* surface of the leaves (Fig. **388**). The larvae (2–3 mm long) are orange-red, and there may be up to five present per gall.

388 Galls of *Macrodiplosis volvens* on leaf of *Quercus*.

Harmandia cavernosa (Rübsaamen)
Galls formed by this species commonly occur in pairs at the base of the leaves of aspen (*Populus tremula*) and white poplar (*Populus alba*). The galls develop as pale green swellings (up to 5 mm across) on the underside of the lamina, with a distinct opening through the upper surface. Each gall contains a single yellowish-red larva, which eventually escapes to pupate in the soil. Mature galls are distinctly blackened; although very noticeable, they cause little or no distortion of the lamina (Fig. **389**).

389 Galls of *Harmandia cavernosa* on leaf of *Populus tremula*.

Harmandia tremulae (Winnertz)
syn. *globuli* (Rübsaamen)
Aspen leaf gall midge

Leaves of aspen (*Populus tremula*) are often disfigured by spherical, red to purplish-red galls; they arise on the upper surface, each from a major vein, but do not cause distortion (Fig. **390**). The galls (up to 4mm across) develop from June onwards, reaching maturity in August or September. There are usually several galls per infested leaf, each containing a single larva. Fully fed larvae vacate the galls in the autumn and eventually pupate in the soil, adults appearing in the spring. There is one generation annually.

390 Galls of aspen leaf gall midge, *Harmandia tremulae*.

Monarthropalpus buxi (Laboulbène)
Box leaf mining midge

An often important pest of box (*Buxus sempervirens*), especially in nurseries. Widely distributed in Europe; also present in North America.

DESCRIPTION **Adult:** 2–3mm long; yellowish-orange; female with abdomen terminating in a long, curved spine. **Larva:** 2.5mm long; white, later orange; flattened and narrowing posteriorly.

LIFE HISTORY Adults are active in the spring, depositing eggs in the young leaves by inserting the ovipositor from below. Eggs hatch in about three weeks; the larvae then mine within the leaf tissue to form distinctive yellowish or brownish blister-like mines on the underside of the leaves (Fig. **391**). Where several larvae are present in the same leaf, the blisters eventually coalesce and may eventually occupy the complete lamina. Larvae overwinter within the mines, completing their development in the following spring. They then pupate, midges appearing 2–3 weeks later.

DAMAGE Larval mines are visible from above as discoloured swellings (Fig. **392**), disfiguring the foliage and spoiling the appearance of both specimen ornamentals and hedges. Heavy infestations weaken host plants, infested leaves dropping prematurely, and persistent attacks will gradually reduce the overall density of the foliage.

CONTROL Chemical treatments should be applied at the early stages of gall formation.

391 Galls of box leaf gall midge, *Monarthropalpus buxi*, viewed from below.

392 Galls of box leaf gall midge, *Monarthropalpus buxi*, viewed from above.

Parallelodiplosis cattleyae (Molliard)
Cattleya gall midge

Infestations of this midge sometimes occur in glasshouses, producing elongate or pea-like galls on the aerial roots of orchids, including various species of *Cattleya* and *Laelia*. The pest is usually of little significance but the growth of heavily infested plants may be retarded. There are several generations annually, several orange-coloured larvae developing within individual cells in each gall. Fully grown larvae pupate in their cells, the pupae eventually protruding from the surface of the gall following emergence of the adults.

CONTROL Galls should be cut off and burnt.

Resseliella oculiperda (Rübsaamen)
larva = red bud borer

A local and usually sporadic pest of budded stock and grafts, attacking rose (*Rosa*) and fruit trees. Widely distributed in Europe.

DESCRIPTION **Adult:** 1.4–2.1 mm long; dark reddish-brown. **Larva:** 3.0–3.5 mm long; salmon-pink to red, with a bilobed spatula.

LIFE HISTORY Adults occur in three generations, in late May to late June, July to early August and late August to early September, depositing eggs in graft slits or cuts in the bark of newly budded stock. Eggs hatch about a week later. Larvae then feed in small groups on sap in the cambium between the scion and stock. They are fully fed in 2–3 weeks, dropping to the ground to pupate in small cocoons a few centimetres below the surface. Larvae of the autumn generation overwinter in their cocoons and pupate in the following spring.

DAMAGE Infestations prevent grafts or buds from taking, so that the scions or buds wither and die. Most damage is caused by larvae which feed from August onwards, when losses on unprotected nursery stock can be considerable.

CONTROL To prevent egg laying, grafted buds or unions should be coated liberally with petroleum jelly. This treatment is ineffective once eggs are laid or larvae have begun to feed.

6. Family **SYRPHIDAE** (drone flies or hover flies)

Small to large, often brightly patterned flies, some of which hover in the air and emit a bee-like hum.

393 Small narcissus fly, *Eumerus tuberculatus*.

394 Larvae of small narcissus fly, *Eumerus tuberculatus*.

Eumerus tuberculatus Rondani
A small narcissus fly

A generally common pest of *Narcissus* bulbs but usually confining its attacks to those which are already diseased (e.g. infected with basal rot, *Fusarium oxysporum* f. sp. *narcissi*) or mechanically damaged. Holarctic. Present throughout Europe.

DESCRIPTION **Adult:** 5–6 mm long; robust-bodied and mainly black; head and thorax with a golden sheen, the thorax also with a pair of whitish longitudinal lines; abdomen with three pairs of white, crescent-shaped marks, and with golden hairs along the sides and towards the tip; stigma of fore wings dark brown or blackish (Fig. **393**); hind femora (especially of male) with a basal projection. **Egg:** 0.7 mm long; white, elongate-oval and tapered at one end. **Larva:** 7–9 mm long; whitish-grey to pale yellowish-white, with an elongate reddish-brown spiracular cone at the hind end, that protrudes beyond two long and two short, fleshy papillae (Fig. **394**). **Puparium:** 6–7 mm long; yellowish-white (Fig. **395**).

LIFE HISTORY Adults appear in the early spring, depositing batches of eggs in association with unhealthy narcissus bulbs. Larvae emerge a few days later; they then enter the bulbs to feed on the inner tissue in groups of five or six (Fig. **396**) but often in much larger numbers. Pupation takes place in or around the neck of the host bulb in June or early July, adults emerging about two weeks later. Larvae of a second generation become fully fed by the autumn. They vacate the remains of the bulb to overwinter in the soil, eventually pupating in the following spring.

DAMAGE Narcissus bulbs are often totally destroyed by the larvae. However, attacks rarely occur on healthy stock and are, therefore, of secondary importance.

CONTROL Infested bulbs should be destroyed, and steps taken to identify and overcome the original problem which predisposed them to attack.

Eumerus strigatus (Fallén)
A small narcissus fly

This fly is essentially similar to *Eumerus tuberculatus* but has a wider host range, including *Colchicum*, hyacinth (*Hyacinthus*), *Iris*, lily (*Lilium*) and *Narcissus*; the larvae will also invade vegetable crops such as carrot, onion, parsnip and potato, increasing damage to previously injured tissue. Adults are distinguishable from those of *Eumerus tuberculatus* by the yellow or light brown stigma on the fore wings and by the absence of a basal projection on the hind femora.

Merodon equestris (Fabricius)
Large narcissus fly

A locally important pest of *Narcissus*; various other cultivated bulbs, including belladonna lily (*Amaryllis belladonna*), *Galtonia*, hyacinth (*Hyacinthus*), *Iris*, snowdrop (*Galanthus nivalis*), snowflake (*Leucojum*), squill (*Scilla*) and *Vallota*, are also attacked. Widely distributed in mainland Europe; in the British Isles most numerous in south-western England. Also present in parts of Australasia, Japan and North America.

DESCRIPTION **Adult:** 12–14 mm long; stout-bodied and very hairy, the body hairs ranging in colour from black to greyish, yellowish, orange or red and often forming a bumblebee-like pattern; each hind leg bears a characteristic tooth-like projection (Fig. **397**). **Egg:** 1.6 mm long; elongate-oval and pearly white. **Larva:** 18 mm long; dirty yellowish-white

395 Puparium of small narcissus fly, *Eumerus tuberculatus*.

396 Small narcissus fly damage to bulb of *Narcissus*.

397 Large narcissus fly, *Merodon equestris*.

and plump, with a short, dark brown respiratory cone at the hind end, bordered on either side by small inconspicuous tubercules (Fig. **398**) (cf. *Eumerus*, p. 170). **Puparium:** 10–12 mm long; brownish.

LIFE HISTORY Adults occur mainly from May to July but individuals may appear as early as February in forcing houses. In cool, inclement weather, they hide in hedgerows and in other suitable situations, but in favourable conditions they visit various flowers in search of nectar and may also be found sunning themselves on nearby banks, posts, tree trunks and leaves. In flight they emit a characteristic, bee-like buzz, often occurring in large numbers in narcissus fields during warm, sunny afternoons. Eggs are deposited singly, either on foliage in the neck region of host bulbs, or directly on the bulb, or in the soil, the female usually crawling into the hole left by the withering foliage and flower stem. Most eggs are laid in June and early July. Eggs hatch in about two weeks, each larva crawling to the base plate of a bulb before burrowing in to begin feeding. The tissue immediately around the larval entry hole soon turns rusty red, visible if the dead tissue around the base plate is scraped away. At first, the larva (typically one per infested bulb, cf. *Eumerus* spp., p. 171) feeds within the base plate but, after a few weeks, it moves into the centre of the bulb, quickly hollowing a large cavity which becomes filled with blackish frass and rotting tissue. Larvae normally complete their development in the original bulb but will move from bulb to bulb if necessary; most individuals become fully grown by the winter. Pupation occurs in the following spring, either in the neck region of the bulb or in the soil, adults emerging 5–6 weeks later.

DAMAGE Infested bulbs become soft, especially in the neck region, and much of the inner tissue is destroyed. Small bulbs are often completely destroyed and, if planted, will fail to grow. Larger ones, however, may appear sound but will produce weak, distorted, yellowish foliage, or merely a ring of small grass-like leaves.

CONTROL Recommended treatments include hot-water treatment of lifted bulbs, the use of a persistent insecticide as a pre-planting treatment of bulbs and the use of a soil insecticide at planting. In early summer, raking soil into any crevices formed around the withering foliage will discourage egg laying. Infested bulbs discovered at lifting or in store should be destroyed.

398 Larva of large narcissus fly, *Merodon equestris*.

399 Drone fly, *Eristalis tenax*.

Eristalis tenax (Linnaeus)
Drone fly

Adults of this widespread and generally common syrphid are often attracted in large numbers to flowers, especially Compositae (e.g. *Aster* and *Chrysanthemum*) and Umbelliferae, where they feed avidly on nectar. They also bask in sunshine, darting into the air and hovering nearby if disturbed. The flies are normally harmless but they sometimes enter glasshouses, especially in the autumn, and may then contaminate the petals of chrysanthemums and other flowers with droplets of excrement. The insects breed in wet, decaying organic matter, the larvae (commonly known as 'rat-tailed maggots') possessing a very long, extensible breathing tube which allows them to breathe whilst submerged well below the surface of stagnant mud or water. The bee-like adults (12–15 mm long) are mainly brownish-black, with the thorax clothed in yellowish to brownish-yellow hairs and the abdomen variably marked with yellowish or yellowish-brown (Fig. **399**). They are distinguished from other closely related syrphids by the dark bands of hairs across the eyes and by the simple, unbranched arista.

7. Family **TEPHRITIDAE**

Small flies, often with large, colourful eyes, patterned wings and, in females, a rigid sheath around the ovipositor.

Trypeta zoe (Meigen)
larva = chrysanthemum blotch miner

A minor pest of *Chrysanthemum*, attacks of significance tending to occur most frequently on autumn-flowering or winter-flowering plants. Certain other cultivated members of the family Compositae are also attacked. Widely distributed in Europe.

DESCRIPTION **Adult:** 4–5 mm long; yellow-bodied, with iridescent wings distinctly marked with brown. **Larva:** 5–7 mm long; yellowish and rather stout-bodied, pointed anteriorly and truncated posteriorly.

LIFE HISTORY Adults occur in the spring, depositing eggs on the leaves of chrysanthemum and certain other composite hosts such as *Aster*, common ragwort (*Senecio jacobaea*), groundsel (*Senecio vulgaris*), mugwort (*Artemesia vulgaris*) and tansey (*Chrysanthemum vulgare*). The larvae then feed within the leaves during May and June, each forming a characteristic mine. This feeding gallery commences as an irregularly rounded blotch but later develops into an expanded, somewhat linear, mine which frequently follows the mid-rib and major veins (Fig. **400**). When fully fed, usually in early July, the larvae enter the soil and pupate, a second generation of adults appearing about two weeks later. Larvae of the second generation feed during the autumn; they eventually pupate to produce adults in the following spring.

DAMAGE When numerous, the yellowish mines seriously disfigure the foliage of chrysanthemum plants, affected tissue eventually turning brown.

CONTROL As for *Phytomyza syngenesiae* (p. 182).

Euleia heraclei (Linnaeus)
Celery fly

Although mainly regarded as a pest of celery, this generally common species also attacks other umbelliferous hosts, including ornamental giant hogweed (*Heracleum mantegazzianum*). Adults (wings 5 mm long) are mainly brown to black, with a yellow scutellum, yellow legs and smoky-patterned wings. They appear in late April and early May, depositing eggs on the underside of leaves. The eggs hatch in about a week, larvae then mining within the lamina of the leaves to form expansive brownish blotches within which black deposits of frass accumulate (Fig. **401**). The larvae (up to 7 mm long) are greenish-white and translucent, with a pointed head

400 Mine of chrysanthemum blotch miner, *Trypeta zoe*.

401 Mine of celery fly in leaf of *Heracleum mantegazzianum*.

402 Larva of celery fly, *Euleia heraclei*.

end and prominent black mouth-hooks (Fig. **402**). They often feed gregariously within the same blotch, development taking about three weeks. Larvae then pupate within the mine or in the soil, each in a yellowish (5 mm long) puparium. Adults emerge about four weeks later, usually in late July and August. Larvae of the second generation complete their development in the autumn, individuals overwintering within puparia; in favourable conditions, there may be a third generation.

8. Family **PSILIDAE**

Small to relatively large, colourful flies; wings with a pale streak or fold extending from the broken costa.

Psila nigricornis Meigen

larva = chrysanthemum stool miner
An infrequent but formerly common pest of *Chrysanthemum*, occurring mainly where clean stock plants are not used. Widely distributed in Europe; also found in Canada.

DESCRIPTION **Adult:** 4–5 mm long; body shiny bluish-black, with brownish-yellow head and legs; antennae mainly black. **Larva:** 5–6 mm long; creamish-white, tough-skinned and slender-bodied. **Puparium:** 4 mm long; dark brown.

LIFE HISTORY Adults of the first generation occur in May and June, laying eggs in the soil close to chrysanthemum plants. The eggs hatch in about two weeks, the larvae then invading the chrysanthemum stools. Larvae are fully fed in 1–2 months. They then return to the soil to pupate, adult flies emerging from late August to early October. Larvae resulting from this second generation feed throughout the winter, usually taking 3–4 months to reach maturity; under glasshouse conditions, however, the rate of development is increased and the first new adults can appear in February or March.

DAMAGE Larvae form long, superficial tunnels in chrysanthemum roots and stools, the outer tissue drying out and then splitting open; shoots, especially of cuttings, are also attacked. Infested plants are weakened and shoot production reduced, while attacked cuttings may be killed; damage is most significant in stool beds during the autumn, especially where early-flowering cultivars (e.g. cv. Favourite) are used for producing cuttings in heated glasshouses. Damage may also be caused to the roots of lettuces planted into infested chrysanthemum beds.

CONTROL The likelihood of attacks will be reduced by using clean stock plants. Where required, an insecticide should be applied to the stools in early May and again in early September; to protect stools used for producing cuttings, chemical treatment before, or immediately after, lifting is recommended.

9. Family **EPHYDRIDAE**

Very small to small flies, found mainly near water or in damp situations.

Scatella spp.

Glasshouse wing-spot flies
Two species, *Scatella stagnalis* (Fallén) and *S. tenuicosta* (Collin), of these small (*c.* 3 mm long), black-bodied flies are commonly present in glasshouses and other protected sites, where they are associated with algal growths in nutrient-film troughs and on potting composts, rockwool growing-media and so on. They are often mistaken for sciarid flies (see p. 154) and, when present in large numbers, may cause concern; however, unlike the larvae of sciarids, those of *Scatella* are harmless and do not attack plant roots, cuttings or seedlings. The adults are characterized by their twice-broken costal veins, and by the slightly darkened wings which have several small, clear patches visible in the membrane (Fig. **403**); differences between species are slight.

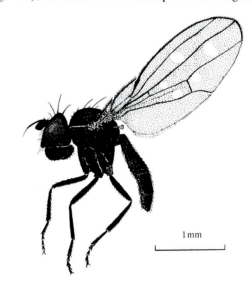

403 A glasshouse wing-spot fly, *Scatella* sp..

10. Family DROSOPHILIDAE

Very small to small flies, much attracted to fermenting juices; often called 'vinegar flies'. Compound eyes bright red; arista of the antennae generally plumose, with a bifid tip.

Scaptomyza flava (Fallén)
syn. *apicalis* Hardy

A widely distributed, often abundant leaf miner which attacks various cruciferous plants, including ornamentals, and nasturtium (*Tropaeolum*) (Fig. **404**). In severe cases infested leaves can be killed, but any effect on plant growth is unimportant. There are several overlapping generations each year, adults occurring from April to September. The larvae feed within the lamina of leaves, forming conspicuous whitish mines that vary from irregular, branched galleries which often follow the major veins. Fully fed larvae are 3–4 mm long, white and translucent, with four prominent tubercules at the hind end. They pupate externally on the ground, each in an elongate (3.0–3.5 mm long) reddish-brown puparium. Adults (wings 2.5 mm long) are pale yellow with grey markings and red eyes.

11. Family AGROMYZIDAE

Very small to small flies. The larvae of most species are host-specific, usually mining leaves to form serpentine or blotch-like mines characterized by double lines of frass (cf. lepidopterous and hymenopterous leaf miners).

Agromyza demeijerei Hendel

A locally abundant pest of *Laburnum*, especially in England, Germany, the Netherlands and Sweden. Widely distributed in Europe.

DESCRIPTION **Adult:** wings 2.4–3.0 mm long; black with mainly yellow legs. **Larva:** 3 mm long; whitish.

LIFE HISTORY This species is bivoltine, adults appearing in May and in August. Larvae feed from June to July and from September to October, fully grown individuals vacating their mine to pupate on the ground. Each mine, which is restricted to the upper leaf surface, commences as a narrow gallery along the leaf margin, but then widens into a substantial blotch (cf. *Phytomyza cytisi*, p. 184; *Leucoptera laburnella*, p. 199); completed mines eventually turn brown (Fig. **405**).

404 Mines of *Scaptomyza flava* in leaves of *Tropaeolum*.

405 Mine of *Agromyza demeijerei* in leaf of *Laburnum*.

DAMAGE The foliage of severely infested trees looks distinctly scorched, attacks leading to premature leaf-fall.

CONTROL Infestations on small trees can be controlled by spraying with a systemic or a leaf-penetrating insecticide at the first signs of damage; treatment of older trees is usually impractical.

Agromyza alnibetulae Hendel

This species occurs widely in central and northern Europe and is often common on young birch (*Betula*) trees in parks, gardens and nurseries. The larvae (2–3 mm long) are whitish to orange-coloured. They feed in very long, serpentine galleries formed on the upper surface of the leaves. The mines, unlike those formed on birch by *Lyonetia clerkella* (p. 201) and certain other lepidopterous pests, are distinctly widened towards their end and contain a double row of black frass (Fig. **406**); the underside of the gallery is also noticeably swollen. The larvae occur in two main broods, from June to July and from August to September, fully grown individuals pupating in the ground. Adults (wings 2.2–2.5 mm long) are mainly greyish-black; they occur in May or June and from late July to early August.

406 Mine of *Agromyza alnibetulae* in leaf of *Betula*.

Agromyza johannae de Meijere

Widespread and locally common in association with broom (*Cytisus scoparius*) and Spanish broom (*Spartium junceum*). The larvae mine the leaves, sometimes causing significant damage to cultivated bushes. Each mine begins as a thin, linear gallery, usually following the leaf margin towards the apex, but then turns abruptly and develops into an elongate, central blotch. Fresh mines are inconspicuous, and infestations often pass unnoticed until mined leaves dry up prematurely and die. Adults (wings 2.2–2.9 mm long) are greyish-black with a reddish frons.

407 Mine of *Amauromyza flavifrons* in leaf of *Dianthus*.

Agromyza potentillae (Kaltenbach)
syn. *spiraeae* Kaltenbach

Irregular, pale brown blotch mines formed by this widely distributed and locally common species sometimes occur on the leaves of ornamental *Geum* and *Potentilla* but are more frequently noted on raspberry and strawberry. Attacks can check the growth of young plants but are rarely important. Although mainly a European pest, this leaf miner is also present in Canada.

Amauromyza flavifrons (Meigen)

Larvae of this widespread but local species mine the leaves of various plants, including carnation and pink (*Dianthus*), *Gypsophila* and *Silene*; outbreaks on cultivated *Dianthus* can be especially serious, and heavily infested plants may be killed. The yellow larvae (up to 3 mm long) feed during the summer, each forming a characteristic white blotch mine, preceded by a narrow gallery (Fig. **407**) (cf. *Delia cardui*, p. 187). Pupation occurs on the ground in a reddish-brown puparium. Adults (wings 2 mm long) are black and yellow, with bright yellow halteres and black legs.

Amauromyza maculosa (Malloch)

A polyphagous, mainly tropical or subtropical, American species found occasionally in northern Europe on imported *Chrysanthemum* cuttings. The larvae feed gregariously within expansive leaf blotches, fully fed individuals pupating externally in reddish puparia; the posterior spiracles each have three pores. There are several generations each year, but infestations are unlikely to become established in northern Europe, except in heated glasshouses.

Amauromyza verbasci (Bouché)

Large, conspicuous blotches formed by larvae of this species are sometimes noted on the leaves of cultivated buddleia (*Buddleja*). Each mine commences as a narrow, contorted gallery but soon widens into a broad feeding gallery (Fig. **408**). Although disfiguring, especially if present in large numbers, damage is not important. Figwort (*Scrophularia nodosa*) and mullein (*Verbascum*) are also hosts. Adults (wings 2.5 mm long) are mainly greyish-black with a yellow frons and bright yellow knees.

Liriomyza trifolii (Burgess in Comstock)

larva = American serpentine leaf miner

A mainly North American species but frequently introduced into Europe on chrysanthemum cuttings imported from America, the Canary Islands, Kenya and Malta; now established in glasshouses in the Netherlands. In addition to *Chrysanthemum*, infestations also occur on many other hosts, including vegetable crops. N.B. This insect is a notifiable pest in several countries.

DESCRIPTION **Adult:** wings 1.2–1.5 mm long; greyish-black with a mainly yellow head and bright yellow scutellum; antennae bright yellow; legs with yellow coxae and femora, and brown tibiae and tarsi; costa reaching vein M1 + 2 (Fig. **409A**) (cf. *Phytomyza*, Fig. **409B**) **Egg:** 0.2×0.1 mm; oval, smooth and translucent. **Larva:** 2 mm long; yellowish-orange; posterior spiracles with three pores, the outer ones elongate (Fig. **410**). **Puparium:** 1.5 mm long; yellowish-brown.

408 Mine of *Amauromyza verbasci* in leaf of *Buddleja*.

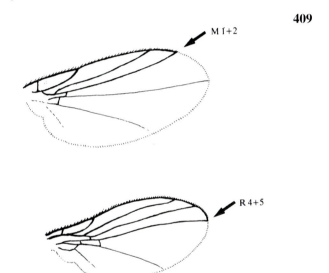

409 Venation of fore wing. A, *Liriomyza*; B, *Phytomyza*.

410 American serpentine leaf miner, *Liriomyza trifolii*.

LIFE HISTORY Eggs are deposited singly in the leaves of host plants, each female capable of depositing several hundred over a period of about a month. Eggs hatch within a few days. The larvae then mine the leaves, each forming a contorted, whitish gallery (Fig. **411**) which contains a meandering double line of dark frass along its length; fully fed larvae vacate the mines about 10–14 days later and pupate in the soil (cf. *Phytomyza syngenesiae*, pp. 181–182), adults emerging a week or two later. Breeding is continuous under favourable conditions, the life-cycle being greatly shortened by high temperatures (but protracted by low temperatures and poor lighting conditions), and populations can build up rapidly following an original infestation. Once established, infestations may also spread to outdoor plants, including common weeds such as bittersweet (*Solanum dulcamara*), common ragwort (*Senecio jacobaea*) and groundsel (*Senecio vulgaris*). In most instances, however, populations in northern Europe are unlikely to survive the winter out-of-doors.

DAMAGE Leaves are disfigured by small, rounded adult feeding punctures (Fig. **412**) (cf. those of *Phytomyza syngenesiae* pp. 181–182); the leaf mines, when numerous, cause considerable damage to the foliage, checking growth and affecting crop quality and marketability.

CONTROL Recommended treatments include the adoption of intensive insecticide treatments (often under the direction of Plant Health authorities) and pre-planting sterilization of glasshouse soil.

Liriomyza congesta (Becker)
larva = a pea leaf miner

A generally common, polyphagous leaf miner; sometimes a minor pest of sweet pea (*Lathyrus odoratus*). The larvae (up to 2mm long) are greenish-white and form narrow linear mines on the upper surface of the leaves, with black frass dispersed characteristically along either side of a green central band (Fig. **413**); the mines, usually no more than one per infested leaf, terminate in an expanded blotch. Fully fed larvae pupate externally on the ground (cf. *Phytomyza horticola*, p. 184). There are two or more generations annually, populations often building up throughout the season; however, the mines cause little or no distortion and are usually insufficiently numerous to affect plant growth. Adults (wings 1.3–1.8mm long) are greyish-black, with the frons, scutellum and sides of both thorax and abdomen, antennae and legs yellow.

411 Mines of American serpentine leaf miner on *Chrysanthemum*.

412 Adult feeding punctures of *Liriomyza trifolii* in leaf of *Chrysanthemum*.

413 Mines of *Liriomyza congesta* in leaf of *Lathyrus*.

Liriomyza huidobrensis (Blanchard)

larva = South American leaf miner

A very polyphagous, mainly South American species occurring occasionally in northern Europe on imported plants, especially *Chrysanthemum*; also well established in California. The larvae (up to 3.25 mm long) feed in distinctive linear leaf mines that often commence their development in association with the mid-rib and other major veins. Later, the mines frequently turn back upon themselves to give the appearance of a broad, blotch-like gallery (Fig. **414**); the larvae may also burrow alongside mines formed in the same leaf by other individuals. Fully grown larvae pupate in the soil or within the feeding gallery, each forming a yellowish-brown or reddish-brown puparium; characteristically, each posterior spiracle has about 6–9 small pores arranged in an arc. Adults are similar in appearance to those of *Liriomyza trifolii* (p. 177) but larger (wings 1.7–2.2 mm long) and slightly darker.

414 Mine of South American leaf miner on *Chrysanthemum*.

Liriomyza sativae (Blanchard)

A mainly American species but sometimes introduced into northern Europe on imported plants, especially *Chrysanthemum* cuttings. The larval mines are sinuous but relatively small; however, large numbers can occur in the same leaf, so that damage becomes extensive and host plants severely weakened or killed. Fully fed larvae pupate externally in pale orange-yellow puparia; the posterior spiracles each bear three stout bulbs. In common with other alien species, infestations may occur on various ornamentals and protected vegetables; larvae may also invade weeds such as bittersweet (*Solanum dulcamara*), common ragwort (*Senecio jacobaea*) and groundsel (*Senecio vulgaris*). Breeding is continuous under suitable conditions and host plant availability.

415 Mine of *Liriomyza strigata* in leaf of *Valeriana*.

Liriomyza strigata (Meigen)

syn. *pumila* (Meigen); *violae* (Curtis)

This common and widespread species attacks a wide variety of herbaceous plants, including ornamentals, but is usually present in only small numbers. The leaf mines, which usually occur only on older plants, tend to follow the mid-rib, with distinctive lateral branches extending into the lamina (Fig. **415**). Although the presence of mines on ornamentals may be disfiguring, they rarely, if ever, cause actual harm to host plants.

Nemorimyza posticata (Meigen)

Widely distributed and locally common on golden rod (*Solidago virgaurea*), the larvae forming characteristic blotch mines on the leaves; the mines are extensive and adorned by wavy lines of frass. Pupation takes place externally. Although usually an unimportant species, significant damage to cultivated hosts has occurred in southern England. Adults (wings 3.0–3.3 mm long) are mainly black, with the knees of the fore legs and (in males) the anterior tergites yellow.

416 Mine of *Paraphytomyza hendeliana* in leaf of *Symphoricarpos*.

Paraphytomyza dianthicola (Venturi)

Mediterranean carnation leaf miner

This primarily Mediterranean leaf miner is a potentially serious pest of carnation (*Dianthus caryophyllus*). The larvae form linear mines, which extend down the leaf blade, usually on the underside, each gallery gradually widening into an elongate blotch (cf. mines of *Amauromyza flavifrons*, p. 176); feeding may also occur in the stems. There are several generations each year. Serious attacks have occurred in Belgium, but reports of this pest in northern Europe are usually limited to the discovery of mines on plants imported from Mediterranean regions, including Crete, Greece, Italy and the South of France.

417 Mines of *Paraphytomyza lonicerae* in leaves of *Lonicera*.

Paraphytomyza hendeliana (Hering)

Infestations of this locally common species occur on honeysuckle (*Lonicera*) and snowberry (*Symphoricarpos rivularis*), the yellowish larvae (up to 3 mm long) forming long, brownish to whitish mines on the leaves (Fig. **416**). Occupied mines occur from late April or May onwards, the larvae pupating externally. There are several generations annually. Adults (wings 2.5–2.75 mm long) are mainly greyish to brownish, with yellow knees.

Paraphytomyza lonicerae (Robineau-Desvoidy)

A widely distributed pest of honeysuckle (*Lonicera*) and snowberry (*Symphoricarpos rivularis*), heavy infestations sometimes developing on cultivated bushes in parks and gardens. The mines are relatively small but distinctly broadened in their later stages, several often occurring on the same leaf (Fig **417**). Unlike the previous species, there is just one generation annually, occupied mines occurring in the early summer. Adults are distinguished from those of *Paraphytomyza hendeliana* by the entirely dark knees on the middle and hind legs.

Phytomyza ilicis Curtis

larva = holly leaf miner

An often abundant pest of holly (*Ilex aquifolium*), occurring on wild and cultivated plants, including variegated varieties. Widely distributed in Europe; also now well established in parts of North America, having been introduced from Europe along with its food plant.

DESCRIPTION **Adult:** wings 2.5–3.0 mm long; body mainly dark brown or black; legs black. **Egg:** 0.4 × 0.2 mm; white. **Larva:** 2.8–3.0 mm long; whitish to yellowish-white. **Puparium:** 2.5 mm long; brown.

LIFE HISTORY Adults occur in May and June, depositing eggs near the base of the mid-rib on the underside of young holly leaves. Eggs soon hatch and the larvae enter the mid-rib to begin their development. They feed slowly until the end of the summer, and then move into the leaf lamina. The larvae usually remain undetected until December, when the first signs of a yellowish to purplish-brown blotch appear on the upper surface of the leaf. The often somewhat linear mine continues to develop throughout the winter, and is completed by the spring, the larva becoming fully fed in March or April. Pupation occurs within the mine, with the anterior tip of the puparium protruding through the upper epidermis, the adult eventually emerging through a pin-head-sized hole. Galled leaves (Fig. **418**) remain on bushes throughout the year and, within each, the remains of the puparium may be found close to the adult emergence hole.

DAMAGE Infested leaves look unsightly and heavy attacks on nursery stock may cause concern; however, plant growth is rarely affected.

CONTROL Although a difficult pest to eliminate, some control can be achieved on small plants by spraying with a systemic or a leaf-penetrating insecticide about 3 weeks after the peak of adult activity. Because of the waxy nature of the leaves, a wetting agent may be needed to improve coverage of the leaves.

418 Mine of holly leaf miner, *Phytomyza ilicis*.

419 Adult of *Phytomyza syngenesiae*.

Phytomyza syngenesiae (Hardy)

larva = a chrysanthemum leaf miner

A generally abundant pest of glasshouse-grown ornamentals, such as *Chrysanthemum*, *Cineraria*, pot marigold (*Calendula*) and sunflower (*Helianthus*). Widespread in Europe; serious infestations have also occurred in Australasia and in North America.

DESCRIPTION **Adult:** wings 2.2–2.6 mm long; body greyish-black, with pale yellow markings on the head and sides; legs mainly black, with yellow knees (Fig. **419**); costa reaching vein R4 + 5 (Fig. **409B**) (cf. *Liriomyza*, Fig. **409A**). **Egg:** 0.35 × 0.15 mm; oval, white and shiny. **Larva:** 3.0–3.5 mm long; greenish-white; yellowish-brown to dark brown; oval but rather flattened (Fig. **420**).

LIFE HISTORY Eggs are laid mainly on the upper surface of the leaves of host plants, especially where foliage is shaded. They hatch in 3–6 days and larvae then mine the leaves, each forming a long white to brownish, serpentine gallery on the upper surface (Fig. **421**), within which grains of dark frass are distributed at irregular intervals. The larvae are fully fed in about 7–10 days; they then burrow through the leaf to the lower surface before pupating, the anterior spiracles of the puparium protruding through the lower epidermis (cf. *Phytomyza horticola*, p. 184, and *Liriomyza trifolii*, pp. 177–178). Adults emerge in 9–12 days at normal glasshouse temperatures. In common with other agromyzids, the adults feed on host leaves by inserting their ovipositor into the leaf tissue and imbibing the exuded sap; they are also relatively inactive, usually making only short, jerky flights. Under protection there are several generations each year and, whilst conditions remain favourable, breeding is continuous. Out-of-doors, when development may be protracted, there are normally two generations annually, adults occurring in May or June and again in late July and August; the winter is passed in the pupal stage. During the summer, glasshouses may be invaded by naturally occurring flies emerging from wild hosts, such as groundsel (*Senecio vulgaris*) and sow-thistle (*Sonchus*), or from those reared on garden plants, especially chrysanthemums.

DAMAGE Foliage is disfigured by the relatively large, wavy-edged (cf. *Liriomyza trifolii*, p. 178) adult feeding and egg-laying punctures; leaves of some hosts also develop wart-like wounds. Larval mines, which may be extensive, spoil the appearance of ornamental plants; heavily infested leaves shrivel and turn brown, weakening host plants and sometimes causing their death. Although infestations on glasshouse-grown plants are often severe, attacks in the open are usually of little or no consequence.

CONTROL Insecticide sprays can give control, but resistance to several well-known pesticides is now widespread. Systemic soil-applied insecticides can be effective; biological control, using parasitic wasps such as *Dacnusa* spp. and *Diglyphus* spp., is also feasible.

420 Puparium of *Phytomyza syngenesiae* in underside of leaf of *Chrysanthemum*.

421 Mines of *Phytomyza syngenesiae* in leaf of *Chrysanthemum*.

Phytomyza aconiti Hendel

Unlike most other phytomyzids, larvae of this species feed gregariously. They form large, irregular, brown blotch mines on the leaves of *Delphinium* and monkshood (*Angelicum*), significant damage often occurring on such plants in gardens in southern England. Up to six larvae occur within each blotch, individuals eventually pupating in dark brown puparia. Adults (wings 2.4–2.7 mm long) are black with the sides of the thorax yellowish.

CONTROL Apply an insecticide at the first signs of damage.

Phytomyza aprilina Goureau

syn. *lonicerella* (Hendel)

The characteristic mines of this widely distributed species occur on honeysuckle (*Lonicera*), causing considerable disfigurement of affected leaves (Fig. **422**). Each mine commences as a stellate gallery but later becomes distinctly linear, with frass deposited in black lines. Infestations are noted most often on wild plants but may occur on cultivated bushes.

422 Mine of *Phytomyza aprilina* in leaf of *Lonicera*.

423 Larva of *Phytomyza aquilegiae*.

Phytomyza aquilegiae Hardy

Severe attacks by this leaf miner commonly occur on wild and cultivated columbine (*Aquilegia vulgaris*). The larvae (Fig. **423**) form large, distinctive, greenish-white blotch mines in the leaves, foliage often being destroyed and plants seriously weakened; galleries are also formed on the stipules (Fig. **424**). Pupation takes place on the ground in reddish-brown puparia (2 mm long). Adults are mainly black with a pale frons, and are larger (wings 2.1–2.5 mm long) than the other *Aquilegia*-mining species *Phytomyza minuscula* (p. 185).

Phytomyza calthophila Hendel

The long, dark, snake-like mines of this widespread but local species occur on the leaves of wild and cultivated marsh marigold (*Caltha palustris*). Although disfiguring, they are not harmful.

424 Mines of *Phytomyza aquilegiae* in stipules of *Aquilegia*.

425 Mine of *Phytomyza cytisi* in leaf of *Laburnum*.

426 Adult feeding punctures of *Phytomyza cytisi* in leaf of *Laburnum*.

Phytomyza cytisi Brischke

Leaf mines of this species are sometimes abundant during the summer and again in the autumn on *Laburnum*. They are, however, less destructive than those of *Agromyza demeijerei* (p. 175) and *Leucoptera laburnella* (p. 198), two other leaf miners on laburnum. The mines occur mainly on the upper leaf surface. They are irregular, white and linear, with obvious black frass scattered along their length (Fig. **425**); those predominantly or entirely on the lower leaf surface appear greenish from above. Larvae pupate on the ground in brown puparia. Adults (wings 1.8–2.2mm long) are mainly black with a pale frons. Adult feeding punctures (Fig. **426**) commonly disfigure the foliage of host trees.

427 Adult of *Phytomyza horticola*.

Phytomyza horticola Goureau

larva = a chrysanthemum leaf miner

A widely distributed and generally common leaf miner. Adults (Fig. **427**) are similar in appearance to those of *Phytomyza syngenesiae* (p. 181). The larvae are very polyphagous, forming galleries in the leaves of various ornamentals, including *Chrysanthemum*, *Petunia*, *Phlox*, poppy (*Papaver*), sweet pea (*Lathyrus odoratus*), tobacco plant (*Nicotiana*) and wallflower (*Cheiranthus cheiri*) (Fig. **428**). In common with *Phytomyza syngenesiae* pupation takes place at the end of the larval mine, but the puparium protrudes through the epidermis on the same side of the leaf as the rest of the gallery (Fig. **429**).

428 Mines of *Phytomyza horticola* in leaf of *Cheiranthus*.

429 Puparium of *Phytomyza horticola* on upper surface of leaf of *Chrysanthemum*.

430 Mines of *Phytomyza minuscula* in leaf of *Aquilegia*.

Phytomyza minuscula Goureau

This species forms short, conspicuous, whitish, irregular, linear leaf mines on columbine (*Aquilegia vulgaris*) (Fig. **430**) and meadow-rue (*Thalictrum*). Attacks are common on columbines in gardens, often occurring in company with the more destructive pest *Phytomyza aquilegiae* (p. 183). Pupation occurs in an orange puparium, frequently attached to the lower surface of a leaf, close to the end of the mine (Fig. **431**). Adults are relatively small (wings 1.7–2.0 mm long) and mainly black with a pale frons. Adult feeding punctures on the leaves of host plants are sometimes very conspicuous.

431 Puparium of *Phytomyza minuscula* on leaf of *Aquilegia*.

Phytomyza primulae Robineau-Desvoidy

Leaves of wild and cultivated species of *Primula* (including *Polyanthus*) are attacked by this common and widespread species. The larvae form long, silver-white linear mines, in which widely scattered lumps of black frass are clearly visible (Fig. **432**). Pupation occurs within the mine in a whitish puparium. Adults (wings 2 mm long) have black legs with yellow knees.

432 Mine of *Phytomyza primulae* in leaf of *Primula*.

433 Mine of *Phytomyza scolopendri* in leaf of *Phyllitis scolopendrium*.

434 Mine of *Phytomyza spondylii* in leaf of *Heracleum*.

Phytomyza scolopendri Robineau-Desvoidy

Associated with ferns, especially common polypody (*Polypodium vulgare*). Cultivated ferns may be attacked, but damage is limited to the presence of mines and, therefore, not important. On hartstongue fern (*Phyllitis scolopendrium*) the yellowish-green mines are extremely long, sometimes exceeding 10 cm, and are very distinctive (Fig. **433**). Pupation occurs within or outside the mine. Adults (wings 2.1–2.6 mm long) are black and yellow.

Phytomyza spondylii Robineau-Desvoidy

A common species on wild hogweed (*Heracleum sphondylium*); also well established on certain introduced ornamental umbellifers, including giant hogweed (*Heracleum mantegazzianum*). The larvae are yellowish; they form broadly linear mines (Fig. **434**), pupating on the ground after escaping through a slit made in the lower surface of the leaf. Adults (wings 2.1–2.4 mm long) are blackish, with the sides of the thorax pale; the legs are black, with just the knees of the fore legs yellow.

435 Mine of *Phytomyza vitalbae* in leaf of *Clematis*.

Phytomyza vitalbae Kaltenbach

This widespread leaf miner is associated with wild traveller's joy (*Clematis vitalba*) and will also attack various kinds of *Clematis* grown in cultivation. The larvae form long, irregular linear mines on the upper surface of leaves, causing noticeable distortion (Fig. **435**). Larvae eventually pupate externally in dark brown puparia. Adults (wings 2.2 mm long) are mainly black with a yellow scutellum.

Cerodontha ireos (Goureau)

larva = iris leaf miner

Conspicuous leaf mines formed by this widely distributed species are often common on both wild and cultivated yellow flag (*Iris pseudacorus*). Each mine is broadly elongate with dark discrete patches of frass clearly visible; the larva pupates in a black

436 Mines of *Cerodontha ireos* in leaves of *Iris pseudacorus*.

puparium orientated lengthwise at the end of the mine (Fig. **436**). Adults (wings 2.0–2.7mm long) are mainly black, with bright yellow knees.

Cerodontha iridis (Hendel)

larva = gregarious iris leaf miner

An abundant, southerly distributed species, associated with *Iris foetidissima*; the larvae feed gregariously, forming relatively large, opaque, greenish to yellowish leaf mines (Fig. **437**). Damage may also occur on other cultivated irises, including *Iris ochroleuca* and *I. spuria*. Pupation occurs at the end of the mine in a stack of reddish-brown or dark brown puparia, each puparium orientated crosswise (Fig. **438**). Adults (wings 2.3–3.2mm long) are mainly black with just the fore knees yellow.

437 Mines of *Cerodontha iridis* in leaves of *Iris foetidissima*.

12. Family **ANTHOMYIIDAE**

Adults are 'house fly'-like. The maggot-like larvae possess distinctive 'mouth hooks', several species attacking plant roots and stems.

Delia cardui (Meigen)

Carnation fly

A common pest of glasshouse and outdoor carnation, pink and sweet william (*Dianthus*). Widespread in Europe.

DESCRIPTION **Adult:** 6mm long; mainly greyish-brown. **Larva:** 8–10mm long; creamish-white.

438 Puparia of *Cerodontha iridis*.

LIFE HISTORY Adults occur throughout the summer but do not become sexually mature until the autumn. Eggs are then laid in the leaf axils of host plants or on the soil close by. The eggs hatch about two weeks later, the larvae then burrowing into the leaves to form elongate blotches (Fig. **439**) (cf. *Amauromyza flavifrons*, p. 176). Larvae may also bore into the pith of the shoots. When fully grown, either in the late autumn or late winter, the larvae enter the soil. They eventually pupate in the spring, adults emerging about two months later.

DAMAGE Affected leaves are extensively discoloured by the mines and may eventually wither and die. Infested shoots may also be killed.

CONTROL Occupied leaves or shoots should be removed and destroyed. Application of an insecticide to control larvae can have some effect; alternatively, spray against adults at regular intervals from late August to the end of October.

439 Mine of carnation fly in leaf of *Dianthus*.

Delia radicum (Linnaeus)

syn. *brassicae* (Bouché); *brassicae* (Wiedemann)
Cabbage root fly

A notorious and generally common pest of vegetable brassicas, but also sometimes a problem on cruciferous ornamentals such as *Alyssum*, stock (*Mathiola*) and wallflower (*Cheiranthus cheiri*). Holarctic. Present throughout Europe.

DESCRIPTION **Adult:** 6–7 mm long; grey to blackish. **Egg:** 1 mm long; elongate-oval, white and ribbed longitudinally. **Larva:** 8–10 mm long; creamish-white with prominent papillae on the anal segment. **Puparium:** 6–7 mm long; elongate-oval, reddish-brown.

LIFE HISTORY Individuals overwinter as pupae within puparia, adults emerging in the following spring from mid-April onwards, the precise timing of their appearance depending on temperature. Eggs are deposited in the soil close to the stems of host plants, the period of egg laying often coinciding with the flowering of cow parsley (*Anthriscus sylvestris*). Eggs hatch after 3–7 days, the larvae immediately attacking the roots of adjacent host plants. They feed for 3–4 weeks and then, when fully grown, move away through the soil for a few centimetres before pupating. Adults of the second generation appear in late June and July, and those of the third from mid-August onwards, the two generations tending to overlap so that subsequent egg laying can occur at virtually any time from July to September.

DAMAGE Seedlings or recent transplants collapse and die, the fibrous roots and much of the tap root being destroyed. Older or less heavily infested plants wilt in warm, dry weather and make poor growth; damaged root systems are also liable to subsequent attack by fungal pathogens.

CONTROL Seedlings raised after mid-April can be protected with a liquid or a granular insecticide, applied after plants have reached the second true-leaf stage. Transplants may require similar treatment, applied within four days of transplanting.

Delia platura (Meigen)

syn. *cilicrura* (Rondani)
Bean seed fly

Although associated mainly with vegetable crops, infestations of this world-wide pest may also occur on ornamentals such as *Anemone*, *Freesia* and hollyhock (*Althaea rosea*); damage is also reported on conifer seedlings. Attacked plants lack vigour and may be killed, and seedlings arising from attacked freesia corms sometimes turn bluish. Adults are active from May onwards, depositing eggs in the soil, especially in the presence of decaying organic matter. Eggs hatch within a few days. The larvae then tunnel inside germinating bean seeds and young stems, and within other suitable plant tissue, feeding for about 1–3 weeks before pupating in the surrounding soil, each in an oval (4–5 mm long), reddish-brown puparium. New adults appear 2–3 weeks later and, after mating, females initiate a further generation, each fly depositing about 50 eggs. There are usually 3–5 generations annually. Adults (6 mm long) are greyish-brown; the larvae (up to 8 mm long) are white and relatively robust, with distinct, curved mouth-hooks and 12 posterior tubercules (and the posterior respiratory stigmata each with 8–10 projections).

CONTROL Drench thoroughly with a soil insecticide as soon as damage is seen.

Order **LEPIDOPTERA** (butterflies and moths)

1. Family **ERIOCRANIIDAE**

A small group of small or very small, primitive, metallic-looking, day-flying moths with reduced mouthparts. The larvae are apodous leaf-miners with a very small head partly shielded by a large prothorax.

Eriocrania semipurpurella (Stephens)

Generally common and often abundant on birch (*Betula*), especially on trees no more than a few metres in height. Present throughout central and northern Europe.

DESCRIPTION **Adult:** 10–16mm wingspan; fore wings purplish-golden; hind wings mainly bronzy-grey. **Larva:** 8mm long; whitish with a pale brown head; legless (Fig. **440**).

LIFE HISTORY Adults occur in March and April, flying in sunshine and often occurring around birch trees in noticeable swarms. Eggs are laid in the leaf buds, the larvae feeding from late March to early May in expansive, brownish-white leaf blotches, each containing quantities of black frass distributed throughout the mine in criss-crossing threads. Fully grown individuals drop to the ground and enter the soil to pupate in tough, silken cocoons. There is one generation annually.

DAMAGE Larval mines occupy a large proportion of the leaf lamina, disfiguring and sometimes totally destroying some of the earliest-expanded leaves (Fig. **441**). Growth of plants, however, is not affected.

Eriocrania subpurpurella (Haworth)

Generally common on oak (*Quercus*) and locally abundant; occasionally troublesome on trees in parks and gardens. Present throughout most of Europe.

DESCRIPTION **Adult:** 9–14mm wingspan; fore wings pale gold and shiny, partly suffused and speckled with purple; hind wings yellowish-grey but purple apically (Fig. **442**). **Larva:** 8mm long; whitish with

440 Larva of *Eriocrania semipurpurella.*

441 Mature mine of *Eriocrania semipurpurella* in leaf of *Betula.*

442 Adult of *Eriocrania subpurpurella.*

443 Larva of *Eriocrania subpurpurella*.

444 Mine of *Eriocrania subpurpurella* in leaf of *Quercus*.

a pale brown head; legless (Fig. 443).

LIFE HISTORY Adults are active in sunny weather during April and May, resting on the bark of oak trees in dull weather. Eggs are laid singly in the buds, larvae later mining the leaves from late May to July. Each mine is an expansive, brownish blotch and, characteristically, contains numerous intertwining threads of blackish frass. When fully fed, larvae vacate their mines and enter the soil to pupate within tough, silken cocoons.

DAMAGE Mined leaves are distorted but infestations have little or no effect on tree growth. However, heavy infestations on ornamental trees are unsightly and sometimes cause concern, the foliage appearing extensively scorched (Fig. 444).

CONTROL On a small scale, infested leaves can be picked off and burnt, but application of an insecticide is rarely, if ever, justified.

445 Larva of *Eriocrania sangii*.

Eriocrania sangii (Wood)
A locally common species associated with birch (*Betula*), the characteristically grey-coloured larvae (8–9mm long) (Fig. 445) feeding from late March to May. Large, brownish blotch mines are formed in the leaves (Fig. 446), heavily infested trees appearing scorched.

Eriocrania sparrmannella (Bosc)
Mines formed by this species are formed in the leaves of birch (*Betula*) in June and July, much later than those of other birch-feeding eriocraniids. They are also characterized by the presence of an initial narrow, linear gallery, which opens abruptly into an expansive, brown blotch (Fig. 447). Adults (10–13mm wingspan) have golden, purple-marked fore wings; they occur in May. The larvae (8–10 mm long) are whitish with a small brown head (Fig. 448).

446 Mine of *Eriocrania sangii* in leaf of *Betula*.

447 Mine of *Eriocrania sparrmannella* in leaf of *Betula*.

448 Larva of *Eriocrania sparrmannella*.

2. Family **HEPIALIDAE** (swift moths)

Medium-sized moths, with vestigial mouthparts and short antennae. The larvae are elongate, with well-developed legs and crochets on the abdominal prolegs arranged into several concentric circles. Larvae are soil-inhabiting and feed on the roots of various wild and cultivated plants.

449 Male garden swift moth, *Hepialus lupulinus*.

Hepialus lupulinus (Linnaeus)
Garden swift moth

A common horticultural pest, especially in the vicinity of grassland, the larvae attacking the roots of various annual and perennial herbaceous plants, and also damaging bulbs, corms, rhizomes and tubers. Infestations are especially serious on ornamentals such as *Anemone*, *Aster*, *Campanula*, *Chrysanthemum*, daffodil (*Narcissus*), *Dahlia*, *Delphinium*, *Gladiolus*, *Iris*, lily of the valley (*Convallaria majalis*), lupin (*Lupinus*), peony (*Paeonia*) and *Phlox*. Widely distributed in Europe.

DESCRIPTION **Adult:** 25–40 mm wingspan; fore wings yellowish-brown, variably marked with white (especially in the male); hind wings yellowish-grey, darker in the male (Fig. **449**). **Egg:** 0.5 mm diameter, almost spherical; whitish when laid but soon becoming black. **Larva:** 35 mm long; white, shiny and translucent, the dark gut contents often clearly visible; head and prothoracic plate light brown (Fig. **450**). **Pupa:** 20 mm long; reddish-brown; abdominal segments with ventral projections and dentate dorsal ridges.

450 Larva of garden swift moth, *Hepialus lupulinus*.

LIFE HISTORY Adults fly at dusk in May and June, and occasionally in August and September, often skimming over grassland, lawns and pastures in considerable numbers. Each female lays up to 300 eggs which she broadcasts at random whilst in flight. Larvae feed in the soil on the roots of grass and many other plants, including those with fleshy or woody underground systems. If disturbed, the larvae retreat rapidly backwards down narrow feeding burrows but they are readily unearthed during soil cultivation. Larval development continues throughout the winter, individuals usually pupating in the following April, each in a loosely woven subterranean tunnel formed among the root system of the host. Adults emerge about six weeks later, often appearing in distinct flushes after rainfall.

DAMAGE Larvae bite off the roots and tunnel into bulbs, corms, rhizomes and tubers of hosts with fleshy underground root systems, retarding growth and often causing plants to wilt; badly damaged plants can be killed. If unchecked, infestations on perennial hosts may persist and increase in importance from year to year. Most serious damage is caused in autumn, winter and early spring, and is often experienced when plants are grown in recently ploughed pasture or grassland.

CONTROL Infested sites or those at special risk should be cultivated and a soil insecticide incorporated before planting. If necessary, perennial plants should be lifted periodically, the soil checked for larvae and an insecticide applied before replanting. Regular cultivation and good weed control will reduce the likelihood of attacks developing.

Hepialus humuli (Linnaeus)
Ghost swift moth
A widely distributed and generally common species, infesting a similar range of hosts to the previous species but usually more damaging to grassland and lawns. The larvae (up to 50 mm long) are whitish, robust and relatively opaque, with a reddish-brown head and prothoracic plate, and prominent, dark brown pinacula (cf. *Hepialus lupulinus*, p.000). When young, they feed on plant rootlets but older individuals attack the larger roots and may also bite into stolons and the lowermost parts of stems. In common with the previous species, individuals construct silk-lined feeding tunnels in the soil, retreating into them or curling up if disturbed. The larvae consume large amounts of food but growth is slow,

451 Male ghost swift moth, *Hepialus humuli*.

the period of development usually extending over two, and occasionally three, years. Damage to plants is particularly severe in the second summer of larval growth, individuals then pupating in the following April or May. Adults occur mainly June and July, and are active at dusk. The females are relatively large (50–70 mm wingspan) with yellowish-ochreous, orange-marked fore wings; males are much smaller (46–50 mm wingspan), with silvery-white wings (Fig. 451).

CONTROL As for *Hepialus lupulinus*.

3. Family NEPTICULIDAE

Minute moths with the first antennal segment forming an 'eye-cap'. The larvae, which feed in leaves and form sinuous mines or blotches, have a wedge-shaped head and are virtually apodous, the thoracic legs being reduced to short, extendible lobes and the abdominal prolegs to fleshy humps without crochets. Pupation usually takes place outside the mine in a small, parchment-like cocoon.

Stigmella anomalella (Goeze)

larva = rose leaf miner

A generally common pest of rose (*Rosa*), often occurring abundantly on both wild and cultivated bushes. Present throughout Europe.

DESCRIPTION **Adult:** 5–6mm wingspan; head orange, often suffused with dark brown; fore wings mainly greenish-bronze to golden, with a partly coppery tinge, the apical region purple; hind wings brownish-grey. **Larva:** 5mm long; yellow with a brown head.

LIFE HISTORY Adults occur in May and in August, eggs being deposited on the underside of leaves, usually close to the mid-rib. The larvae form long, contorted mines which become filled with greenish-grey to blackish frass; each gallery widens considerably in its later stages to leave a clear marginal line along both sides of the central band of frass (Fig. **452**). Occupied mines occur mainly in July and October, fully grown larvae then pupating in brownish or reddish-brown cocoons spun at the base of a petiole, in the angle between two shoots or on the surface of a leaf.

DAMAGE Infested leaves are unsightly but attacks have little or no effect on plant growth.

CONTROL Leaves with occupied mines can be picked off and destroyed; application of an insecticide is seldom, if ever, justified.

Stigmella hemargyrella (Kollar)

Widely distributed and locally common on beech (*Fagus sylvatica*), the larvae feeding in elongate galleries formed in the leaves. Infestations are often present on beech hedges, the mines disfiguring the leaves but not causing significant damage (Fig. **453**). There are two generations annually, larvae feeding in June and from August to September; adults (5–6mm wingspan, the fore wings bronzy-brown with a distinct silver or golden crossband) occur from April to May and from late July to early August.

452 Mine of rose leaf miner on *Rosa*.

453 Mine of *Stigmella hemargyrella* in leaf of *Fagus*.

193

Stigmella lapponica (Wocke)

Generally common on birch (*Betula*), including young amenity and nursery trees. Adults occur in May, eggs being laid on the underside of the leaves, usually close to a major vein. The larvae feed from mid- or late June onwards, forming long, angular galleries which often follow but may also cross the leaf veins (Fig. **454**). Development is completed in early July, there being normally just one generation annually. The first quarter of the mine is filled with greenish frass; there is then an abrupt change to a central black line of frass which continues throughout the rest of the gallery. There may be several mines per leaf but the lamina is not distorted and shoot growth is unaffected.

454 Mine of *Stigmella lapponica* in leaf of *Betula*.

Stigmella obliquella (Heinemann)

Widely distributed on smooth-leaved willows, including ornamentals such as weeping willow (*Salix vitellina* var. *pendula*). The larval mines appear as narrow, frass-filled galleries which end in a blotch with frass accumulated in the centre (Fig. **455**). Adults appear in May and in August, tenented mines occurring from June to July, and from September to October. Larvae (up to 3mm long) are orange-yellow with a brown head and, on the ventral surface, a line of dark, pear-shaped spots. When fully grown, they pupate externally in brownish-orange cocoons surrounded by strands of silk. Adults (4–5 mm wingspan) have mainly dark brown fore wings, each marked by a narrow, yellowish crossband.

455 Mine of *Stigmella obliquella* in leaf of *Salix vitellina*.

Stigmella roborella (Haworth)

A widely distributed and generally common leaf miner on oak (*Quercus*) but its true status uncertain due to confusion with close relatives, especially *Stigmella atricapitella* (Haworth) and *S. ruficapitella* (Haworth). The larvae form long galleries on the leaves, characterized by the clearly defined central line of black frass (Fig. **456**). Adults (5.0–6.5mm wingspan, with dark bronzy-brown, purplish-tinged fore wings) occur in May and June, with a second generation appearing in August and September; occupied mines occur mainly in late June, July, September and October.

456 Mine of *Stigmella roborella* in leaf of *Quercus*.

Stigmella suberivora (Stainton)

This local species is reported in southern England in association with evergreen oak (*Quercus ilex*), and is gradually extending its range; in Mediterranean areas, including France and Italy, it is found on cork oak (*Quercus suber*). Adults occur in May and in September, depositing eggs on the upperside of the leaves; the yellow-bodied larvae feed in July and August, and from November onwards, those of the second generation pupating in the late winter or early spring. Although not affecting plant growth, the elongate, serpentine mines disfigure the foliage and are especially conspicuous as they age and turn brown (Fig. **457**).

457 Mine of *Stigmella suberivora* in leaf of *Quercus ilex*.

4. Family **TISCHERIIDAE**

A small family of very small moths with narrow, pointed wings. The leaf-mining larvae are very flat-bodied, with vestigial abdominal prolegs.

Tischeria ekebladella (Bjerkander)

A locally common leaf miner on oak (*Quercus*) and sweet chestnut (*Castanea sativa*). Present throughout Europe.

458 Larva of *Tischeria ekebladella*.

DESCRIPTION **Adult:** 8–11 mm wingspan; fore wings deep ochreous-yellow, speckled with blackish scales apically; hind wings grey. **Larva:** 7 mm long; pale yellow with the gut partly visible; head pale brown; prothoracic plate brown (Fig. **458**).

LIFE HISTORY Adults occur in June, depositing eggs on the leaves of oak and sweet chestnut. The larvae form whitish, initially shell-shaped mines on the upper surface (Fig. **459**), feeding from about September to November; unlike many leaf miners, a slit is formed at the edge of the mine, through which frass is ejected. When fully fed, each larva forms a circular chamber in the middle of its mine, within which the winter is passed and pupation eventually occurs.

DAMAGE Although several mines may occur on a leaf the tissue is not distorted; also, the lateness of attacks lessens their importance.

459 Young mines of *Tischeria ekebladella* in leaf of *Castanea*.

Tischeria angusticollella (Duponchel)

Minor infestations of this widely distributed but local species occur on rose (*Rosa*) including cultivated bushes, the small, pale green larvae forming characteristic blotch mines on the upperside of the leaves (Fig. **460**). Although causing distortion, and affecting the appearance of foliage, the mines have no effect on plant growth. Occupied mines occur from June to July and again in September and October. The insignificant, dark purplish-brown adults (8–9 mm wingspan) appear in the spring, with a second generation emerging in the late summer.

460 Young mines of *Tischeria angusticolella* in leaf of *Rosa*.

5. Family **INCURVARIIDAE**

Small, metallic-looking, mainly day-flying moths with well-developed antennae. The larvae commence feeding as leaf miners but later become case-dwellers; they have a single transverse band of crochets on each abdominal proleg.

Incurvaria pectinea Haworth

syn. *zinckenii* (Zeller); *pectinella* (Fabricius)

A minor pest of birch (*Betula*) and hazel (*Corylus avellana*). Widely distributed and locally common in central and northern Europe; in mainland Europe alder (*Alnus*), dogwood (*Cornus*), hornbeam (*Carpinus betulus*), maple (*Acer*) and rowan (*Sorbus aucuparia*) are also attacked.

461 Larval cases of *Incurvaria pectinea* on leaf of *Betula*.

DESCRIPTION **Adult:** 12–16 mm wingspan; head yellow (male with strongly bipectinated antennae); fore wings light brownish-bronze, each marked on the dorsal margin with two whitish spots; hind wings greyish-bronze. **Larva:** 6–7 mm long; whitish with a brown head and yellowish-grey thoracic plates. **Case:** 8 × 5 mm; oval and flattened.

LIFE HISTORY Adults occur in April and May, depositing eggs on the leaves of host plants. The larvae commence feeding in June, at first mining the leaves and forming small, circular blotches. Later, they form portable cases by cutting out and spinning together oval pieces from each leaf surface (Fig. **461**). The larvae then wander or fall away to feed on dead leaves and other vegetable debris, completing their development in the autumn. The cases are then attached to upright surfaces, including fences, posts or tree trunks, the larvae pupating in the winter and adult moths emerging in the following spring.

DAMAGE Larval mines disfigure the foliage but are less noticeable than the series of holes left by the case-forming larvae. Damage caused is not significant.

6. Family COSSIDAE

Large or very large moths. The larvae are wood-borers, feeding in the trunks and branches of trees or shrubs.

Zeuzera pyrina (Linnaeus)
Leopard moth

A sporadically important pest of various trees and shrubs, including ash (*Fraxinus excelsior*), birch (*Betula*), *Cotoneaster*, crab apple (*Malus*), hawthorn (*Crataegus*), honeysuckle (*Lonicera*), horse chestnut (*Aesculus hippocastanum*), lilac (*Syringa vulgaris*), maple (*Acer*), ornamental cherry (*Prunus*), *Rhododendron* and sycamore (*Acer pseudoplatanus*). Widespread in mainland Europe; in Britain restricted mainly to the southern half of England and eastern Wales. Also introduced into North America.

DESCRIPTION **Adult:** 45–65 mm wingspan; wings white and translucent, with black or blue-black spots; body similarly coloured and rather velvety; male considerably smaller than female and with the antennae strongly bipectinated basally (Fig. 462). **Egg:** 1 mm long; oval and pinkish-orange. **Larva:** 50–60 mm long; yellowish with prominent black pinacula; head and prothoracic plate brownish-black, the latter with a characteristically scalloped hind margin; young larva at first pinkish, readily distinguished from those of other wood-boring species by the characteristic prothoracic plate (Fig. 463). **Pupa:** 25–35 mm long; reddish-brown (Fig. 464).

LIFE HISTORY Adults occur in June or July, depositing eggs in groups in wounds or cracks in the bark of host plants. The newly emerged larvae often attack the leaf petioles and major leaf veins, buds and shoots of host plants but then enter the larger twigs and branches to feed in the heart wood. Larval galleries extend for 40 cm or more, a larva taking two or three years to complete its development. Pupation occurs in the feeding gallery in a slight cocoon into which particles of wood are incorporated. In early summer the pupa wriggles out of the cocoon and breaks through the surface of the branch where it remains protruding after emergence of the adult.

462 Male leopard moth, *Zeuzera pyrina*.

463 Larva of leopard moth, *Zeuzera pyrina*.

464 Pupa of leopard moth, *Zeuzera pyrina*.

DAMAGE Attacks usually occur in branches, stems or trunks less than 10 cm in diameter. The presence of a larva is indicated by the accumulation of frass and particles of wood which are forced out of the entry holes and, later, by the withering and die-back of the leaves and shoots. Infested branches are weakened and may snap off in a strong wind.

CONTROL Young larvae can be removed or killed by piercing them with a piece of wire inserted into the galleries; wounds on the tree should then be treated with canker paint. Insertion of certain fumigant pesticides into the tunnels and then sealing the entry holes with clay, putty or grafting wax is also recommended. If infestations are more advanced, dying branches should be cut back below the damaged area and burnt.

465 Goat moth damage to trunk of *Quercus*.

466 Adult of *Leucoptera laburnella*.

Cossus cossus (Linnaeus)
Goat moth
Infestations of this widely distributed but generally uncommon, wood-boring species occur in various kinds of mature tree, including ash (*Fraxinus excelsior*), birch (*Betula*), cherry (*Prunus*), oak (*Quercus*) and willow (*Salix*). The large (70–100 mm wingspan), dull, greyish-brown moths occur in June and July, and may sometimes be found resting on tree trunks during the daytime. Eggs are deposited in crevices in the bark, usually in groups of about fifty. After hatching, the pinkish larvae immediately burrow into the trunks to feed within the sap and heart wood. Larval development is greatly protracted, lasting up to three or four years, fully grown individuals (80–100 mm long) eventually pupating just below the bark, or in the ground, in strong, silken cocoons. Larval feeding galleries are very extensive and the larvae often cause the death of their hosts. Infested trees are sometimes recognized by the presence of large emergence holes in the bark (Fig. **465**), from which sap is sometimes exuded, the larval galleries also emitting an unpleasant, goat-like smell.

CONTROL With slight attacks, measures suggested for controlling *Zeuzera pyrina* (pp. 197–198) should be adequate but severely infested trees should be felled and burnt.

7. Family LYONETIIDAE

Small moths with narrow wings fringed with long cilia. The larvae have a complete circle of crochets on each abdominal proleg.

Leucoptera laburnella (Stainton)
larva = laburnum leaf miner
An often common pest of *Laburnum*; also found on lupin (*Lupinus*). Present throughout Europe; also present in North America.

DESCRIPTION **Adult:** 7–9 mm wingspan; fore wings shiny white, marked towards the apex with brownish and yellowish-orange; hind wings white; body mainly white (Fig. **466**). **Larva:** 6 mm long; greyish-white with a distinct greenish gut; head light brownish; prothoracic plate broad and greyish-brown; abdominal prolegs reduced but with crochets forming a complete circle (Fig. **467**). **Pupa:** 4mm long; yellowish-brown.

LIFE HISTORY Adults occur in May and in August with, in favourable conditions, a third generation in the autumn. Eggs are deposited on the underside of leaves of laburnum and, less frequently, on lupin. Larvae feed singly within the leaves, each burrowing to the upperside and then forming a large, pale blotch; blackish frass within the mine forms a distinctive pattern (Fig. **468**) (cf. mines of *Agromyza demeijerei*, p. 175; *Phytomyza cytisi*, p. 184). Occupied mines are most common from June to July and in August. When fully grown, each larva bites its way through the upper surface of the mine and then spins a white web on the underside of the same or an adjacent leaf (Fig. **469**). Pupation then takes place, pupae occurring from July to August and from September to May.

DAMAGE Mines distort the foliage and disfigure host plants; heavy infestations can reduce the vigour of young hosts.

CONTROL Attacks on small trees can be controlled by spraying with a systemic or a leaf-penetrating insecticide.

467 Laburnum leaf miner, *Leucoptera laburnella*.

468 Mine of laburnum leaf miner on *Laburnum*.

Lyonetia clerkella (Linnaeus)
larva = apple leaf miner
A generally common pest of trees and shrubs, including birch (*Betula*), cherry (*Prunus*), cherry laurel (*Prunus laurocerasus*), *Cotoneaster*, crab apple (*Malus*), hawthorn (*Crataegus*) and rowan (*Sorbus aucuparia*); often present on nursery stock and ornamentals. Palearctic. Widely distributed throughout Europe.

469 Pupation site of laburnum leaf miner on *Laburnum*.

470 Adult of apple leaf miner, *Lyonetia clerkella*.

471 Apple leaf miner, *Lyonetia clerkella*.

DESCRIPTION **Adult:** 8–9mm wingspan; fore wings shiny white (often partly or entirely suffused with brown), marked apically with a dark spot and by several black streaks which extend through the fringe of cilia; hind wings dark grey (Fig. **470**). **Larva:** 7–8mm long; green, translucent and moniliform; head and legs brown (Fig. **471**). **Pupa:** 3.5 mm long; pale green with yellowish-brown wing cases.

LIFE HISTORY Adults occur in June and August, and from October onwards, individuals of the autumn generation hibernating under loose bark, amongst thatch and in out-buildings, and reappearing in the following April. Eggs are usually laid singly in the underside of a leaf. They hatch about two weeks later. Each larva then commences to mine towards the upper surface and eventually forms a very long, pale-coloured gallery which widens gradually throughout its length and terminates in a distinct, elongate chamber (Fig. **472**). Feeding is completed in three or four weeks. The larva then emerges and wanders about on the foliage for a few hours before beginning to spin a slight, 6–7mm long, hammock-like cocoon (Fig. **473**), attached by strands of silk to a leaf or rough bark. Pupation then takes place, the adult moth appearing about two weeks later. There are usually three generations each year.

472 Mine of apple leaf miner on *Amelanchier laevis*.

473 Pupal cocoon of apple leaf miner, *Lyonetia clerkella*.

474 Apple leaf miner damage to leaf of *Amelanchier laevis.*

475 Mines of apple leaf miner on *Betula.*

DAMAGE The mines disfigure the foliage and, if numerous, cause distortion and premature death of infested leaves. On some hosts, notably snowy mespilus (*Amelanchier laevis*), infested leaves become greatly discoloured, especially where the gallery isolates a section of the lamina (Fig. **474**). On cherry laurel the leaf tissue often splits along the length of the mine; portions of the lamina may also fall away to leave rounded or irregular holes. N.B. Birch (Fig. **475**) is attacked only by the later generations.

CONTROL Infested leaves on young trees and nursery stock can be picked off and burnt; if required, apply an insecticide recommended for control of leaf-mining lepidoptera.

476 Larva of *Bucculatrix thoracella.*

Bucculatrix thoracella (Thunberg)

syn. *hippocastanella* (Duponchel)

Locally common on lime (*Tilia*), and of increasing importance as a pest of amenity or shade trees in towns and cities, especially in the Netherlands and Western Germany. Widely distributed in Europe.

DESCRIPTION **Adult:** 8 mm wingspan; fore wings pale ochreous-yellow with blackish markings; hind wings grey. **Larva:** 7 mm long; pale creamish-white with an orange tinge (Fig. **476**).

LIFE HISTORY Adults of the first generation occur in May, depositing eggs on the leaves of lime. At first the larvae mine within the leaves but they later feed externally on the undersurface, the change between instars occurring beneath opaque, silken webs (Fig. **477**). When fully grown the larvae

477 Larval cocoon of *Bucculatrix thoracella* on underside of leaf of *Tilia.*

descend on silken threads to pupate in ribbed cocoons (*c.* 5mm long) (Fig. **478**) formed on the trunks of trees or on fallen leaves or amongst other debris on the ground. A second flight of adults occurs in about July, larvae of this second generation completing their development in late August or September. The pupal cocoons can persist on the trunks of infested trees for several seasons, some with a split through which the pupa burst on emergence of the adult moth, and others with a rounded hole through which a parasitic wasp would have emerged.

DAMAGE Infested leaves are extensively disfigured by whitish or brownish patches, visible from above and from below. In public places the larvae descending on their silken threads can also be a nuisance, especially on trees growing close to market stalls and open-fronted food shops.

CONTROL Application of an insecticide in May can be effective, so long as the undersides of the leaves are adequately covered, but chemical treatment is often impractical.

478 Pupal cocoon of *Bucculatrix thoracella*.

8. Family **HIEROXESTIDAE**

Opogona sacchari (Bojer)
larva = sugar cane borer
This tropical species is a native of various African offshore islands, including the Canary Islands, Mauritius, St. Helena and the Seychelles, where it is
a pest of banana and sugar-cane. Infestations also occur on various plants imported into Europe as ornamentals, including *Dracaena*, *Ficus elastica*, *Hibiscus* and *Yucca elephantipes*.

DESCRIPTION **Adult:** 20–28mm wingspan; fore wings lanceolate, pale ochreous suffused with brown and blackish-brown; hind wings shiny brownish. **Larva:** 25mm long; whitish with small yellowish pinacula; head chestnut-brown; prothoracic and anal plates brown.

LIFE HISTORY Larvae burrow into the stems of host plants, forming extensive, silk-lined, frass-filled galleries. Fully grown individuals pupate close to the surface, the pupae protruding from the galleries following emergence of the adult moths (Fig. **479**). Infested plants imported into glasshouses in northern Europe usually produce moths in the winter months; the insect is unable to survive the winter out-of-doors.

DAMAGE The stems of infested plants are riddled with galleries, spoiling the appearance of ornamen-

479 Sugar cane borer damage to stem of *Dracaena*.

tals, reducing vigour and, sometimes, causing death of plants.

CONTROL This pest is difficult to control and it is usually necessary to destroy affected plants. Suspected infestations should be reported to Plant Health authorities.

480 Adult of azalea leaf miner, *Caloptilia azaleella*.

481 Azalea leaf miner, *Caloptilia azaleella*.

9. Family GRACILLARIIDAE

Very small moths with narrow wings bearing long fringes of cilia; adults adopt a typical resting posture, with the head end raised up by the long, widely spaced front legs. The larvae are 14-legged (prolegs absent on the sixth abdominal segment); they are mainly leaf miners and, in their young stages, are apodous and have a peculiar flattened head. The pupa typically protrudes from the cocoon following emergence of the adult.

482 Mine of azalea leaf miner, *Caloptilia azaleella*.

Caloptilia azaleella (Brants)

syn. *azaleae* (Busck)

larva = azalea leaf miner

A native of eastern Asia but now a well-established pest of indoor azalea (*Rhododendron*) in various European countries, including Belgium, southern England, France, Germany and the Netherlands; survives out-of-doors in favourable areas. Also introduced into New Zealand and North America.

DESCRIPTION **Adult:** 10–12 mm wingspan; fore wings violet-brown, with a large, elongate, golden blotch and fine brownish-black markings; hind wings grey (Fig. **480**). **Larva:** 10–12 mm long; pale greenish-yellow; head pale brownish-yellow; early instars legless (Fig. **481**).

LIFE HISTORY Eggs are laid singly, close to the mid-rib on the underside of the leaves of azaleas. Immediately after egg-hatch each larva forms a small, narrow mine which is then extended into an expansive leaf blotch (Fig. **482**), the latter often occurring without a preliminary mine. After completion of the blotch, the larva moves to the leaf tip; this is then rolled back and secured with silk to form a distinctive cone-like shelter (Fig. **483**),

483 Habitation of azalea leaf miner, *Caloptilia azaleella*.

within which further feeding occurs; a second such shelter is constructed before the larva is fully grown. Pupation then takes place within a white cocoon on the underside of a longitudinally folded leaf. Outdoors, this species is bivoltine (adults occurring in May and August, and larvae in June and September), the winter being spent in the pupal stage. Under glass, however, there are several generations and, if conditions remain favourable, breeding (and, hence, damage) may be continuous.

DAMAGE Heavily infested shrubs are seriously disfigured, both by the larval shelters and by the blotch mines; the latter commonly turn brown and infested leaves often shrivel and fall prematurely.

CONTROL At the first signs of damage (ideally before formation of leaf-tip cones), apply an insecticide recommended for control of leaf-mining lepidoptera; several sprays may be needed to eliminate an infestation.

484 Adult of *Caloptilia stigmatella*.

Caloptilia stigmatella (Fabricius)
Generally common on aspen (*Populus tremula*), poplar (*Populus*), sallow and willow (*Salix*); also reported on birch (*Betula*). Holarctic. Present throughout Europe.

DESCRIPTION **Adult:** 12–14 mm wingspan; fore wings chestnut-brown with a whitish, triangular, costal blotch (Fig. **484**). **Larva:** 8 mm long; head pale yellowish-brown; body whitish-yellow to whitish-green (Fig. **485**).

LIFE HISTORY Adults overwinter, emerging in the spring and eventually depositing eggs on the underside of leaves of willows or poplars. Eggs hatch in late June or July. The larvae each form an elongate gallery which eventually terminates in a small blotch. Each then feeds externally, forming a characteristic cone-like shelter, typically at the leaf tip (Fig. **486**). When the surface tissue within the feeding habitation is consumed, the larva moves to another leaf and forms a second shelter; rarely, feeding may continue on a further leaf. Fully fed individuals pupate in silken cocoons (Fig. **487**) formed on the underside of an adjacent leaf. Larvae develop from July to September, adults appearing from September onwards.

485 Larva of *Caloptilia stigmatella*.

486 Larval habitation of *Caloptilia stigmatella* on leaf of *Salix*.

DAMAGE Damage is of no significance but the presence of larval cones on young ornamental trees is disfiguring, affected tissue often turning black.

CONTROL On a small scale, affected leaves can be picked off and burnt, but application of an insecticide is rarely, if ever, justified.

Caloptilia syringella (Fabricius)

larva = lilac leaf miner

An often abundant pest of lilac (*Syringa vulgaris*) and privet (*Ligustrum vulgare*); ash (*Fraxinus excelsior*) and, less frequently, mock privet (*Phillyrea latifolia*) and white jasmine (*Jasminum officinale*) are also attacked. Eurasiatic. Widespread in Europe; also present in North America.

DESCRIPTION **Adult:** 10–13 mm wingspan; fore wings golden brown, with various dark-edged, whitish-yellow markings; hind wings dark brown (Fig. 488). **Larva:** 7 mm long; translucent greenish-white to yellowish-white, with a darker green gut; head pale brownish-yellow (Fig. 489). **Pupa:** 4 mm long; brownish-yellow.

LIFE HISTORY Adults occur in two distinct generations, from April to May and in July. They may be seen at rest on host plants, adopting a typical position with the wings and body held at an angle to the leaf surface. Eggs are laid singly or in rows on the mid-rib of host plants. Young larvae mine in the leaves, the galleries commencing as narrow mines but soon widening into expansive blotches, several larvae usually feeding together. Later in their development the larvae feed within distinctive, frass-filled shelters formed from webbed-down leaves. Occupied larval mines or tents are most common in June and from August to September. Fully grown individuals pupate within greyish-white cocoons constructed amongst litter on the ground or on the underside of a leaf. Individuals of the first generation usually pupate in June or July, adults emerging shortly afterwards, and those of the second in October, adults appearing in the following spring.

DAMAGE Feeding by young larvae leads to distinctive leaf discoloration and distortion. The

487 Pupal cocoon of *Caloptilia stigmatella*.

488 Adult of lilac leaf miner, *Caloptilia syringella*.

489 Lilac leaf miner, *Caloptilia syringella*.

prominent larval tents, formed either from a single leaf (as on lilac, Fig. **490**) or from several webbed-down leaves or leaflets (as on ash, Fig. **491**), disfigure host plants and affect the growth of young shoots. The leaves of affected sections of privet hedges appear scorched.

CONTROL On a small scale, affected leaves can be picked off and burnt; if required, apply an insecticide recommended for control of leaf-mining lepidoptera at the first signs of damage.

Acrocercops brongniardella (Fabricius)

An often abundant pest of oak, including evergreen oak (*Quercus ilex*). Widely distributed in mainland Europe; very local in the British Isles, and found mainly in the southern half of England, south-east Wales and southern Ireland. Also present in parts of Asia and North Africa.

DESCRIPTION **Adult:** 8–10 mm wingspan; fore wings ochreous-brown to brown, marked with white; hind wings grey (Fig. **492**). **Larva:** 6–7 mm long; translucent, pale yellowish-brown, with a blackish gut; pinacula and a series of oval dorsal plates, shiny greyish-brown; head and prothoracic plate pale yellowish-brown (Fig. **493**).

LIFE HISTORY Eggs are laid in the spring on the upperside of oak leaves. Larvae then mine the leaves, several individuals commonly feeding on the same leaf. The mines commence as silvery galleries and, later, develop into blotches; these eventually join to form an expansive mine that becomes distinctly bloated and may occupy most of the lamina (Fig. **494**). The blotches often become partly filled with moisture but this does not appear to harm the occupants. Larvae are fully fed by late June or early July. They then pupate in cocoons formed amongst debris on the ground. Adults appear in late July or early August and then overwinter, to reappear in the following April and May. There is usually just one generation annually but, in favourable situations, there may be a partial second.

DAMAGE The larval mines disfigure host plants and are especially noticeable and important on hedges of evergreen oak.

490 Habitation of lilac leaf miner on *Syringa vulgaris*.

491 Habitation of lilac leaf miner on *Fraxinus*.

492 Adult of *Acrocercops brongniardella*.

CONTROL On a small scale, affected leaves can be picked off and burnt. If required, at the first signs of damage apply an insecticide recommended for control of leaf-mining lepidoptera; on evergreen oak, additional wetter may be needed to ensure adequate coverage of the leaves.

Parectopa robiniella (Clemens)

This North American leaf miner is associated with false acacia (*Robinia pseudoacacia*) and has become established in Europe, infestations having been reported first in Italy in 1970; the pest has since appeared in Czechoslovakia, France, Hungary, Switzerland and Yugoslavia. The general popularity and abundance of false acacia as an ornamental or shade tree suggests that spread of the pest to more northerly regions could also be expected. Moths appear in May or June, eggs being deposited singly on the underside of the leaves of host trees. The larvae form large, irregular blotch mines, frass being ejected through a small hole in the lower surface of the gallery. Fully grown larvae are 4.0–4.5 mm long and mainly green, individuals pupating in white cocoons spun at the edges of leaves; moths emerge shortly afterwards. The adults (8–9 mm wingspan) are mainly dark brown, the fore wings being ornamented with several narrow, white, wedge-shaped marks. In favourable areas there are three generations annually, full-grown larvae of the final brood overwintering within cocoons on fallen leaves.

Phyllonorycter maestingella (Müller)
syn. *faginella* (Zeller)

Generally common on beech (*Fagus sylvatica*), and sometimes a minor pest of ornamental and nursery plants. Present throughout central and northern Europe.

DESCRIPTION **Adult:** 7.5–9.0 mm wingspan; fore wings dark brown with white, wedge-shaped markings; hind wings greyish-brown. **Larva:** 4–5 mm long; pale greenish-yellow, with a darker green gut; head light brown. **Pupa:** 3 mm long; light brown.

LIFE HISTORY Adults occur in two distinct generations, from May to early June and in August. Eggs are deposited on the underside of beech leaves. The larvae then develop in brownish blotch mines formed on the underside of the leaves between two lateral veins. The mines extend from the mid-rib to, or almost to, the leaf edge (Fig. **495**), their

493 Larva of *Acrocercops brongniardella*.

494 Mine of *Acrocercops brongniardella* in leaf of *Quercus ilex*.

495 Mine of *Phyllonorycter maestingella* in leaf of *Fagus*.

elongate form distinguishing them from those formed on beech by *Phyllonorycter messaniella* (see below). Larvae may be found in July and again from September to October. Pupation takes place in the mine within a white, silken cocoon; pupae occur in July and August, and from November to May.

DAMAGE Mines cause slight distortion of the leaves but infestations, unless heavy, have little effect on the general appearance of hosts; they do not affect plant growth.

CONTROL On a small scale, infested leaves can be picked off and burnt, but application of an insecticide is rarely, if ever, justified.

496 Mine of *Phyllonorycter messaniella* in leaf of *Quercus ilex*.

Phyllonorycter messaniella (Zeller)
Zeller's midget moth
Generally common on evergreen oak (*Quercus ilex*), but also found on deciduous oak, and occasionally an important pest; will also attack certain other trees, including beech (*Fagus sylvatica*), hornbeam (*Carpinus betulus*) and sweet chestnut (*Castanea sativa*). Widespread throughout much of Europe, except for more northerly areas; an introduced pest of ornamentals in Australia and New Zealand.

DESCRIPTION **Adult:** 7–9 mm wingspan; fore wings shiny pale golden-ochreous, with whitish, slightly brownish-tinged markings partly edged with dark brown; hind wings greyish-brown. **Larva:** 4.5 mm long; yellow to whitish-yellow; head brown. **Pupa:** 3.0–3.5 mm long; dark brown; a pair of large spines on the first three abdominal segments.

LIFE HISTORY Adults occur mainly from April to May, in August and from late October to late November, females depositing eggs on the underside of oak leaves. Larvae then mine the underside of the leaves, each forming a pale brown, oval or elongate blotch. Fully grown larvae pupate in a slight cocoon spun to one side of the gallery. Adults emerge two to three weeks later. Larvae are most frequent in July and October but, on evergreen oak, they also occur from December to March.

DAMAGE On evergreen oak, mines develop a strong elongate crease on their lower surface, which distorts the leaf so the lamina folds downwards. The upper surface over the mine becomes slightly mottled and eventually turns brown (Fig. **496**), heavy infestations significantly disfiguring host plants. On deciduous oak, mines are relatively small but will cause some distortion of the leaf, a distinct crease developing on the upper surface of the lamina above the mine. Mines on beech are also tightly folded (cf. *Phyllonorycter maestingella* p. 207).

CONTROL On a small scale, infested leaves can be picked off and burnt. If required on small trees, apply an insecticide recommended for control of leaf-mining lepidoptera; on evergreen oak, additional wetter may be needed to ensure adequate coverage of the leaves.

208

Phyllonorycter platani (Staudinger)

A minor pest of plane (*Platanus*), originating in Asia Minor but now widely distributed in Mediterranean regions; also well established further north, occurring throughout France and in various other countries, including Belgium, Germany and the Netherlands. Not found in the British Isles.

DESCRIPTION **Adult:** 7–9 mm wingspan; fore wings golden-brown, with white, dark-edged markings.

LIFE HISTORY Adults appear in May or June, eggs being laid on the underside of the leaves, especially on those on the lower branches. Larvae feed within the leaves, forming distinctive blotch mines visible from below (Fig. **497**). There are often several mines per leaf, the blotches varying considerably in size and sometimes exceeding 20 mm in length. Fully fed larvae pupate within the feeding gallery, adults of a second generation emerging in August. Larvae of the second brood complete their development in the autumn, pupae overwintering in the fallen leaves.

DAMAGE Mines disfigure the foliage and, if numerous, will cause considerable deformation of the lamina. Attacks on mature trees are of minor importance but heavy infestations on nursery trees can be troublesome.

CONTROL In nurseries, fallen leaves with occupied mines should be collected and burnt.

497 Mines of *Phyllonorycter platani* in leaf of *Platanus*.

498 Mine of *Phyllonorycter comparella* in leaf of *Populus tremula*.

Phyllonorycter comparella (Duponchel)

A local, southerly distributed and usually uncommon species, associated with poplar (*Populus*); not of importance as a pest but sometimes noted on young amenity trees, especially in mainland Europe. The larvae mine the underside of the leaves, forming relatively small blotches that cause noticeable discolouration of the upper surface (Fig. **498**). Occupied mines occur in mid-summer and in the autumn. Unusually for a species of *Phyllonorycter*, the adults are mainly white and grey (Fig. **499**), those of the second generation overwintering.

499 Adult of *Phyllonorycter comparella*.

Phyllonorycter corylifoliella (Hübner)

Widely distributed and often common on rosaceous trees, including crab apple (*Malus*), hawthorn (*Crataegus*), rowan (*Sorbus aucuparia*) and other kinds of *Sorbus*; less frequently, birch (*Betula*) is also attacked. The larvae feed throughout July and again in September and October, each forming a large, silvery-white, russet-flecked blotch on the upperside of leaves (Fig. 500); the mines commonly overlap the mid-rib or a major vein and eventually cause infested leaves to crinkle. Larvae are 5–6mm long, with a brown head and the body dirty whitish to pale yellowish, but appearing greenish intersegmentally. Adults occur in May and June, and in August. They are relatively large (8–9mm wingspan); the fore wings are chestnut-brown, suffused with blackish scales and marked with two or three narrow, whitish striae and a long, white basal streak.

500 Mine of *Phyllonorycter corylifoliella* in leaf of *Sorbus aucuparia*.

Phyllonorycter leucographella (Zeller)

An important pest of firethorn (*Pyracantha*), originating in Mediterranean areas where it also feeds on *Calycotome spinosa*; now extending its range into more northern parts of Europe, including Austria, France, Germany, the Netherlands, south-eastern England and Switzerland. Adults (8–9mm wingspan) are mainly yellowish-orange, the fore wings each marked with distinctive white striae. They occur from May to June and from August to mid-September, eggs being deposited singly on the leaves of firethorn. The larvae mine within the leaves, each forming a large, silvery blotch on the upper surface along the mid-rib (Fig. 501). The mine often extends over the whole of the upper surface, affected leaves eventually folding longitudinally at the completion of larval development (Fig. 502). Mines can also occur on other rosaceous plants [including *Chaenomeles*, *Cotoneaster*, crab apple (*Malus*), hawthorn (*Crataegus*), ornamental cherry (*Prunus*) and *Sorbus*] growing in the immediate vicinity of infested firethorn bushes; however, mines on such hosts tend to split open and larval development is then aborted. The larvae occur during the summer and from September onwards; individuals continue to feed throughout the winter, except in very cold weather, but are unable to survive in fallen leaves. Badly affected leaves tend to drop during the spring and summer, persistent attacks on bushes leading to noticeable defoliation.

CONTROL On nursery plants, apply an insecticide at the first signs of damage. Sprays are usually unnecessary on established bushes in parks and gardens, because of high levels of natural parasitism.

501 Mine of *Phyllonorycter leucographella* in leaf of *Pyracantha*.

502 *Phyllonorycter leucographella* damage to leaf of *Pyracantha*.

Phyllonorycter oxyacanthae (Frey)

Generally common on hawthorn (*Crataegus*), the larvae forming blotch mines on the underside of the leaves. The mines occur most frequently on the leaf lobes closest to the petiole, and cause considerable distortion of the lamina (Fig. **503**). Infestations are often present on nursery stock but do not affect plant growth. The larvae are pale yellowish-green; they feed in July and from September to October. Adults (6–8mm wingspan) occur in May and August; the fore wings are yellowish-brown, marked with dark brown scales and white strigulae.

503 Mine of *Phyllonorycter oxyacanthae* in leaf of *Crataegus*.

Phyllonorycter quercifoliella (Zeller)

Generally common on oak (*Quercus*), including young trees, the larvae forming mainly oval mines on the underside of the leaves. The mines usually occur between the major veins, their upper surface being characterized by a central patch of unconsumed tissue (Fig. **504**). The larvae feed in July and from September to November, the golden-brown, white-marked adults (7–9mm wingspan) emerging in late April or May and in August.

504 Mines of *Phyllonorycter quercifoliella* in leaf of *Quercus*.

Phyllonorycter rajella (Linnaeus)
syn. *alnifoliella* (Hübner)

A widespread and often abundant leaf miner on alder (*Alnus*), including amenity and nursery trees. There are two generations annually, larvae mining the underside of the leaves during the summer and again in the autumn. There are often several mines in each infested leaf but they cause only slight distortion. Adults are brown, marked with white (Fig. **505**); they occur in May and again in August.

505 Adult of *Phyllonorycter rajella*.

Phyllonorycter salicicolella (Sircom)

Blotch mines of this generally common species occur on the underside of leaves of common sallow (*Salix atrocinerea*) and other broad-leaved willows. They are often present on young shrubs and, if formed close to the leaf edge, cause noticeable distortion of the lamina (Fig. **506**). There are two generations annually, larvae feeding in July and from September to October; they are whitish-green with an orange spot on the sixth abdominal segment. Adults (7.0–8.5 mm wingspan) have light brown, white-strigulated fore wings. They occur in May and from late July to August.

506 Mine of *Phyllonorycter salicicolella* in leaf of *Salix atrocinerea*.

Phyllonorycter trifasciella (Haworth)

Generally common on honeysuckle (*Lonicera*) and snowberry (*Symphoricarpos rivularis*), and sometimes a minor pest of cultivated bushes. The larvae feed in distinctive blotches formed on the underside of the leaves, causing considerable distortion of the lamina (Fig. **507**). There are usually three generations, larvae feeding in March to April, July and October. The mainly orange-brown adults (7–8 mm wingspan) appear in May, August and November.

507 Mine of *Phyllonorycter trifasciella* in leaf of *Symphoricarpos rivularis*.

Phyllonorycter ulmifoliella (Hübner)

A generally common species on birch (*Betula*), the larvae developing in blotch mines formed on the underside of the leaves. The mines are characterized by several longitudinal folds which coalesce across the middle (Fig. **508**); the upper surface appears slightly mottled. Although relatively small, they can cause noticeable distortion of leaves and often disfigure young trees, including nursery stock. There are two generations annually, occupied mines occurring in July and again in the autumn. The larvae (up to 5 mm long) are pale yellowish-green with a yellow spot on the sixth abdominal segment. Adults (7–9 mm wingspan) occur in May and August; the fore wings are golden-brown, marked with several white, inwardly blackish-bordered strigulae and a white basal streak.

508 Mines of *Phyllonorycter ulmifoliella* in leaf of *Betula*.

10. Family PHYLLOCNISTIDAE

Very small moths, with an 'eye-cap'. The larvae are apodous, sap-feeding leaf miners.

509 Mines of poplar leaf miner on *Populus*.

Phyllocnistis unipunctella (Stephens)
syn. *suffusella* (Zeller)
larva = poplar leaf miner

An often abundant leaf miner on black poplar (*Populus nigra*) and Lombardy poplar (*Populus nigra* 'Italica'), including trees growing in gardens and nurseries. Although present throughout much of mainland Europe, in Britain it is found mainly in England and south-east Wales. Adults occur in July and from September to April, those of the second generation hibernating and eventually depositing eggs in the spring. The larvae feed in June and in August, each forming a very long slightly raised, but inconspicuous, gallery on the upper surface of the leaves. The mines are most noticeable when the leaf is viewed at an angle, then appearing as silvery, slug-like trails (Fig. **509**); they are usually noticed only when foliage of trees is examined closely for other disorders. The larvae (up to 5 mm long) are pale greenish-white with a transparent head and a blackish prothoracic plate (Fig. **510**); unlike members of the Gracillariidae they are sap feeders throughout their development, excreting a clear liquid instead of faecal pellets. Fully grown larvae pupate in whitish cocoons, each usually spun externally under a fold of the leaf margin at the end of the larval gallery (Fig. **511**). The adults (7–8 mm wingspan) are superficially similar to those of various species of *Phyllonorycter* (p. 207 *et seq.*) with a *Leucoptera*-like (p. 198) wing pattern. Damage caused by this insect is unimportant and usually noticed only when foliage of trees is examined closely for other disorders; in some situations mined leaves are reduced in size, suggesting that persistent attacks on young hosts might have an adverse effect on tree growth.

510 Poplar leaf miner, *Phyllocnistis unipunctella*.

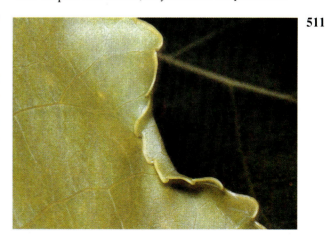

511 Pupal habitation of poplar leaf miner, *Phyllocnistis unipunctella*.

11. Family SESIIDAE (clearwing moths)

Unusual, wasp-like moths with partly clear (scaleless) wings and a distinct fan-like anal tuft of scales. The larvae are stem-borers in shrubs and trees; crochets on the abdominal prolegs are arranged into two transverse bands.

Sesia apiformis (Clerck)
Hornet moth

Associated mainly with poplar (*Populus*), and sometimes damaging to mature specimen trees. Palearctic. Widely distributed in mainland Europe; in the British Isles most numerous in central, eastern and southern England. Also introduced into North America.

DESCRIPTION **Adult:** 35–45 mm wingspan; wings mainly clear with brownish scales along the costal margin, brownish veins and cilia; body black and yellow, tinged with brown. **Larva:** 30–40 mm long; yellowish-white with a large reddish-brown head and a yellowish prothoracic plate. **Pupa:** 20–30 mm long; dark reddish-brown; very plump, with rows of backwardly directed spines on the abdominal tergites.

Conopia formicaeformis (Esper)
Red-tipped clearwing moth

An occasional but minor problem in willow (*Salix*) stool beds, the larvae mining within the stumps. The adults (18–20 mm wingspan) are mainly black with a bright red abdominal band and the wings tipped with red scales (Fig. **512**). The moths are very active, flying away rapidly if disturbed; they occur in July and August.

512 Red-tipped clearwing moth, *Conopia formicaeformis.*

LIFE HISTORY Adults occur mainly in June and July, freshly emerged individuals often basking in sunlight on the trunks of host trees. Eggs are laid in bark crevices or holes, close to the base of the trunks of poplar trees. Eggs hatch in about two weeks. The larvae then burrow into the trees to feed, forming extensive, frass-filled galleries between the bark and heart wood; the feeding galleries may also extend into the roots. Larvae continue to feed for at least two years, fully fed individuals then overwintering in tough cocoons formed just below the bark at the base of infested trees. Pupation occurs in the following spring, adults emerging several weeks later.

DAMAGE Infested trees can be distinguished by the presence at the base of the trunks of adult emergence holes, each about 8 mm in diameter.

12. Family CHOREUTIDAE

Small moths, the wings held flat over the body in repose. The larvae have long, pencil-like abdominal prolegs.

Eutromula pariana (Clerck)
larva = apple leaf skeletonizer

Although associated mainly with apple, including crab apple (*Malus*), this species is occasionally damaging to the foliage of other rosaceous plants, including hawthorn (*Crataegus*), ornamental cherry (*Prunus*) and ornamental pear (*Pyrus*). The pale green to yellowish, black-spotted larvae (up to 14 mm long) (Fig. **513**) cause extensive damage to the leaves, typically removing tissue from the upperside but leaving the lower surface intact. They feed beneath silken webs from May onwards, individuals eventually pupating in white, boat-shaped cocoons spun beneath a leaf or among plant debris. The greyish-brown adults (11–13mm wingspan) (Fig. **514**) appear in July and early August. Larvae of a second generation complete their development in the autumn, producing adults which then overwinter.

13. Family YPONOMEUTIDAE

A large family of small moths with elongate, relatively broad wings and well-developed, projecting maxillary palps. The larvae are of variable form; crochets on the abdominal prolegs are often arranged into several concentric circles.

Argyresthia dilectella Zeller
Juniper shoot moth
larva = juniper shoot borer

A locally common pest of juniper (*Juniperus*), and sometimes troublesome on ornamentals in gardens and nurseries; also reported on other members of the Cupressaceae. Restricted to central Europe and to the more southerly parts of northern Europe.

DESCRIPTION **Adult:** 8–10mm wingspan; fore wings violet-whitish, suffused and ornamented with golden-brown; hind wings pale grey (Fig. **515**). **Larva:** 5mm long; yellowish-green, sometimes reddish

513 Apple leaf skeletonizer, *Eutromula pariana*.

514 Adult of apple leaf skeletonizer, *Eutromula pariana*.

515 Juniper shoot moth, *Argyresthia dilectella*.

intersegmentally, with a brownish-black head and blackish prothoracic and anal plates (Fig. **516**).

LIFE HISTORY Moths occur in July, depositing eggs on the shoots of juniper. After egg hatch the larvae tunnel within the shoots, feeding throughout the winter months and completing their development by the end of May. They then vacate the feeding gallery to pupate externally in white cocoons.

DAMAGE Affected shoots turn purplish and then brown (Fig. **517**), spoiling the appearance and affecting the development of infested plants.

CONTROL Apply an insecticide in the summer to prevent larvae tunnelling into the shoots. Damaged shoots discovered during the winter should be picked off and burnt.

516 Juniper shoot borer, *Argyresthia dilectella.*

517 Juniper shoot borer damage to shoot of *Juniperus.*

Argyresthia thuiella (Packard)
American thuja shoot moth
larva = American thuja leaf miner
A North American pest of cedars and cypresses, especially Lawson cypress (*Chamaecyparis lawsoniana*) and white cedar (*Thuja occidentalis*); accidentally introduced in the 1970s into parts of mainland Europe, including Austria, Germany and the Netherlands, where it is now well established. Not recorded in the British Isles. N.B. Reports of this species attacking juniper (*Juniperus*) are due to its confusion with *Argyresthia trifasciata* (p. 217).

DESCRIPTION **Adult:** 8mm wingspan; head and thorax whitish; fore wings white to greyish-white, with silver-grey markings (Fig. **518**). **Egg:** greenish. **Larva:** 5mm long; body green or brownish, often

518 American thuja shoot moth, *Argyresthia thuiella.*

with a reddish tinge at the posterior margin of each segment; head and prothoracic plate shiny black or brown. **Pupa:** 3.5 mm long; green to brown.

LIFE HISTORY Adults occur mainly in late June and July, when they may be found settled on the foliage of host plants; if disturbed, they fly a short distance before resettling. Eggs are laid between the tips of leaf scales and the base of adjoining scales, hatching in 2–3 weeks. The larvae then bore into the shoots to begin feeding, each mining away from the tip. Larval development continues from July onwards, fully grown individuals pupating in the following May or June.

DAMAGE Tips of mined shoots become discoloured, and eventually turn brown and die (Fig. **519**).

CONTROL As for *Argyresthia dilectella* (pp. 215–216).

Argyresthia trifasciata (Staudinger)

A common pest of juniper (*Juniperus*) in southern Europe, also becoming established further north in parts of France, Germany, the Netherlands and Switzerland. Infestations also occur on *Chamaecyparis*, *Cupressus* and *Thuja* (cf. *Argyresthia thuiella*, p. 216).

DESCRIPTION **Adult:** 10 mm wingspan; fore wings blackish, each marked with whitish crosslines; hind wings dark grey. **Larva:** 5 mm long; greenish with a blackish head.

LIFE HISTORY Adults occur from mid-May to mid-June, eggs being deposited on the shoots of juniper. The larvae mine within the shoots from late June onwards, completing their development in the following spring. They then pupate on a main shoot, each in a small cocoon hidden beneath a flake of bark.

DAMAGE Infested shoots turn brown (Fig. **520**), damage to hosts often being extensive. Especially severe infestations have occurred on *Juniperus virginiana* 'Sky Rocket'.

CONTROL Spraying with an insecticide in the first half of June has proved effective.

519 American thuja leaf miner damage to shoot of *Thuja*.

520 *Argyresthia trifasciata* damage to shoot of *Juniperus*.

521 *Argyresthia sorbiella* damage to shoot of *Sorbus aucuparia*.

522 Larva of *Argyresthia sorbiella*.

Argyresthia sorbiella (Treitschke)

A locally common species associated with rowan (*Sorbus aucuparia*). The larvae tunnel in the young shoots, causing the tips to wilt and turn black (Fig. **521**); infestations are found occasionally on young amenity trees but are rarely extensive. Larvae are whitish (5–6 mm long) with a brownish-black head, prothoracic plate and anal plate (Fig. **522**); they feed mainly in May, and can be found if an infested shoot is split open. Adults (*c.* 12 mm wingspan) are mainly whitish, with golden-brown markings on the fore wings; they occur in June and July.

523 Common small ermine moth, *Yponomeuta padella*.

Yponomeuta padella (Linnaeus)
Common small ermine moth

Often abundant on hawthorn (*Crataegus*) and *Prunus*, including blackthorn (*Prunus spinosa*) and wild cherry (*Prunus avium*) but not bird cherry (*Prunus padus*) (cf. *Yponomeuta evonymella* p. 220), and sometimes injurious to hedges and ornamentals. A locally distributed race affecting rowan (*Sorbus aucuparia*) also occurs; this is sometimes a pest of rowans planted as shade or ornamental trees on alpine camp sites. Widely distributed in Europe.

DESCRIPTION **Adult:** 20–22 mm wingspan; fore wings whitish to whitish-grey, with light grey cilia and four longitudinal rows of black dots (including a row of 4–7 dots towards the lower margin); hind wings grey (Fig. **523**). **Egg:** flat, yellowish to dark reddish. **Larva:** 15–20 mm long; dirty yellowish-grey to greenish-grey, marked with black spots and a blackish dorsal stripe; head, prothoracic plate and anal plate black (Fig. **524**). **Pupa:** 10 mm long; yellowish with a blackish head, thorax, wing cases and tip; cremaster with six filaments (Fig. **525**).

LIFE HISTORY Adults occur in July and August, commonly resting in numbers on the original food-plant or on nearby hosts. Eggs are laid in overlapping tile-like rows in flat, scale-like batches on the bark of shoots and branches of host plants. They are then coated with a gelatinous secretion which soon hardens to form a protective cap. Eggs hatch two to three weeks later, the larvae biting their way through the lower surface of the egg shells but leaving the upper surface of the egg cluster intact. The young larvae remain *in situ* throughout the winter; they are very resistant to low temperatures, surviving without apparent difficulty even in exposed mountainous regions. Activity is resumed in the following May, the larvae venturing on to the foliage to feed in communal webs which eventually cover the shoots and branches of host trees. Larvae are fully grown in June. They then pupate in pendulous, semi-transparent, silken cocoons dispersed throughout the communal web (Fig. **526**) (cf. *Yponomeuta evonymella*, p. 221), adults emerging a few weeks later.

DAMAGE Host plants are disfigured by the presence of webbing and loss or death of foliage; they are also weakened following severe attacks. Large numbers of larvae and pupae are often killed by parasites (especially various kinds of parasitic wasp) but natural control is often insufficient to prevent infestations from reaching damaging levels.

CONTROL Where practicable, occupied webs should be cut out and burnt, ideally in the early spring before they become too extensive. If necessary apply an insecticide in the early spring.

524 Larvae of common small ermine moth, *Yponomeuta padella.*

525 Pupae of common small ermine moth, *Yponomeuta padella.*

526 Pupal cocoons of common small ermine moth, *Yponomeuta padella.*

527 Web of *Yponomeuta cagnagella* on *Euonymus*.

528 Adult of *Yponomeuta cagnagella*.

Yponomeuta cagnagella (Hübner)

syn. *cognatella* Treitschke; *evonymi* Zeller

Infestations of this relatively common species occur on spindle (*Euonymus europaeus* and *E. japonica*), and are noted occasionally on cultivated bushes. The life-cycle is similar to that of *Yponomeuta padella* (pp. 218–219) but larval activity commences slightly earlier in the spring, the larvae at first attacking the unfurling leaves and later forming conspicuous webs on the shoots or branches (Fig. **527**). Adults (24–26mm wingspan) are mainly white, the fore wings with white cilia and four longitudinal rows of black dots (including a row of 4–7 dots towards the lower margin) (Fig. **528**). The larvae (18–22mm long when fully grown) are yellowish-grey to greenish-grey, marked with black spots; head, prothoracic and anal plates black (Fig. **529**); younger larvae are darker-bodied. Fully fed larvae pupate in groups in white, opaque cocoons (Fig. **530**) (cf. *Yponomeuta plumbella*, p. 221), adults emerging in July.

529 Larva of *Yponomeuta cagnagella*.

Yponomeuta evonymella (Linnaeus)

syn. *padi* Zeller

A widely distributed and often abundant species on bird cherry (*Prunus padus*), the greenish-grey, black-spotted larvae commonly causing complete defoliation of host trees and coating the trunks and branches in dense, polythene-like sheets of webbing (Fig. **531**). Infestations often occur in gardens and hedgerows, causing extensive damage. Adults (23–25mm wingspan) are larger than those of *Ypono-*

530 Pupal cocoons of *Yponomeuta cagnagella*.

531 Web of *Yponomeuta evonymella* on *Prunus padus*.

meuta padella (p. 218); also, the fore wings are pure white with white cilia and five or six longitudinal rows of black dots (including a row of 9–11 dots towards the lower margin); the hind wings are dark grey (Fig. **532**). The biology of both species is similar but individuals of *Yponomeuta evonymella* pupate in white, opaque cocoons which are formed in distinct clusters within the larval web (Fig. **532**).

CONTROL As for *Yponomeuta padella* (p. 219), but chemical treatments are often impractical.

Yponomeuta plumbella (Denis & Schiffermüller)

This uncommon, relatively small species also occurs on spindle (*Euonymus*), the larvae at first feeding within the shoots causing them to wilt, and later forming webs on the foliage. Adults (16–18 mm wingspan) have greyish-white fore wings (each with four longitudinal rows of black dots) and dark grey hind wings. The larvae are yellowish-grey with black spots, and the pupae light brown; the latter are formed within the larval web in spatially separated cocoons.

532 Adult and pupal cocoons of *Yponomeuta evonymella*.

533

533 Web of *Yponomeuta rorrella* on *Salix fragilis*.

534

534 Larvae of *Yponomeuta rorrella*.

Yponomeuta rorrella (Hübner)
syn. *rorella* (Hübner)

This uncommon, local, southerly-distributed species is essentially similar to *Yponomeuta padella* (p.000) but the larvae feed on white willow (*Salix alba*) and sometimes on other species of *Salix*, including grey sallow (*Salix cinerea*), osier (*Salix viminalis*) and pussy willow (*Salix caprea*). They form conspicuous webs on the branches (Fig. **533**) and can be damaging to ornamental trees. Individuals (18–20 mm long when fully grown) are greyish-green, marked with black or dark grey; the head, prothoracic plate and anal plate are dark brown (Fig. **534**). The adults (22–25 mm wingspan) are greyish-white, with four longitudinal rows of black dots on each fore wing and dark grey hind wings (Fig. **535**). The pupae occur amongst the webbing, with little protection (Fig. **536**).

CONTROL Occupied webs should be cut out and burnt.

535

535 Adult of *Yponomeuta rorrella*.

Yponomeuta vigintipunctata (Retzius)
syn. *sedella* Treitschke

This moth is a local, mainly southerly-distributed pest of cultivated *Sedum*, occurring most commonly on *Sedum telephium*; in France, also reported on *Sedum album* and *S. purpurescens* and, in southern England, on *Sedum cauticolum* × *telephium* 'Ruby Glow' and *S. telephium maximum* 'Atropurpureum'.

536

536 Pupae of *Yponomeuta rorrella*.

The yellowish-grey, black-spotted larvae (up to 25 mm long) feed in small groups in June to July and in September to October, forming silken webs on the foliage. When fully grown, they pupate in opaque, silvery-white cocoons spun amongst withered leaves at the base of the host plant. The adults (22–25 mm wingspan) are mainly grey, with three longitudinal rows of black spots on each fore wing. They occur in April and May, and again in August.

Prays fraxinella (Bjerkander)
syn. *curtisella* (Donovan)
Ash bud moth
A generally common pest of ash (*Fraxinus excelsior*), sometimes causing extensive damage to young trees. Present throughout much of Europe.

DESCRIPTION **Adult:** 15 mm wingspan; fore wings white with brownish-grey cilia and irregular blackish markings, including a distinct triangular patch; head white, antennae black (Fig. **537**); melanic forms, largely suffused with black, also occur. **Larva:** 12 mm long; dirty reddish-green with paler sides, more-or-less marked with grey; head, prothoracic and anal plates black.

LIFE HISTORY Adults occur in June and July, depositing eggs on the shoots of ash. Larvae commence feeding in the autumn, mining within the buds or leaves before overwintering beneath the bark close to a terminal bud. In the following spring they attack the terminal shoots, boring into the pith. In May, growth from infested shoots wilts (Fig. **538**), affected tissue soon turning black. If examined, a distinct hole may be seen at the base of each wilted shoot and a larva found feeding inside. Larvae complete their development by the end of May; they then pupate, adults emerging a few weeks later. In northern Europe there is just one generation annually but two generations are produced in southern Europe.

DAMAGE Growth from affected buds is aborted, development becoming forked as paired lateral shoots become dominant.

CONTROL Some control can be achieved by applying an insecticide immediately before bud invasion.

537 Ash bud moth, *Prays fraxinella*.

538 Ash bud moth damage to shoot of *Fraxinus*.

223

Scythropia crataegella (Linnaeus)
Hawthorn webber moth

A locally important pest of hawthorn (*Crataegus*) also associated with *Cotoneaster*. Widespread in central and southern Europe; also found in the more southerly parts of northern Europe.

DESCRIPTION **Adult:** 13–14 mm wingspan; fore wings silvery-white, mottled with ochreous and brown (Fig. **539**). **Larva:** 12–15 mm long; reddish-brown, the thoracic segments marked dorsally with yellowish-orange; body hairs whitish and relatively long; head black. **Pupa:** 7–8 mm long; mainly black, suffused with dirty creamish-white dorsally (Fig. **540**).

LIFE HISTORY Infestations are usually first noticed during the spring, the overwintered larvae feeding gregariously on the foliage of host plants, protected by a thin but expansive web (Fig. **541**). Larvae are fully grown from mid-June onwards. They then pupate within the web, moths emerging in late June and July. Eggs deposited during the summer hatch in the late summer. The young larvae then mine briefly within the leaves before hibernating, the mined leaves remaining attached to host plants throughout the winter.

DAMAGE Larvae cause significant defoliation and their webs disfigure host plants (Fig. **542**), affecting the development of the new shoots.

CONTROL Large numbers of the immature stages fall victim to hymenopterous parasites, which reduces the likelihood of infestations reaching epidemic proportions. However, where necessary, webs on host plants should be cut out and burnt, preferably in the early spring. If necessary, apply an insecticide but do not spray during the flowering stage.

Ypsolopha dentella (Fabricius)
syn. *harpella* (Denis & Schiffermüller)

Generally common on honeysuckle (*Lonicera*) and occasionally a minor pest of cultivated plants. Widely distributed in Europe.

DESCRIPTION **Adult:** 17–21 mm wingspan; fore wings elongate with a dentate tip, chocolate-brown

539 Hawthorn webber moth, *Scythropia crataegella*.

540 Pupa of hawthorn webber moth, *Scythropia crataegella*.

541 Web of hawthorn webber moth on *Crataegus*.

542 Hawthorn webber moth damage to foliage of *Crataegus*.

543 Adult of *Ypsolopha dentella*.

544 Larva of *Ypsolopha dentella*.

marked with yellow and white; hind wings grey (Fig. **543**). **Larva:** 18mm long; yellowish-green with a broad, reddish, black-marked stripe down the back; head yellowish, marked with brown (Fig. **544**).

LIFE HISTORY Larvae feed on honeysuckle during May and June, each sheltering in a rolled leaf. They are extremely active if disturbed, wriggling backwards out of their habitation and dropping immediately to the ground. When fully fed, the larvae pupate in pale brown, boat-shaped cocoons formed either on the foodplant or amongst debris on the ground. The adult moths appear in July and August.

DAMAGE Larvae cause loss of leaf tissue and slight distortion of new growth but damage is not important.

14. Family COLEOPHORIDAE (casebearer moths)

A distinctive family of small moths with long, narrow, pointed wings. The larvae commence feeding as leaf miners but then inhabit characteristic portable cases formed from silk and a cut-out portion of leaf or other plant tissue; crochets on the abdominal prolegs are arranged into transverse bands.

545 Larch casebearer, *Coleophora laricella*.

Coleophora laricella (Hübner)

Larch casebearer moth
larva = larch casebearer
An often common pest of larch (*Larix*), including ornamentals, but mainly of importance in forestry. Widespread in central and northern Europe; introduced into North America.

DESCRIPTION **Adult:** 9–11 mm wingspan; fore wings shiny grey; hind wings grey; head grey, with light grey antennae. **Larva:** 4.0–4.5 mm long; dark brown with a black head and prothoracic plate, and a pair of small, black plates on the second thoracic segment; 16-legged. **Case:** 4.0–4.5 mm long; pale straw-coloured to greyish-white (Fig. **545**).

LIFE HISTORY Adults occur in June and July, depositing eggs on the shoots of larch. The larvae feed on the needles from September onwards, hibernating during the winter. In spring the larvae attack the young needles, entering a short distance from the tip of a needle and eating out the contents. Individuals are fully fed by the end of May or early June. They then pupate, adults emerging a few weeks later. There is just one generation annually.

546 Larch casebearer damage to needles of *Larix*.

DAMAGE The tips of damaged needles appear whitish and, if examined under a lens, the rounded entry hole (characteristic of casebearer feeding blotches) can be seen (Fig. **546**); the tips of damaged needles often break off.

Coleophora potentillae Elisha

Small blotch mines formed by this locally common species are sometimes found on the leaves of cultivated rose (*Rosa*) (Fig. **547**), disfiguring the foliage but usually not causing distortion; they also occur on ornamental cinquefoil (*Potentilla*), goat's beard (*Aruncus dioicus*) and *Rubus*. The larval cases are blackish-brown and they may be found on the

547 Mines of *Coleophora potentillae* in leaf of *Rosa*.

underside of the leaves from August to September. Pupation takes place in the case, attached to fallen leaves or other debris on the ground. Adults (10 mm wingspan) are mainly greyish-bronze; they occur in June.

Coleophora serratella (Linnaeus)
syn. *fuscedinella* Zeller

An abundant species on alder (*Alnus*), birch (*Betula*), crab apple (*Malus*), elm (*Ulmus*) and hazel (*Corylus avellana*); often associated with ornamentals and young amenity trees, infested foliage developing numerous pale blotches visible from above. The larvae inhabit short, cylindrical cases, attached to the underside of the leaves (Fig. 548), the blotch mines often causing noticeable distortion (Fig. 549). Cases may be found on host plants from September onwards, but are most obvious in spring and early summer, the brownish and ochreous-tinged adults (12 mm wingspan) appearing in June and July.

548 Case of *Coleophora serratella* on leaf of *Ulmus*.

549 Mines of *Coleophora serratella* in leaf of *Betula*.

15. Family OECOPHORIDAE

A group of relatively small moths with prominent labial palps and broadly elongate fore wings. The larvae have a well-developed prothoracic plate and are very active.

Carcina quercana (Fabricius)

A minor pest of deciduous trees and shrubs, including beech (*Fagus sylvatica*), crab apple (*Malus*), oak (*Quercus*) and ornamental cherry (*Prunus*). The bright, apple-green, somewhat flattened larvae (up to 15 mm long) feed on the underside of leaves during May and June, each sheltered beneath a slight web. Pupation occurs beneath an opaque web from mid-June onwards, the bright yellow to greyish-pink adults (15–20 mm wingspan) (Fig. 550) appearing in July and August.

550 Adult of *Carcina quercana*.

Diurnea fagella (Denis & Schiffermüller)
A generally abundant woodland species, attacking various ornamental trees and shrubs, including beech (*Fagus sylvatica*), birch (*Betula*), crab apple (*Malus*), hazel (*Corylus avellana*), oak (*Quercus*) and sallow (*Salix*). Widely distributed in Europe; also present in Asia Minor.

551 Female of *Diurnea fagella*.

DESCRIPTION **Adult female:** 18–20 mm wingspan, but wings much reduced and pointed; fore wings whitish, suffused to a greater or lesser degree with brownish-black and bearing blackish and buff scale tufts; hind wings greyish to brownish-black; palps long (Fig. **551**). **Adult male:** 25–28 mm wingspan; coloration similar to female (Fig. **552**). **Larva:** 15–18 mm long; pale, dull yellowish-green, with yellowish intersegmental markings; head light brown; prothoracic plate mainly yellowish with a pair of brown lateral markings; third pair of thoracic legs fleshy and projecting well beyond width of body (Fig. **553**). **Pupa:** 10 mm long; brown; cremaster with a cluster of long, hooked setae.

LIFE HISTORY Adults appear in March and April. The females, although incapable of flight, are very active and crawl with considerable speed if disturbed. Eggs are laid on various hosts, the larvae then feeding from June to October in spun or, occasionally, rolled leaves. Fully fed larvae spin cocoons amongst debris on the ground, passing the winter in the pupal stage.

DAMAGE Damage is rarely extensive and usually limited to loss of a few leaves.

552 Male of *Diurnea fagella*.

553 Larva of *Diurnea fagella*.

Agonopterix conterminella (Zeller)
Willow shoot moth

A generally common but minor pest of willow (*Salix*). Present throughout central Europe; also found in the more southerly parts of northern Europe.

DESCRIPTION **Adult:** 18–19 mm wingspan; fore wings reddish-brown, suffused with blackish and yellowish-white scales; hind wings brownish-white (Fig. **554**). **Larva:** 15–18 mm long; light green with black pinacula; head yellowish-brown; young larva pale with the head, prothoracic plate and pinacula black (Fig. **555**).

LIFE HISTORY Larvae feed on willow trees in May and June, spinning the leaves of the new shoots together to form a shelter. When fully fed they drop to the ground and eventually pupate, each in a subterranean, silken cocoon. Adults occur from July to September.

DAMAGE The larvae cause noticeable distortion of the shoot tips but infestations are rarely extensive.

554 Willow shoot moth, *Agonopterix conterminella.*

555 Larva of willow shoot moth, *Agonopterix conterminella.*

Agonopterix ocellana (Fabricius)

This widely distributed species is also associated with willow (*Salix*), the larvae feeding within spun leaves or shoots and sometimes causing slight damage to cultivated plants. Adults (23 mm wingspan) are characterized by the elongate, pale greyish-buff fore wings, each suffused with black and marked with a prominent red streak (Fig. **556**); they occur from August to April, hibernating throughout the winter. The larvae (up to 20 mm

556 Adult of *Agonopterix ocellana.*

long) are green, with yellowish intersegmental divisions and small, but distinct, black pinacula (Fig. **557**).

557 Larva of *Agonopterix ocellana*.

16. Family **GELECHIIDAE**

Anacampsis blattariella (Hübner)
syn. *betulinella* Vári
A common but minor pest of birch (*Betula*) including, occasionally, nursery and ornamental trees. Widely distributed in Europe.

DESCRIPTION **Adult:** 16–18 mm wingspan; fore wings pale grey, marked extensively with black or dark grey, and with several black scale tufts and black terminal spots; hind wings dark grey (Fig. **558**). **Larva:** 14 mm long; greyish-green with large, black pinacula; head and prothoracic plate shiny black (Fig. **559**).

LIFE HISTORY Larvae of this single-brooded species feed in webbed leaves of birch during May and June, and are often mistaken for tortricids (see p. 233 *et seq.*). When fully grown they pupate in a rolled leaf, the adults occurring from July to September.

DAMAGE Larvae cause slight distortion of the new growth but attacks are not important.

558 Adult of *Anacampsis blattariella*.

559 Larva of *Anacampsis blattariella*.

560 Larva of *Anacampsis populella*.

561 Adult of *Anacampsis populella*.

Anacampsis populella (Clerck)

This species is essentially similar to *Anacampsis blattariella* but is associated with poplar (*Populus*) and willow (*Salix*). The larvae (up to 14 mm long) are greyish to yellowish-grey, with prominent black pinacula and a black head and prothoracic plate (Fig. **560**). They feed within rolled leaves during the late spring and early summer, the dark grey adults (Fig. **561**) appearing in July and August.

Hypatima rhomboidella (Linnaeus)
syn. *conscriptella* (Hübner)

A minor pest of birch (*Betula*) and hazel (*Corylus avellana*); occasionally found on nursery trees. Widely distributed in central Europe; also found in the more southerly parts of northern Europe.

562 Adult of *Hypatima rhomboidella*.

DESCRIPTION **Adult:** 13–16 mm wingspan; fore wings elongate, pale grey, each with a conspicuous blackish, triangular blotch and a terminal streak; hind wings light grey (Fig. **562**). **Larva:** 12 mm long; green or pinkish-brown, with small, black pinacula; head and prothoracic plate shiny black (Fig. **563**).

LIFE HISTORY Adults occur in July, August and September. They often hide amongst dead leaves during the day, scurrying away if disturbed. The larvae feed on the leaves of birch and hazel in the following May and June, each sheltering within a rolled leaf or leaf edge. Fully grown individuals usually pupate amongst debris on the ground, the adults emerging shortly afterwards. There is one generation annually.

563 Larvae of *Hypatima rhomboidella*.

DAMAGE Infested leaves attract attention but damage caused is of no importance.

Dichomeris marginella (Fabricius)
Juniper webber moth
larva = juniper webworm

A locally common pest of juniper (*Juniperus*), but also found on conifers such as *Cryptomeria*, often causing damage to ornamental plants in gardens and nurseries. Widespread in Europe; also present in parts of North America.

DESCRIPTION **Adult:** 15–16 mm wingspan; fore wings dark brown with a white stripe along the anterior and hind margins; head and thorax white (Fig. **564**). **Larva:** 10 mm long; light brown with darker dorsal and subdorsal stripes, the area between often slightly whitish; head and prothoracic plate brown to brownish-black (Fig. **565**).

LIFE HISTORY Adults occur in July and August, depositing eggs in groups on host plants. The larvae feed gregariously in May and June, webbing the shoots together with silken threads. Fully grown larvae pupate in cocoons spun amongst the webbing, moths appearing a few weeks later.

DAMAGE Infested shoots are disfigured by webbing, much of the foliage dying and turning brown.

CONTROL Cut out and burn infested shoots as soon as attacks are seen. If necessary on nursery stock, apply a contact insecticide.

564 Juniper webber moth, *Dichomeris marginella.*

565 Juniper webworms, *Dichomeris marginella.*

17. Family **BLASTOBASIDAE**

A small family of small moths with elongate wings. The larvae have relatively small abdominal spiracles.

Blastobasis lignea (Wollaston)

A minor pest of yew (*Taxus baccata*); also reported attacking Norway spruce (*Picea abies*). Of subtropical origin but now widely distributed in Europe.

DESCRIPTION **Adult:** 16–18 mm wingspan; fore wings elongate, yellowish-brown, marked with black (Fig. **566**). **Larva:** 8–10 mm long; purplish-brown and rather plump, with a blackish head and prothoracic plate, and small, inconspicuous pinacula (Fig. **567**).

LIFE HISTORY Adults occur during the summer, depositing eggs in August on yew and, less com-

566 Adult of *Blastobasis lignea.*

monly, on other evergreen hosts such as Norway spruce. Larvae feed from September onwards, sheltering on the shoots within a cluster of dead or decaying leaves held together with silk. The larvae survive the winter in these habitations, and complete their development in the following spring or early summer. Although grazing directly on living tissue, the larvae also feed on decaying leaves and other dead vegetative matter.

DAMAGE Larvae cause only slight damage to the leaves, and do not affect the growth of plants.

Blastobasis decolorella (Wollaston)

A subtropical species, now established in parts of Europe where it sometimes causes damage in orchards, attacking tree bark and the maturing fruits; the larvae also feed on various other trees and shrubs, including ornamentals such as *Rhododendron*. The larvae are similar in appearance to those of *Blastobasis lignea* but the adults are more uniformly coloured, the wings being mainly pale ochreous-yellow.

18. Family **TORTRICIDAE** (tortrix moths)

A large and important group of small, relatively broad-winged moths, including several well-known pests. The larvae have well-developed thoracic legs and five pairs of abdominal prolegs, each proleg bearing a complete circle of similarly sized crochets; many species have an anal comb with which particles of frass are ejected from the anus. Larvae often feed in spun or rolled leaves, wriggling backwards rapidly if disturbed. Pupation takes place in the larval habitation or elsewhere, the pupa protruding from the cocoon following emergence of the adult.

Pandemis cerasana (Hübner)
Barred fruit tree tortrix moth

An often common but minor pest of trees and shrubs in gardens and nurseries established close to woodlands, including alder (*Alnus*), birch (*Betula*), crab apple (*Malus*), elm (*Ulmus*), hazel (*Corylus avellana*), lime (*Tilia*), oak (*Quercus*), ornamental

567 Larva of *Blastobasis lignea*.

568 Female barred fruit tree tortrix moth, *Pandemis cerasana*.

cherry (*Prunus*), rowan (*Sorbus aucuparia*), sycamore (*Acer pseudoplatanus*) and willow (*Salix*). Eurasiatic. Widely distributed in Europe.

DESCRIPTION **Adult:** 16–24 mm wingspan; fore wings pale yellowish to yellowish-brown, with pale brown, chestnut-edged markings; hind wings greyish-brown; antennae of male with a basal notch (Fig. **568**). **Egg:** flat and oval; laid in an oval, raft-like batch. **Larva:** 20 mm long; rather thin and flattened; whitish-green, but darker above, with pale green pinacula; head light green to brownish-green; prothoracic plate pale green or yellowish-green, with dark sides and hind edge; anal plate green,

233

dotted with black; anal comb yellowish, with 6–8 teeth; first and last spiracles elliptical and distinctly larger than the others (unlike *Pandemis heparana* and *P. corylana*) (Fig. **569**). **Pupa:** 8–13 mm long; brown or blackish-brown; cremaster longer than broad.

LIFE HISTORY Adults occur from June to August, eggs being deposited on the leaves or branches of various trees and shrubs. Some eggs hatch after a few weeks but others not until the following spring. Young summer larvae feed on the foliage for a short time and then, whilst still small, spin silken retreats on the twigs in which they overwinter. Activity recommences at bud-burst, when over-wintered eggs also hatch. Larvae feed in a rolled or folded leaf until May or early June and then pupate in whitish cocoons, each spun in the larval habitation or in a folded leaf.

DAMAGE Leaf damage is usually unimportant as larvae rarely feed together in large numbers.

CONTROL Application of an insecticide in the spring can be effective but is rarely, if ever, justified.

Pandemis heparana (Denis & Schiffermüller)
Dark fruit tree tortrix moth

Often common in woodland areas, nurseries and gardens, attacking various trees and shrubs, including birch (*Betula*), crab apple (*Malus*), *Forsythia*, honeysuckle (*Lonicera*), lime (*Tilia*), ornamental cherry (*Prunus*) and sallow (*Salix*). Eurasiatic. Widely distributed in Europe.

DESCRIPTION **Adult:** 16–24 mm wingspan; fore wings reddish-ochreous to reddish-brown, with darker markings; hind wings dark brownish-grey (Fig. **570**); antennae of male with a basal notch. **Egg:** flat and oval; laid in a small, raft-like batch. **Larva:** 25 mm long; bright green with pale sides; head and prothoracic plate green to brown; anal plate green; anal comb whitish, usually with 6–8 teeth (Fig. **571**). **Pupa:** 10–12 mm long; brownish-black; cremaster about as long as wide.

LIFE HISTORY Adults occur from late June to August or September. Eggs are laid on the upper surface of leaves of various trees and shrubs,

569 Larva of barred fruit tree tortrix moth, *Pandemis cerasana.*

570 Female dark fruit tree tortrix moth, *Pandemis heparana.*

571 Larva of dark fruit tree tortrix moth, *Pandemis heparana.*

usually in batches of 30–50. They hatch two to three weeks later, the larvae feeding for a short time before hibernating in silken retreats spun on the twigs. Larvae resume feeding in the following May and June, each sheltering within rolled leaves on the young shoots or under webs spun on the underside of expanded leaves. Pupation occurs in the larval habitation or in spun leaves near the tips of infested shoots.

DAMAGE In addition to feeding on leaves, larvae sometimes destroy blossoms, but infestations are usually unimportant.

CONTROL Application of a pre-blossom spray in spring can be effective, but is rarely, if ever, justified.

572 Hazel tortrix moth, *Pandemis corylana*.

Pandemis corylana (Fabricius)
Hazel tortrix moth

A locally common but minor pest of trees and shrubs, including ash (*Fraxinus excelsior*), dogwood (*Cornus*), hazel (*Corylus avellana*) and oak (*Quercus*). Adults occur from July to September, eggs being laid on the bark of host plants where they overwinter until the following spring. The larvae feed on the foliage from May to July, inhabiting either a longitudinally folded leaf or a bunch of spun leaves. Pupation occurs in June within the larval habitation. Adults (18–24 mm wingspan) are similar in appearance to those of *Pandemis cerasana* (p. 233) but the fore wings are paler and more strongly reticulated (Fig. 572). The larvae (up to 25 mm long) are green and slender-bodied, with a well-developed anal comb.

573 Female of *Archips crataegana*.

Archips crataegana (Hübner)

A minor pest of various trees and shrubs, including ash (*Fraxinus excelsior*), birch (*Betula*), crab apple (*Malus*), elm (*Ulmus*), lime (*Tilia*), oak (*Quercus*), ornamental cherry (*Prunus*) and sallow (*Salix*). Eurasiatic. Widely distributed in central and much of northern Europe.

DESCRIPTION **Adult female:** 23–27 mm wingspan; fore wings brown with darker, chocolate-coloured markings; hind wings brownish-grey (Fig. 573). **Adult male:** 19–22 mm wingspan; fore wings lighter than in female. **Egg:** 0.6 × 0.4 mm; laid in a white mass. **Larva:** 23 mm long; dull greenish-black with black pinacula; head, prothoracic plate and anal plate shiny black; anal comb black with 6–8 teeth (Fig. 574). **Pupa:** 9–12 mm long; dull black; cremaster elongate.

574 Larva of *Archips crataegana*.

LIFE HISTORY Adults of this single-brooded species occur from late June to August. Eggs are laid in conspicuous batches of about 30 on the trunks and main branches, and then coated with a white substance which quickly hardens and disguises them as bird-droppings. Eggs hatch in April or early May; the tiny larvae are very active and rapidly climb the tree to begin feeding on the underside of the leaves. Later, each feeds inside a tightly rolled leaf edge, usually on fully expanded foliage at the shoot tips. Pupation occurs in a rolled leaf or between two spun leaves, adults emerging a few weeks later.

DAMAGE Feeding is restricted mainly to the expanded leaves and, unless larvae are very numerous, damage is insignificant.

CONTROL Minor infestations can be eliminated by picking off and burning occupied tied leaves; alternatively, spray with an insecticide in the spring as soon as larval attacks commence.

575 Male fruit tree tortrix moth, *Archips podana*.

Archips podana (Scopoli)
Fruit tree tortrix moth

A generally common, polyphagous pest of trees and shrubs, including birch (*Betula*), crab apple (*Malus*), hawthorn (*Crataegus*), ornamental cherry (*Prunus*), ornamental pear (*Pyrus*), ornamental spindle (*Euonymus japonica*) and *Sorbus*; often present on garden trees and nursery stock. Eurasiatic; also present in North America, probably as an introduced species. Widely distributed in Europe.

DESCRIPTION **Adult female:** 20–28 mm wingspan; fore wings purplish-ochreous, each with a brown, reticulated pattern and darker markings, and a dark spot at the tip; hind wings brownish-grey, suffused with orange apically. **Adult male:** 19–23 mm wingspan; fore wings purplish to purplish-ochreous, each with dark reddish-brown, velvety markings and a dark spot at the tip; hind wings greyish, tinged with orange apically (Fig. **575**). **Egg:** 0.6-0.7 mm across; flat and almost circular; green and laid in a large raft-like batch. **Larva:** 14–22 mm long; light green to greyish-green, with pale pinacula; head chestnut-brown or black; prothoracic plate chestnut-brown with darker lateral and hind margins, a pale anterior margin and a pale, narrow mid-line; anal plate green or grey; thoracic legs brownish-black or black; prothoracic spiracle elliptical and last body spiracle distinctly larger than the rest (Fig. **576**). **Pupa:** 9–14 mm long; dark yellowish-brown to blackish-brown.

576 Larva of fruit tree tortrix moth, *Archips podana*.

LIFE HISTORY Adults occur from June to September but are usually most abundant in July. Eggs are deposited on the leaves in flat, oval batches of about 50, with the shells overlapping like roofing tiles. The eggs, which are extremely difficult to find as they closely match the colour of the leaves, hatch in about three weeks. The larvae feed on the foliage for a few weeks and then enter hibernation. Overwintered larvae become active in late March or April, burrowing into the opening buds. Larvae later attack the foliage, each webbing two or more leaves together and sheltering between them, or forming a retreat by spinning a dead leaf to a healthy one or to a twig. Larval development is completed in May or June. Individuals then pupate within the larval habitation or within freshly spun leaves nearby, adults emerging three or more weeks later. In favourable situations, when adult emergence and egg laying is particularly advanced, some larvae may feed up and pupate to produce a partial second generation of moths in the late summer or early autumn.

DAMAGE Attacks on buds can be serious, the overwintered larvae often totally destroying them; feeding by larvae or the presence of their webbing disfigures and may also affect the development of young shoots but damage to fully expanded foliage is usually unimportant.

CONTROL Minor infestations can be eliminated by picking off and burning occupied tied leaves. Alternatively, spray with an insecticide as soon as larval attacks develop; summer sprays are more effective than spring treatments.

577 Male rose tortrix moth, *Archips rosana*.

Archips rosana (Linnaeus)
Rose tortrix moth

An often abundant but minor garden pest of trees and shrubs, including ornamentals such as cherry (*Prunus*), crab apple (*Malus*), rose (*Rosa*) and certain conifers. Eurasiatic; also occurs in North America. Widely distributed in Europe.

DESCRIPTION **Adult female:** 17–24 mm wingspan; fore wings darker brown and markings more diffuse than in male; hind wings grey, suffused apically with orange-yellow. **Adult male:** 15–18 mm wingspan; fore wings light brown to purplish-brown, with dark brown, often pinkish-tinged, markings; hind wings grey (Fig. **577**). **Egg:** 0.9 × 0.7 mm; flat, oval and greyish-green; laid in a large raft. **Larva:** 22 mm long; light green to dark green, with pale pinacula; head and prothoracic plate light brown to black; anal plate green or pale brown (Fig. **578**). **Pupa:** 9–11 mm long; dark brown.

LIFE HISTORY Eggs are laid during August and September on the bark of host plants. They hatch in the following April. The larvae then feed in the buds; later, each inhabits a rolled leaf or spun

578 Larva of rose tortrix moth, *Archips rosana*.

shoot. Pupation occurs in the larval habitation from June onwards, adults appearing from July to early September.

DAMAGE Infested leaves or shoots are disfiguring but of little or no importance.

CONTROL As for *Archips crataegana* (pp. 235–236).

Archips xylosteana (Linnaeus)
Brown oak tortrix moth

Generally common in woodland habitats on trees, shrubs and certain herbaceous plants; minor infestations occur occasionally in gardens and nurseries on ash (*Fraxinus excelsior*), birch (*Betula*), crab apple (*Malus*), elm (*Ulmus*), fir (*Abies*), hazel (*Corylus avellana*), honeysuckle (*Lonicera*), *Hypericum*, lime (*Tilia*), oak (*Quercus*) and certain other hosts. Eurasiatic. Present throughout much of Europe.

DESCRIPTION **Adult:** 15–23 mm wingspan; fore wings whitish-ochreous with reddish-brown, variegated markings; hind wings greyish (Fig. **579**). **Larva:** 16–22 mm long; whitish-grey, sometimes grey or dark bluish-grey, with black pinacula; head shiny black; prothoracic plate black or dark brown, with a whitish mid-line and collar; anal comb present (Fig. **580**). **Pupa:** 9–12 mm long; dark brown or black; cremaster elongate.

LIFE HISTORY Adults occur in July, eggs being laid in batches on the trunks or branches of various trees and shrubs. The eggs are then coated with a brownish secretion which camouflages them against the bark. They hatch in the following April or early May, each larva attacking the foliage and inhabiting a rolled leaf (Fig. **581**). Individuals are fully grown in June; they then pupate in the larval habitation, adults emerging a few weeks later.

DAMAGE Larval habitations may disfigure host plants and cause concern but feeding is confined mainly to fully expanded leaves and is, therefore, of little or no significance.

579 Female brown oak tortrix moth, *Archips xylosteana*.

580 Larva of brown oak tortrix moth, *Archips xylosteana*.

581 Larval habitation of brown oak tortrix moth on *Betula*.

Cacoecimorpha pronubana (Hübner)
Carnation tortrix moth

Originally from Africa and nowadays a common glasshouse pest in many parts of Europe, including the British Isles where it was first reported in 1905. Ornamental hosts include bay laurel (*Laurus nobilis*), broom (*Cytisus*), carnation and pink (*Dianthus*), cypress (*Cupressus*), *Daphne*, *Fuchsia*, honeysuckle (*Lonicera*), *Hypericum*, ivy (*Hedera helix*), false acacia (*Robinia pseudoacacia*), ornamental spindle (*Euonymus japonica*), privet (*Ligustrum vulgare*), silk bark oak (*Grevillea*) and many others. In favourable districts infestations occur out-of-doors. Also present in Asia Minor and North America.

DESCRIPTION **Adult female:** 18–22 mm wingspan; fore wings pale orange-brown, reticulated with darker brown; hind wings mainly orange (Fig. **582**). **Adult male:** 12–17 mm wingspan; fore wings orange-brown, with variable reddish-brown and blackish markings; hind wings bright orange, each with a blackish border (Fig. **583**). **Egg:** flat and oval; light green, laid in a large, scale-like batch. **Larva:** 15–20 mm long; olive-green to bright green, paler below, and with slightly paler pinacula; head greenish yellow or yellowish brown, marked with dark brown; prothoracic and anal plates green, marked with dark brown; anal comb green, usually with six teeth (Fig. **584**). **Pupa:** 9–12 mm long; brownish black to black; cremaster elongate and tapered, with eight strong, hooked spines.

LIFE HISTORY Adults appear mainly from April to October, but are most numerous from May to June, and in late August and September. The males have a characteristic, erratic flight and are very active in sunny weather. Eggs are laid on leaves in large groups of up to 200, the shells overlapping like roofing tiles. They hatch in two to three weeks. Larvae at first graze the surface of leaves, sheltering beneath a slight web. Later they feed in spun leaves, shoots or blossom trusses. Pupation occurs within the larval habitation, in a freshly folded leaf or amongst webbed foliage, adults emerging shortly afterwards. The winter is usually passed as young larvae sheltering on the host in silken webs.

DAMAGE The larvae are voracious feeders, causing considerable harm to foliage, buds and flowers.

CONTROL Spray with an insecticide as soon as damage is seen, and repeat as necessary.

582 Female carnation tortrix moth, *Cacoecimorpha pronubana.*

583 Male carnation tortrix moth, *Cacoecimorpha pronubana.*

584 Larva of carnation tortrix moth, *Cacoecimorpha pronubana.*

Syndemis musculana (Hübner)

A generally common but minor pest of birch (*Betula*), oak (*Quercus*), *Rubus* and various other trees, shrubs and herbaceous plants; infestations sometimes occur on nursery plants, including conifer seedlings. Eurasiatic. Widely distributed in Europe.

DESCRIPTION **Adult:** 15–22 mm wingspan; fore wings white to greyish, suffused with black and marked conspicuously with dark brown (Fig. **585**). **Larva:** 18–22 mm long; olive-green to blackish-brown but paler below, with distinct, pale pinacula; head yellowish-brown to brownish-orange; prothoracic plate greyish-brown to yellowish-brown, characteristically marked posteriorly with black; anal plate yellowish-brown or greenish (Fig. **586**).

LIFE HISTORY Adults occur mainly in May and June, depositing eggs on the leaves of various plants. The larvae feed from July onwards, typically inhabiting a folded leaf or tightly rolled tube of spun leaves. Feeding is completed in the autumn, individuals hibernating in the larval habitation or in a silken cocoon formed amongst fallen leaves; pupation occurs in the following spring.

DAMAGE Larval habitations cause noticeable distortion of terminal shoots and can affect the quality of young trees and shrubs.

CONTROL Spray with an insecticide in the summer as soon as damage is seen.

585 Adult of *Syndemis musculana.*

586 Larva of *Syndemis musculana.*

Ptycholomoides aeriferanus (Herrich-Schäffer)

Recorded in England in 1951 and since then widely reported in the southern counties; also occurs in central and south-eastern Europe, parts of Russia and in Japan. The larvae feed in May and June on larch (*Larix*), including garden trees, spinning a few needles together as a shelter. Fully grown individuals are green with a light brown head and a yellowish-brown prothoracic plate. Adults (17–21 mm wingspan) occur from late June to August; the fore wings are light golden-yellow suffused with dark brown and marked with brownish-black; the hind wings are dark brown (Fig. **587**).

587 Adult of *Ptycholomoides aeriferanus.*

Clepsis spectrana (Treitschke)
syn. *costana* sensu Fabricius
Straw-coloured tortrix moth

Especially common in fenland and coastal habitats on herbaceous plants, including various horticultural crops; often a pest in glasshouses and nurseries on ornamentals such as *Cyclamen, Iris,* rose (*Rosa*) and various conifers. Widely distributed in central and northern Europe.

DESCRIPTION **Adult:** 15–24 mm wingspan; (male usually noticeably smaller than female); fore wings pale ochreous to yellowish, with variable dark brown to blackish markings; hind wings pale grey (Fig. **588**). **Egg:** orange, flat and oval. **Larva:** 18–25 mm long; brown to greyish olive-green, paler dorsally, with whitish pinacula; head and prothoracic plate shiny black or blackish-brown; anal plate whitish brown, marked with black or brown; anal comb with 6–8 teeth (Fig. **589**). **Pupa:** 10–14 mm long; dull black; cremaster stout and elongate.

LIFE HISTORY In the open, first-generation adults of this double-brooded species occur from early June to July, sometimes earlier. Eggs are laid in small groups on the food plant, hatching 2–3 weeks later. The larvae then feed beneath a web or in young, webbed leaves and 'capped' flowers. When fully grown each pupates in a white, silken cocoon in the larval habitation, in webbed leaves or amongst dead leaves. A second generation of adults appears in August and September. Larvae from these moths feed for a short time before hibernating. The larvae re-appear in early spring to continue feeding, fully grown individuals eventually pupating in May or June. Some populations in glasshouses breed continuously, having evolved the ability to develop without a diapause phase.

DAMAGE The larvae destroy flowers and flower buds, and cause considerable damage to the foliage, especially on the young shoots.

CONTROL Spray with an insecticide as soon as damage is seen, and repeat as necessary.

588

588 Female straw-coloured tortrix moth, *Clepsis spectrana.*

589

589 Larva of straw-coloured tortrix moth, *Clepsis spectrana.*

Epichoristodes acerbella (Walker)
syn. *ionephela* (Meyrick)
African carnation tortrix moth

A native of Africa, sometimes introduced into Europe on imported carnation (*Dianthus caryophyllus*) cuttings. Although infestations are usually intercepted by Plant Health authorities, the pest has sometimes established itself in glasshouses in parts of northern Europe, including Denmark and Norway; also introduced into parts of southern Europe, such as Italy. Although mainly a pest of carnation, the larvae also feed on *Chrysanthemum*, rose (*Rosa*) and various other plants.

DESCRIPTION **Adult:** 17–20 mm wingspan; fore wings reddish-brown to yellowish, merging into dark reddish-brown or blackish on the hind margin; lighter areas noticeably speckled with brownish-black; hind wings light grey. **Larva:** 15 mm long; green to yellowish-green, with dark green dorsal and subdorsal lines and whitish pinacula; head greenish-brown, marked with brownish-black; prothoracic plate green marked with black along the lateral margin and above the prothoracic spiracle (cf. *Cacoecimorpha pronubana*, p. 239).

LIFE HISTORY Eggs are laid on the leaves of carnation in elongate clusters of about 25, hatching in about ten days. Larvae then feed for 4–8 weeks, sheltering within rolled leaves which they spin together with silk. Pupation occurs within a silken cocoon spun between the leaves, adults emerging 2–4 weeks later. Breeding is continuous so long as conditions remain favourable.

DAMAGE Larvae eat the leaves and flowers, and also burrow into the buds and stems, often causing considerable destruction; they also disfigure and distort plants by spinning the leaves together.

CONTROL Suspected infestations should be reported to Plant Health authorities.

590 Female light brown apple moth, *Epiphyas postvittana*.

591 Male light brown apple moth, *Epiphyas postvittana*.

Epiphyas postvittana (Walker)
Light brown apple moth

An Australasian species, unknown in Europe until 1936 when breeding colonies were found on ornamental spindle (*Euonymus japonica*) in south-western England. The pest is now well established in this area on various outdoor and glasshouse ornamentals including azalea (*Rhododendron*), *Azara serrata*, bay laurel (*Laurus nobilis*), *Camellia*, *Ceanothus*, Chilean fire-bush (*Embothrium coccineum* var. *longifolium*), daisy-bush (*Olearia*), *Drimys*, *Fothergilla major*, honeysuckle (*Lonicera*), *Leptospermum scoparium*, *Phygelius*, *Pittosporum*, firethorn (*Pyracantha*) and *Skimmia japonica*.

DESCRIPTION **Adult female:** 17–25 mm wingspan, fore wings pale yellowish-brown, speckled with black, but sometimes noticeably darkened apically (Fig. **590**). **Adult male:** 16–21 mm wingspan; fore wings subdivided into a pale yellow to orange-yellow basal area and a dark brown to reddish-brown apical section; hind wing grey (Fig. **591**).

Larva: 18 mm long; yellowish-green with pale pinacula; head light brown; prothoracic plate light greenish-brown; anal plate brownish-green; anal comb light green; thoracic legs brown (Fig. **592**).
Pupa: 8–10 mm long; dark reddish-brown.

LIFE HISTORY In England there are two overlapping generations annually, the adults occurring form May to October with peaks of activity in June and from August to September. Eggs are laid in small batches on the leaves of various host plants, hatching in about ten days. The larvae inhabit spun leaves at the tips of the new shoots, feeding in June and July, and from September to April. Pupation occurs in the larval habitation in May and in August. Under protected conditions the life-cycle varies and adults can appear in late autumn or during the winter.

DAMAGE Infestations affect the growth of new shoots and also spoil the appearance of ornamentals, attacks being especially damaging on young container-grown and pot plants.

CONTROL Spray with an insecticide as soon as damage is seen, and repeat as necessary.

592 Larva of light brown apple moth, *Epiphyas postvittana*.

593 Larva of summer fruit tortrix moth, *Adoxophyes orana*.

Adoxophyes orana (Fischer von Röslerstamm)
Summer fruit tortrix moth
This species is widely distributed in Europe and since the 1950s, following a noticeable westerly extension of its range, has also become established in south-eastern England. Although associated mainly with fruit trees, in mainland Europe the larvae also occur on various other hosts, including alder (*Alnus*), birch (*Betula*), hazel (*Corylus avellana*), honeysuckle (*Lonicera*), poplar (*Populus*), rose (*Rosa*) and willow (*Salix*). They feed on buds and also web the foliage of the young shoots, but damage caused is unimportant. The larvae (up to 20 mm long) are yellowish-green or greyish-green to dark green, with small pale pinacula and a yellowish-brown head and prothoracic plate (Fig. **593**). They occur in two main broods, from June to August, and from September to May. The greyish-brown adults (Fig. **594**) (15–22 mm wingspan) occur in June, with a larger flight extending from mid-August to September or October.

594 Summer fruit tortrix moth, *Adoxophyes orana*.

Ptycholoma lecheana (Linnaeus)

Leche's twist moth

A generally common, polyphagous species but associated mainly with oak woodlands; a minor pest of trees and shrubs including crab apple (*Malus*), hawthorn (*Crataegus*), hazel (*Corylus avellana*), oak (*Quercus*), poplar (*Populus*), sycamore (*Acer pseudoplatanus*), willow (*Salix*) and various conifers, including fir (*Abies*), larch (*Larix*) and spruce (*Picea*). Eurasiatic. Widely distributed in Europe.

DESCRIPTION **Adult:** 16–22 mm wingspan; fore wings blackish-brown suffused with greenish-yellow, especially basally, and with silvery-metallic markings; hind wings blackish-brown with pale cilia (Fig. **595**). **Egg:** 1.0 × 0.7 mm, oval, flat and pale green. **Larva:** 18–20 mm long; bluish-green with a darker line along the back and another along each side; pale yellowish-green below; pinacula yellowish; head yellowish-brown to black; prothoracic plate pale yellow to yellowish-brown, marked on sides with black; anal comb small; thoracic legs black (Fig. **596**). **Pupa:** 9–11 mm long; black; cremaster with apical hook-like setae forming a strong projection.

LIFE HISTORY Adults occur in June and July, eggs being deposited on various host plants. Eggs hatch in the summer, the young larvae feeding on the foliage for a few weeks and then hibernating, each in a silken cocoon spun on a twig or spur. Larvae reappear early in the following spring to feed on the opening buds and young foliage. Later they feed in rolled leaves, becoming fully grown in May or early June. Pupation occurs in a dense, white cocoon formed in the larval habitation or in a freshly rolled leaf.

DAMAGE Larvae cause defoliation and damage to young shoots but are rarely sufficiently numerous to be a major problem.

Lozotaenia forsterana (Fabricius)

A minor pest of various cultivated plants, including *Campanula*, cherry laurel (*Prunus laurocerasus*), honeysuckle (*Lonicera*), ivy (*Hedera*) and various conifers. Widely distributed in Europe.

DESCRIPTION **Adult:** 20–29 mm wingspan, female larger than male; fore wings pale greyish-brown, with dark brown markings; hind wings grey (Fig. **597**). **Larva:** 20–25 mm long; dull grey-green, darker above, with a brown or black head; prothoracic plate green or brown; anal plate greenish, marked with black on each side; anal comb present; thoracic

595 Female Leche's twist moth, *Ptycholoma lecheana*.

596 Larva of Leche's twist moth, *Ptycholoma lecheana*.

597 Female of *Lozotaenia forsterana*.

legs brown (Fig. **598**). **Pupa:** 12–14mm long; dark brown.

LIFE HISTORY Adults occur in late June and July. Eggs are laid on the foliage. They hatch in September, the larvae then feed during the autumn before hibernating. Activity is resumed in the following April, each larva living between two or more leaves spun together strongly with silk. Pupation occurs in June, between spun leaves.

DAMAGE If numerous, larvae can cause considerable defoliation but serious attacks rarely develop.

CONTROL Apply an insecticide at the early stages of larval development.

598 Larva of *Lozotaenia forsterana*.

Ditula angustiorana (Haworth)

A locally common and polyphagous pest of various trees, shrubs and herbaceous plants including ornamentals such as bay laurel (*Laurus nobilis*), beech (*Fagus sylvatica*), box (*Buxus sempervirens*), cherry (*Prunus*), juniper (*Juniperus*), larch (*Larix*), pine (*Pinus*), *Rhododendron* and yew (*Taxus baccata*). Widely distributed in Europe; also present in parts of North Africa and North America.

DESCRIPTION **Adult female:** 14–18mm wingspan; fore wings pale ochreous-brown to whitish ochreous, with brown markings; hind wings brown. **Adult male:** 12–16mm wingspan; fore wings greyish-brown to ochreous-brown, marked distinctly with dark purplish-brown and black; hind wings dark brown (Fig. **599**). **Egg:** pale yellow, flat and almost circular; laid in a scale-like batch. **Larva:** 12–18mm long; slender, pale yellowish-green to brownish-green or greyish-green, but darker above, with light green pinacula; head greenish-yellow or yellowish-brown, marked with blackish-brown; prothoracic plate yellowish-green, light brown or dark brown; anal comb greenish or brownish, with four teeth; thoracic legs green, tipped with blackish-brown; spiracles small, the last twice the diameter of the others (Fig. **600**). **Pupa:** 8mm long; light brown; cremaster elongate with eight tightly hooked setae.

LIFE HISTORY Adults occur in June and July, the males often flying in sunshine. Eggs are laid on the leaves in moderately large batches. The newly emerged larvae feed on the foliage during the late summer. In the autumn, whilst still small, they spin silken hibernacula on the buds or spurs. The larvae reappear early in the following spring. They then

599 Male of *Ditula angustiorana*.

600 Larva of *Ditula angustiorana*.

attack the buds and young leaves, spinning the tissue together with silk to form a shelter. The larvae are very active if disturbed, wriggling backwards and dropping to the ground. Pupation occurs in May or June in a cocoon spun in a folded leaf, in webbed foliage or amongst dead leaves on the ground.

DAMAGE Attacked leaves are either grazed on one surface or bitten right through; attacked young growth is also distorted but attacks are rarely sufficiently numerous to cause economic injury.

601 Flax tortrix moth, *Cnephasia asseclana*.

Cnephasia asseclana (Denis & Schiffermüller)
syn. *interjectana* (Haworth); *virgaureana* (Treitschke)
Flax tortrix moth
Polyphagous on herbaceous plants, and sometimes a pest of garden and glasshouse-grown ornamentals such as *Chrysanthemum*, golden rod (*Solidago*), *Helenium*, *Pelargonium*, *Phlox*, primrose (*Primula*), *Rudbeckia* and sweet pea (*Lathyrus odoratus*); also damaging to young spruce (*Picea*) trees. Widely distributed in Europe; also recorded in the Canary Islands and in Newfoundland.

DESCRIPTION **Adult:** 15–18 mm wingspan; fore wings whitish-grey suffused with black and dark yellowish-brown, blackish-edged markings; hind wings greyish-brown (Fig. **601**). **Egg:** greenish-yellow; flat and oval. **Larva:** 12–14 mm long; grey or bluish-white to dark cream or greyish-green, with black pinacula; head pale or yellowish-brown, marked with black; prothoracic plate light brownish to dark brown, marked with black; anal comb with about six long teeth; thoracic legs light to dark brown; prolegs marked with grey or black (Fig. **602**). **Pupa:** 7–8 mm long; light brown; setae on cremaster terminating in stout hooks.

LIFE HISTORY Moths occur from June to August. Eggs are deposited, either singly or in small batches, on herbaceous plants, tree trunks, posts or other supports. They hatch in about three weeks and the larvae then spin small hibernacular cocoons in suitable shelter nearby, having fed only on their egg shells. Activity is resumed early in the following spring, the first-instar larvae mining leaves to feed in irregular, usually blotch-like mines. Later each larva feeds amongst spun leaves or on a flower, spinning the petals down with silk to form a 'capped' blossom. If disturbed, the larvae roll into a tight 'C' and drop to the ground. Pupation takes place in

602 Larva of flax tortrix moth, *Cnephasia asseclana*.

June in the folded edge of a leaf or amongst debris on the ground.

DAMAGE Attacks are usually of only minor importance but infestations, especially on flowers, may affect the marketability of commercial pot plants.

CONTROL On a small scale, infested leaves can be picked off and burnt; alternatively, apply an insecticide in the early spring as soon as damage is seen.

Cnephasia incertana (Treitschke)
Allied shade moth

Larva of this generally distributed and common species are also polyphagous on herbaceous plants and are sometimes found on ornamentals, including *Chrysanthemum, Geranium*, saxifrage (*Saxifraga*), vines and young conifers. They occur from early spring onwards, pupating in late May or early June, adults appearing in June and July. The larvae are similar to those of *Cnephasia asseclana* but usually darker and have no anal comb; the adults of both species are also similar but the fore wings of *Cnephasia incertana* are slightly narrower (Fig. **603**).

603

603 Allied shade moth, *Cnephasia incertana.*

Cnephasia longana (Haworth)
larva = omnivorous leaf tier

This species occurs mainly in coastal and chalkland sites, and is reported occasionally as a minor pest of glasshouse and outdoor flowers. The larvae are polyphagous but are most often associated with members of the Compositae, including catsear (*Hypochoeris*), *Chrysanthemum*, Michaelmas daisy (*Aster*) and *Pyrethrum*; they attack the young shoots and flower heads, often feeding under a canopy of webbed-down petals. Larvae are rather plump (up to 18 mm long) and greenish-grey or yellowish-grey, with pale longitudinal lines along the back and sides, a light brown head, prothoracic plate and anal plate. They occur from early spring onwards, pupating in June, the whitish-ochreous to brownish-ochreous adults (15–22 mm wingspan) emerging in July.

604

604 Larval habitation of *Cnephasia stephensiana* on leaf of *Geranium.*

Cnephasia stephensiana (Doubleday)

Larvae of this widely distributed species are polyphagous on herbaceous plants, and are often pests of cultivated plants in gardens and glasshouses, including various members of the Compositae. The larvae feed from April to June, at first mining the leaves but later living in spun or folded leaves (Fig. **604**) beneath slight webs; flowers are also attacked. Fully grown individuals are 15–18 mm long and shiny grey or bluish-grey to greenish-grey, with large, black pinacula, a brown or black head, a mainly black prothoracic plate, black thoracic legs and a blackish-brown anal plate; there is no anal comb. Adults (18–22 mm wingspan) are mainly light grey with blackish-edged brownish-grey markings (Fig. **605**); they occur in July and August.

605

605 Adult of *Cnephasia stephensiana.*

Tortricodes alternella (Denis & Schiffermüller)
syn. *tortricella* (Hübner)
Minor infestations of this widely distributed and locally common species sometimes occur on ornamental trees growing in the vicinity of mixed deciduous woodlands. The larvae (up to 15mm long) are reddish-brown, with mottled, yellowish-brown prothoracic and anal plates, and prominent white or yellowish-white pinacula (Fig. **606**). They feed in May and June on the leaves of hornbeam (*Carpinus betulus*) and oak (*Quercus*), but will also attack the foliage of various other trees including birch (*Betula*), hawthorn (*Crataegus*), hazel (*Corylus avellana*) and lime (*Tilia*), each sheltered by a folded leaf edge or within spun leaves. Fully grown larvae pupate in tough, silken cocoons formed in the soil or amongst debris on the ground, remaining *in situ* throughout the winter. The rather drab, brownish-coloured adults (19–23mm wingspan) (Fig. **607**), emerge in the following February or March, but later in more northerly areas. The males fly rapidly in sunny weather, the females usually remaining at rest on tree trunks, but both sexes are active at night.

606 Larva of *Tortricodes alternella*.

Aleimma loeflingiana (Linnaeus)
This widely distributed and generally common woodland species is associated mainly with oak (*Quercus*) but will also attack hornbeam (*Carpinus betulus*) and maple (*Acer*). The larvae feed within folded or rolled leaves throughout May, usually pupating in June. Attacks are sometimes noted on young cultivated trees but are of no economic importance. The larvae (up to 15mm long) are green to blackish-green, with blackish-brown to black pinacula, head, prothoracic and anal plates (Fig. **608**). Adults, which occur in June and July, are mainly whitish-yellow, with the fore wings marked with yellowish-brown and dark brown, including a dark margin at the base of the cilia (Fig. **609**).

607 Adult of *Tortricodes alternella*

608 Larva of *Aleimma loeflingiana*.

Tortrix viridana (Linnaeus)
Green oak tortrix moth

An important pest of oak, especially English oak (*Quercus robur*), in some years appearing in considerable numbers. Although primarily a forest pest, damage is also caused to trees in parks, gardens and nurseries. Widely distributed in Europe; also present in North Africa and parts of Asia.

DESCRIPTION **Adult:** 17–24 mm wingspan; head, thorax and fore wings light green; hind wings light grey (Fig. **610**). **Egg:** flat and oval, pale yellow to orange-brown. **Larva:** 15–18 mm long; greenish-grey to light olive-green, with prominent blackish-brown or black pinacula; head blackish-brown to black; prothoracic plate greenish-brown to greyish, marked with blackish-brown, often edged anteriorly with white and with a pale median line; anal plate green or black; anal comb with eight prongs; thoracic legs black (Fig. **611**). **Pupa:** 10–12 mm long; brownish-black to black.

LIFE HISTORY Adults are most numerous in June and July, and are especially abundant in oak woodlands. They rest on host trees during the daytime but are readily disturbed. Eggs are laid in pairs on the bark, especially close to the leaf bases and where the shoots and small branches divide. They hatch in the following spring at about bud-burst. The larvae feed from late April to June, at first entering the opening· buds but later inhabiting rolled or folded leaves. Pupation usually occurs in a folded leaf on the host tree, but larvae may also descend on a silken thread and pupate on underlying plants. When larvae are present in large numbers they may defoliate the trees before becoming fully fed; in such circumstances the larvae will attack the foliage of various other trees, shrubs and nearby herbage.

DAMAGE Larvae often defoliate host trees, but damage is redressed by the appearance of new leaves on the Lammas shoots which develop in the summer.

CONTROL On nursery trees, apply an insecticide in the spring as soon as damage is seen.

609

609 Male of *Aleimma loeflingiana.*

610

610 Female green oak tree tortrix moth, *Tortrix viridana.*

611

611 Larva of green oak tree tortrix moth, *Tortrix viridana.*

Croesia bergmanniana (Linnaeus)

Generally common on rose (*Rosa*), including cultivated varieties; in mainland Europe also reported on alder buckthorn (*Frangula alnus*) and common buckthorn (*Rhamnus carthartica*). Present throughout central and northern Europe; also occurs in North America.

DESCRIPTION **Adult:** 12–15 mm wingspan; head, thorax and fore wings bright yellow, the latter suffused with brownish-orange and bearing often dark-edged silvery-grey markings; hind wings brownish-grey (Fig. **612**). **Larva:** 10 mm long; greenish-grey to yellowish-white or yellow, with indistinct pinacula; head, thoracic legs, prothoracic and anal plates brownish-black or black; anal comb present (Fig. **613**). **Pupa:** 7–8 mm long; dark brown with a yellowish-brown or light brown abdomen.

LIFE HISTORY Moths occur in June and July, often flying during the afternoon as well as at night. Eggs are laid on the stems of rose bushes, where they remain until the following spring. Larvae feed from April onwards, inhabiting tightly folded leaves or spun shoots and sometimes spinning a shoot tip to an adjacent flower bud. Pupation occurs in the larval habitation or in a folded leaf, adults appearing a few weeks later.

DAMAGE If numerous, the larvae cause noticeable distortion and loss of young shoots, affecting growth and reducing flowering potential of the bushes.

CONTROL Apply an insecticide in the spring as soon as damage is seen.

612 Adult of *Croesia bergmanniana*.

613 Larva of *Croesia bergmanniana*.

Croesia holmiana (Linnaeus)

An often common but minor pest of ornamental trees and shrubs, including blackthorn (*Prunus spinosa*), *Chaenomeles*, crab apple (*Malus*), hawthorn (*Crataegus*) and rose (*Rosa*). Eurasiatic. Widely distributed in central and northern Europe.

DESCRIPTION **Adult:** 12–15 mm wingspan; fore wings orange-yellow, suffused with dark reddish-brown, each with a distinctive white triangular mark; hind wings dark grey. **Larva:** 10 mm long; yellowish-green; head light brown; prothoracic plate brown to black; thoracic legs brown. **Pupa:** 7–8 mm long; reddish-yellow or dark yellow.

LIFE HISTORY Adults occur in July and August, eggs being deposited on the bark of the shoots and small branches of host plants. Eggs hatch in the following spring, the larvae feeding from May to June, each in a shelter formed from two leaves spun together. When fully fed, individuals pupate in the larval habitation or in a folded leaf. There is just one generation annually.

DAMAGE Larvae cause minor damage to the leaves but infestations on cultivated plants are unimportant.

Croesia forsskaleana (Linnaeus)
syn. *forskaliana* (Haworth)

Larvae of this widely distributed species feed from September to May or June on field maple (*Acer campestre*) and sycamore (*Acer pseudoplatanus*); after hibernation individuals attack the unfurling leaves and flowers and, later, inhabit a leaf rolled longitudinally. Although infestations are often common in parks and gardens, damage caused to cultivated plants is unimportant. Larvae (up to 10 mm long) are yellow with small, pale pinacula, a yellowish-green head and anal plate and a greenish prothoracic plate. Adults (12–15 mm wingspan) are mainly yellow, the fore wings reticulated with brownish-orange, partly bordered with black and more-or-less suffused centrally with dark grey (Fig. 614). They occur in July and August.

614 Adult of *Croesia forsskaleana*.

Acleris rhombana (Denis & Schiffermüller)
Rhomboid tortrix moth

An often common pest of trees and shrubs, including crab apple (*Malus*), hawthorn (*Crataegus*), ornamental cherry (*Prunus*), ornamental pear (*Pyrus*) and rose (*Rosa*); often present in gardens and nurseries. Widely distributed in central and northern Europe; also present in North America.

DESCRIPTION **Adult:** 13–19 mm wingspan; fore wings with a sub-falcate tip, dark reddish-brown to ochreous, with darker markings, a reticulate pattern and, sometimes, a distinct central scale tuft; hind wings pale grey, darker in female (Fig. 615). **Egg:** 0.8 × 0.5 mm; flat, oval and yellowish-green. **Larva:** 12–14 mm long; greyish-green or yellowish-green, with pale inconspicuous pinacula; prothoracic plate reddish-brown or green; anal plate pale green; anal comb yellowish; thoracic legs black (Fig. 616). **Pupa:** 7–9 mm long: reddish-brown or dark brown; cremaster short and broad.

615 Rhomboid tortrix moth, *Acleris rhombana*.

LIFE HISTORY Adults occur from August to October. Eggs are laid singly or in small batches on the bark of trunks and branches. They hatch in spring and the larvae invade the opening buds. Later they feed in webbed leaves, usually at the tips of the young shoots. They may also feed on blossom trusses. Pupation takes place in June or early July, usually in a cocoon in the soil. The pupal stage is protracted, lasting for six to eight weeks or more.

DAMAGE Restricted mainly to the loss or distortion of younger leaf tissue and blossoms but with little or no effect on tree growth.

616 Larva of rhomboid tortrix moth, *Acleris rhombana*.

Acleris schalleriana (Linnaeus)

Locally common in association with *Viburnum*, infestations sometimes occurring on ornamental shrubs. Holarctic. Widely distributed in mainland Europe; in the British Isles most frequent in southern England.

DESCRIPTION **Adult:** 15–19 mm wingspan; fore wings greyish-white, greyish-brown or dark ochreous, with a more-or-less conspicuous dark costal blotch (Fig. **617**). **Larva:** 12–14 mm long; green to yellowish-green, with a yellowish or brownish-greenish head; prothoracic plate yellowish-green or green (Fig. **618**). **Pupa:** 8–9 mm long; light brown.

LIFE HISTORY Adults appear from late August or September onwards, hibernating during the winter and reappearing in the following spring. Eggs are laid on viburnum, hatching in late May or June. The larvae then feed within spun leaves until August. Pupation occurs in the larval habitation.

DAMAGE Infested shoots are distorted but infestations are of little importance.

617 Adult of *Acleris schalleriana*.

618 Larva of *Acleris schalleriana*.

Acleris sparsana (Denis & Schiffermüller)

syn. *fagana* (Curtis); *reticulana* (Haworth)

A generally common but minor pest of beech (*Fagus sylvatica*) and sycamore (*Acer pseudoplatanus*), and often found on such hosts in parks and gardens; in mainland Europe additional hosts include birch (*Betula*), oak (*Quercus*), poplar (*Populus*) and rowan (*Sorbus aucuparia*). Widespread in Europe.

DESCRIPTION **Adult:** 18–22 mm wingspan; fore wings grey to white, marked with reddish-brown and yellowish, often with an expanded costal blotch; hind wings grey. **Larva:** 14–16 mm long; light green with inconspicuous, shiny pinacula; head brownish-green or light green; prothoracic plate green with a black mark on each side; anal plate green; thoracic legs green. **Pupa:** 8–10 mm long; light brown.

LIFE HISTORY Adults appear in the late summer and autumn, hibernating in the winter and reappearing in the following spring. Larvae occur from June to late July or early August. Young larvae spin flimsy webs on the underside of expanded leaves. Older individuals occur between two overlapping leaves which they spin together with silk, sheltering in a folded portion of the upper leaf. Fully fed individuals pupate in July or August, either within the larval habitation or amongst debris on the ground.

DAMAGE Larval feeding is restricted to the leaves and, although sometimes noticeable, is of no economic importance.

Acleris variegana (Denis & Schiffermüller)
Garden rose tortrix moth

A generally common pest of rosaceous plants, including ornamental cherry (*Prunus*), -but especially common on rose (*Rosa*); also occurs on barberry (*Berberis*). Palearctic. Widely distributed in Europe.

DESCRIPTION **Adult:** 14–18 mm wingspan; fore wings whitish-ochreous to purplish, variably suffused with grey, frequently with much of the basal half white and with distinct, often black, scale tufts; hind wings grey (Fig. **619**). **Egg:** 0.6 × 0.5 mm: pale yellowish to reddish. **Larva:** 12–14 mm long; pale green or yellowish-green; head and prothoracic plate yellowish-brown or greenish-brown; anal plate green; thoracic legs yellowish-brown. **Pupa:** 7–8 mm long; light brown.

LIFE HISTORY Adults appear from July to September. Eggs are then laid singly or in small batches on either side of leaves, usually along the mid-vein. They hatch in the following spring. The larvae feed on the young shoots from May to late June or early July, sheltering within loosely spun leaves or in folded leaf edges. Pupation occurs in the larval habitation or amongst fallen leaves.

DAMAGE Larval habitations are unsightly and can cause distortion.

CONTROL Apply an insecticide in the spring as soon as damage is seen.

619 Female garden rose tortrix moth, *Acleris variegana*.

620 Larva of *Acleris emargana*.

Acleris emargana (Fabricius)

A generally common species, associated with birch (*Betula*), poplar (*Populus*), pussy willow (*Salix caprea*) and other broad leaved willows. Minor infestations are sometimes noted on nursery trees, the mainly green larvae (up to 15 mm long) (Fig. **620**) feeding within folded leaves or spun shoots during May and June. Damage can also occur on alder (*Alnus*) and hazel (*Corylus avellana*). Adults (16–22 mm wingspan) are mainly brown, with a characteristic emargination on the costa of each fore wing (Fig. **621**). They appear from July onwards.

621 Adult of *Acleris emargana*.

Acleris laternana (Fabricius)
syn. *latifasciana* (Haworth)
Broad-barred button moth
This widely distributed and generally common species is occasionally a minor pest of rose (*Rosa*), and certain other trees and shrubs, but is more common on raspberry (*Rubus idaeus*). The larvae (up to 15 mm long) are whitish-green, with a pale yellowish-brown head (Fig. **622**); they feed between spun leaves in May and June, causing conspicuous but usually minor damage. The adults (15–20 mm wingspan) are mainly whitish to ochreous, marked with black, brown or red (Fig. **623**); they fly in August and September, and are often common in gardens.

622 Larva of broad-barred button moth, *Acleris laternana*.

623 Broad-barred button moth, *Acleris laternana*.

Olethreutes lacunana (Denis & Schiffermüller)
Dark strawberry tortrix moth
Larvae of this widely distributed species are associated mainly with low-growing plants in damp habitats and are often important pests of strawberry crops. They are sometimes noted on garden ornamentals, especially marsh marigold (*Caltha palustris*); minor infestations may also occur on trees and shrubs, including birch (*Betula*), crab apple (*Malus*), larch (*Larix*), privet (*Ligustrum vulgare*), sallow (*Salix*) and spruce (*Picea*). The larvae (up to 14 mm long) are mainly dark purplish-brown to blackish-brown (Fig. **624**) and extremely active if disturbed; they occur from August onwards, usually overwintering whilst still small and completing their development in the following spring. The greyish-ochreous to greenish-black adults (15–18 mm wingspan) (Fig. **625**) appear from May onwards, often with a second generation in August.

624 Larva of dark strawberry tortrix moth, *Olethreutes lacunana*.

Hedya dimidioalba (Retzius)
syn. *nubiferana* (Haworth); *variegana* (Hübner)
An often common pest of various trees and shrubs, including alder (*Alnus*), ash (*Fraxinus excelsior*), ornamental cherry (*Prunus*), rose (*Rosa*) and *Sorbus*, but especially common on hawthorn (*Crataegus*) and often abundant on garden hedges. Eurasiatic; also occurs in North America.

DESCRIPTION **Adult:** 15–21 mm wingspan; fore wings ochreous-white apically, suffused with silver and ochreous grey, the remainder marbled with dark brown, bluish-grey and black; hind wings brownish-grey (Fig. **626**). **Egg:** 0.85 × 0.65 mm; oval, flat and iridescent. **Larva:** 18–20 mm long; olive-green to dark green, with black pinacula; head, prothoracic plate, anal plate and anal comb dark brown or black; thoracic legs black (Fig. **627**). **Pupa:** 8–10 mm long; dull black; cremaster tapered, with an apical tuft of eight hooked bristles.

LIFE HISTORY Adults occur in June and July, closely resembling bird-droppings when at rest. Eggs are laid singly or in small groups, mainly on the underside of leaves. They hatch in about two weeks, the larvae then feeding for several weeks before hibernating, each in a dense cocoon spun within a bark crevice or below old bud scales. Activity is resumed early in the following spring, the larvae attacking the opening buds, blossom trusses, foliage and young shoots, often sheltering between two leaves webbed together with silk. Fully grown individuals pupate in spun leaves in late May or June, adults emerging three or four weeks later.

DAMAGE The larvae contribute to damage done to leaves and blossoms by other species of caterpillar; they may also tunnel into the young shoots and cause wilting or death of the tips.

CONTROL On young trees, apply an insecticide as soon as larval attacks commence; summer treatments are more effective than spring sprays.

625 Dark strawberry tortrix moth, *Olethreutes lacunana*.

626 Adult of *Hedya dimidioalba*.

627 Larva of *Hedya dimidioalba*.

Hedya ochroleucana (Frölich)

Locally common on wild rose (*Rosa*) but also associated with cultivated bushes; adults commonly rest exposed on the foliage. The larvae feed mainly in the spring, from April to June, each inhabiting a bunch of spun leaves. Fully fed specimens (16–18 mm long) are dull greyish-green to olive-green, with inconspicuous pinacula and a brownish-black to black head and prothoracic plate (Fig. **628**). They pupate within the larval habitation in May or June. Adults (16–23 mm wingspan) are mainly blackish to bluish-grey and brownish-grey, with the fore wings partly creamish-white and often tinged with pink (Fig. **629**). They occur in June and July.

628 Larva of *Hedya ochroleucana.*

Orthotaenia undulana (Denis & Schiffermüller)

A locally common but minor pest of trees and shrubs, including alder (*Alnus*), birch (*Betula*), elm (*Ulmus*), honeysuckle (*Lonicera*), juniper (*Juniperus*), maple (*Acer*), pine (*Pinus*), sea buckthorn (*Hippophaë rhamnoides*) and willow (*Salix*); certain herbaceous plants are also attacked. Widely distributed in Europe; also present in North America.

DESCRIPTION **Adult:** 15–20 mm wingspan; fore wings creamish-white, marked with silvery-grey, greyish-yellow, olive-yellow and brownish-black; hind wings greyish (Fig. **630**). **Larva:** 15–18 mm long; reddish-brown to blackish-brown, with brown or reddish-brown, inconspicuous pinacula; head, prothoracic plate and anal plate black (Fig. **631**).

629 Adult of *Hedya ochroleucana.*

LIFE HISTORY Adults occur mainly in June and July. The larvae feed within spun leaves from April onwards, most becoming fully grown by late May or early June. They then pupate, either in a spun leaf or in the larval habitation, adults appearing shortly afterwards. There is just one generation annually. Adults and larvae of this species are often mistaken for those of *Olethreutes lacunana* (p. 254).

DAMAGE The larvae cause slight foliage damage and disfigurement but infestations are unimportant.

630 Adult of *Orthotaenia undulana.*

Lobesia littoralis (Humphreys & Westwood)
Thrift tortrix moth

Locally common and often abundant in coastal habitats where thrift (*Armeria maritima*) is established; also occurs commonly on thrift cultivated in gardens and nurseries. Widespread in Europe.

DESCRIPTION **Adult:** 11–16mm wingspan; fore wings white, suffused with grey and variably marked with brown and reddish-brown; hind wings whitish-grey (Fig. **632**). **Larva:** 10mm long; greenish-grey to yellowish-grey, with inconspicuous pinacula; head and prothoracic plate brownish-black or black; anal comb present; thoracic legs black (Fig. **633**). **Pupa:** 6–8mm long; greenish-brown.

LIFE HISTORY This species is bivoltine, adults occurring in June and July, and in the autumn. The moths are especially active in sunny weather, flying rapidly over the foodplant. The larvae feed amongst the vegetative shoots, each inhabiting a silken tube; they will also attack the flower heads to feed on the unripe seeds. Larvae of the first generation occur from April to May or early June, and those of the second in August. Pupation takes place in a dense silken cocoon spun amongst the leaves or secreted within a flower head.

DAMAGE Larvae cause death of shoots and slight distortion of flower heads, but damage is of importance only on nursery plants.

CONTROL An insecticide aimed to kill larvae before they form protective shelters might be effective.

631

631 Larva of *Orthotaenia undulana.*

632

632 Thrift tortrix moth, *Lobesia littoralis.*

633

633 Larva of thrift tortrix moth, *Lobesia littoralis.*

Ancylis mitterbacheriana (Denis & Schiffermüller)

A locally common pest of beech (*Fagus sylvatica*) and oak (*Quercus*). Eurasiatic. Widely distributed in Europe.

DESCRIPTION **Adult:** 12–16mm wingspan; fore wings mainly brownish-orange to whitish, with a prominent reddish-brown or chocolate-brown dorsal patch, and a curved blackish streak towards the apex. **Larva:** 12mm long; greyish-green to yellowish-green, with pale and relatively large pinacula; head yellowish-brown to greenish-brown, marked with black; prothoracic plate pale green, with a large lateral blackish patch towards the hind margin and blackish markings dorsally; anal plate pale green, marked with blackish-brown (Fig. 634). **Pupa:** 6–8 mm long; light reddish-brown.

LIFE HISTORY Adults occur in May and June, and are especially common in or close to beech and oak woodlands. Larvae inhabit individual pod-like shelters, formed out of a folded leaf (Fig. 635), grazing upon the innermost surface. They feed from July to September, overwintering in the larval habitation and pupating in the following spring.

DAMAGE Larval habitations attract attention when present on nursery or specimen trees, but feeding damage is superficial and of little or no importance.

Ancylis upupana (Treitschke)

A locally distributed species, associated mainly with birch (*Betula*) but also attacking elm (*Ulmus*). Although generally uncommon the larvae are found, occasionally, on nursery or specimen trees, sheltering between spun leaves and causing minor damage to the expanded foliage. They feed from July to September, overwintering when fully fed and pupating in the spring. Larvae (up to 12mm long) are greenish-grey to dark olive-brown, with dark pinacula (those on the thoracic segments especially prominent) and a black or brownish head and prothoracic plate (Fig. 636). Adults (12–18mm wingspan) are mainly brown, marked with silvery-grey and brownish-orange (Fig. 637); they occur in May and June.

634 Larva of *Ancylis mitterbacheriana*.

635 Larval habitation of *Ancylis mitterbacheriana* on *Quercus*.

636 Larva of *Ancylis upupana*.

637 Adult of *Ancylis upupana*.

638 Adult of *Epinotia brunnichana*.

639 Larva of *Epinotia brunnichana*.

Epinotia brunnichana (Linnaeus)

A minor pest of birch (*Betula*), hazel (*Corylus avellana*) and sallow (*Salix*), including young cultivated trees. Widespread in Europe.

DESCRIPTION **Adult:** 18–22 mm wingspan; fore wings white, each more-or-less suffused with brownish-orange and usually marked with a pale blackish-edged dorsal patch; hind wings pale brownish (Fig. **638**). **Larva:** 16–18 mm long; greenish-grey with black pinacula; head brownish-black; prothoracic plate greenish-brown, paler anteriorly; anal plate pale greenish; thoracic legs black (Fig. **639**).

LIFE HISTORY Larvae occur most commonly on birch trees. They feed during May and June, each inhabiting a transversely rolled leaf (Fig. **640**) (cf. *Epinota solandriana*, p. 260). When fully fed the larvae enter the soil to pupate in silken cocoons. The moths appear in July and August.

DAMAGE The characteristic larval habitations attract attention, especially if present on young trees, but damage caused is of no importance.

640 Larval habitation of *Epinotia brunnichana* on *Betula*.

Epinotia immundana (Fischer von Röslerstamm)

A locally common species; associated with alder (*Alnus*) and birch (*Betula*) including, occasionally, cultivated trees. The larvae feed from the autumn to the following spring, attacking the buds and catkins; they pupate in silken cocoons formed amongst leaf debris or in the ground, adults appearing from April to June. In favourable districts a small number of larvae occur during the summer, inhabiting characteristically rolled leaves (Fig. **641**); the larvae eventually pupate to produce a partial second generation of adults in August and September. Adults (12–14mm wingspan) are whitish to greyish-brown, heavily marked with black, with a pale angular patch dorsally on each fore wing (Fig. **642**). The larvae (up to 10mm long) are translucent, greenish-grey to yellowish-grey, suffused above with red (Fig. **643**).

641 Larval habitation of *Epinotia immundana* on *Alnus*.

642 Adult of *Epinotia immundana.*

Epinotia solandriana (Linnaeus)

Larvae of this widely distributed species feed in May and June on birch (*Betula*), hazel (*Corylus avellana*) and pussy willow (*Salix caprea*), each inhabiting a longitudinally folded leaf (Fig. **644**) (cf. *Epinotia brunnichana*, p. 259). Fully fed individuals are 15–18mm long and dull grey to greenish-grey, with grey or blackish-grey pinacula and a yellowish-brown, darker-mottled head and a yellowish-brown to dark brown prothoracic plate. The adults occur in July and August; there are various colour forms, the fore wings commonly being creamish-white, suffused with brown and with a dark dorsal blotch (in some forms the wings are mainly reddish-brown and the blotch light brown or white); the hind wings are grey.

643 Larva of *Epinotia immundana.*

Rhopobota naevana (Hübner)
Holly tortrix moth
larva = holly leaf tier

A generally common pest of various trees and shrubs, but especially abundant on holly (*Ilex aquifolium*); often present in parks and gardens. Eurasiatic; also present in North America. Widely distributed in Europe.

DESCRIPTION **Adult:** 12–15mm wingspan; fore wings notched below the apex, dark grey, marked with blackish and dark reddish-brown, and with several whitish striae on the front margin; hind wings grey (Fig. **645**). **Egg:** 0.7 × 0.5mm; flat and oval; translucent-whitish, becoming yellowish to reddish. **Larva:** 10–12mm long; oily, yellowish-green to greenish-brown; head, prothoracic plate and anal plate brown; anal comb usually with two dark teeth; thoracic legs brown (Fig. **646**). **Pupa:** 5–7mm long; yellowish-brown; tip with four thorn-like spines and a small bump behind the anal slit.

LIFE HISTORY This species overwinters as eggs laid singly on the smooth bark of trunks and branches. They hatch in the spring. The young larvae then feed within a tightly webbed shelter of young, tender leaves; on certain hosts, such as hawthorn (*Crataegus*), they may also feed within spun blossom trusses. Feeding is completed in June, each larva pupating in a white cocoon formed in a folded leaf or amongst dead leaves or debris on the ground. Adults emerge about three weeks later, depositing eggs in July and August.

DAMAGE Larvae destroy young leaves and are especially damaging to the new growth on clipped holly hedges.

CONTROL Apply an insecticide in the spring as soon as damage is seen; application of additional wetter will help improve coverage of the foliage.

644 Larval habitation of *Epinotia solandriana* on *Betula*.

645 Holly tortrix moth, *Rhopobota naevana*.

646 Holly leaf tier, *Rhopobota naevana*.

Epiblema cynosbatella (Linnaeus)

A common species on wild and cultivated rose (*Rosa*). In mainland Europe various other plants, including ornamentals such as *Chaenomeles* are also attacked. Widespread in Europe.

DESCRIPTION **Adult female:** 16–22 mm wingspan; fore wings creamish-white, marked with brownish-black, bluish-grey and pale brown, and a large blackish basal patch; hind wings dark grey; labial palps yellow (Fig. **647**). **Adult male:** similar to female but hind wings light grey. **Larva:** 18 mm long; reddish-brown with inconspicuous pinacula. **Pupa:** 9–11 mm long; blackish-green.

LIFE HISTORY Larvae attack the young shoots and flower buds of rose bushes, feeding mainly in April and May. They then pupate in the larval habitation, usually between two spun leaves or in a spun shoot, adults appearing in May and June.

DAMAGE Damage to foliage is of little consequence, but infested shoots can be distorted and flower buds destroyed.

CONTROL Apply an insecticide in the early spring as soon as damage is seen.

Epiblema roborana (Denis & Schiffermüller)

Generally common on wild and cultivated rose (*Rosa*). Eurasiatic. Widely distributed in Europe.

DESCRIPTION **Adult:** 16–22 mm wingspan; fore wings ochreous-white, suffused or marked with buff, dark brown, bluish-grey and black, and each with a prominent brownish-black basal patch which extends along the front margin; hind wings whitish-grey (Fig. **648**). **Larva:** 18 mm long; reddish-brown to dark brown, with inconspicuous brown pinacula; head yellowish-brown; prothoracic plate brownish-black or black (Fig. **649**). **Pupa:** 9–11 mm long; light brown.

LIFE HISTORY Adults occur from late June to August, often occurring abundantly in the vicinity of rose bushes. Eggs are deposited on the bushes; they hatch in the following spring. The larvae feed throughout May and early June, each sheltering in a tightly spun shoot and devouring the innermost tissue. Pupation usually occurs in the larval habitation.

DAMAGE Infested bushes are disfigured and new shoots distorted or destroyed. The larvae can also damage flower beds.

647 Adult of *Epiblema cynosbatella*.

648 Adult of *Epiblema roborana*.

649 Larva of *Epiblema roborana*.

CONTROL Spray with an insecticide as soon as damage is seen. If necessary, larval habitations can be picked off by hand and burnt.

Epiblema rosaecolana (Doubleday)

Associated with sweet briar (*Rosa rubiginosa*) and frequently a pest of cultivated species of rose. Eurasiatic; also established, as an introduced pest, in North America. Widely distributed in Europe.

DESCRIPTION **Adult:** 16–20 mm wingspan; fore wings white, partly suffused and striated with brown and bluish-grey, and marked with several prominent black spots; hind wings grey (Fig. **650**). **Larva:** 18 mm long; reddish-brown or purplish-brown above, but creamish along the sides and below, with small blackish pinacula; head yellowish-brown; prothoracic plate blackish-brown or black; anal plate brown; thoracic legs blackish-brown (Fig. **651**). **Pupa:** 8–10 mm long; light brown.

LIFE HISTORY Moths fly in June and July; they are often common in parks and gardens. Eggs laid on the foodplant hatch in the following spring. The larvae then feed within spun shoots during May and June; they may also bore into the young shoots. Pupation occurs in the larval habitation or in a slight cocoon spun amongst debris on the ground.

DAMAGE Infestations cause loss of young shoots and lead to a reduction in the number of flowers; they also disfigure bushes.

CONTROL Spray with an insecticide in the spring, when damage is first seen.

Epiblema trimaculana (Haworth)

syn. *suffusana* (Duponchel)

A minor pest of hawthorn (*Crataegus*), and sometimes present on nursery stock. Palearctic. Widely distributed in Europe.

DESCRIPTION **Adult:** 15–17 mm wingspan; fore wings white, suffused and striated with brown and bluish-grey, and marked with black; hind wings grey (Fig. **652**). **Larva:** 16 mm long; brown to reddish-brown, with small inconspicuous pinacula; head brown; prothoracic plate brownish-black to

650

650 Adult of *Epiblema rosaecolana*.

651

651 Larva of *Epiblema rosaecolana*.

652

652 Adult of *Epiblema trimaculana*.

black; anal plate brown; thoracic legs blackish-brown (Fig. **653**). **Pupa:** 8–9 mm long; brown.

LIFE HISTORY Larvae shelter and feed within spun shoots of hawthorn from April to May. Fully fed individuals pupate in loose cocoons spun within the larval habitation, adults emerging in June.

DAMAGE Larvae cause slight distortion of young terminal shoots but infestations are unimportant.

653 Larva of *Epiblema trimaculana*.

Spilonota ocellana (Denis & Schiffermüller)
Bud moth

An often common pet of rosaceous trees and shrubs, including ornamentals; various other trees, including alder (*Alnus*), hazel (*Corylus avellana*), larch (*Larix*) and oak (*Quercus*), are also attacked, the last by a distinct form — *Spilonota ocellana laricana* (Heinermann). Eurasiatic; also occurs in North America. Widely distributed in Europe.

DESCRIPTION **Adult:** 12–16 mm wingspan; fore wings whitish, more-or-less suffused with grey, marked towards the apex with metallic bluish-grey and black, and each with a dark, triangular dorsal spot and a blackish, angular basal patch; hind wings dark grey (Fig. **654**). **Egg:** flat and more-or-less circular; pale yellowish-white. **Larva:** 9–12 mm long; dark purplish-brown with lighter pinacula (larch-feeding form light greyish-brown with indistinct pinacula); head, prothoracic plate and anal plate shiny black or blackish-brown (Fig. **655**). **Pupa:** 6–7 mm long; brown, with outline of wing cases distinctly darker than abdomen; tip blunt.

654 Bud moth, *Spilonota ocellana*

LIFE HISTORY Adults occur from June to August. Eggs are deposited singly or in small batches on the leaves of host plants. The larvae feed from August onwards, each inhabiting a silken tube and eventually hibernating in a silken retreat spun on the bark, often close to a bud. Activity is resumed early in the following spring, the larvae invading the buds and, later, each feeding within a tightly woven tent of young unfurling leaves. Pupation occurs in the larval habitation, usually in June.

DAMAGE Direct damage to leaves and shoots is relatively unimportant, but attacked buds wither and die.

CONTROL Application of an insecticide in the early spring will give some control but spraying in August is more effective.

655 Larva of bud moth, *Spilonota ocellana*.

Petrova resinella (Linnaeus)
Pine resin-gall moth

A minor and locally common pest of pine (*Pinus*), the characteristic larval habitations (Fig. **656**) occasionally attracting attention on young amenity trees. Adults occur in June, eventually depositing eggs on the young shoots. After egg hatch each larva bores into the base of a needle and also damages the surface of the stem, forming an elongate groove from which resin exudes; the larva then feeds close to a bud whorl, protected by silk and a coating of resin which, at this stage, forms a pea-like gall. Feeding continues throughout the following year, the protective gall becoming greatly enlarged (*c.* 20–30 mm across) and obvious. Pupation occurs in the following spring, almost two years after eggs were laid. Adults (16–22 mm wingspan) have whitish fore wings, mottled with grey and blackish-brown; the hind wings are brown, with white cilia.

656 Larval habitation of pine resin-gall moth on *Pinus*.

Enarmonia formosana (Scopoli)
syn. *woeberiana* (Denis & Schiffermüller)
Cherry bark tortrix moth

A locally common pest of ornamental cherry (*Prunus*); sometimes also damaging to crab apple (*Malus*), ornamental pear (*Pyrus*) and *Sorbus*. Widely distributed in Europe.

DESCRIPTION **Adult:** 15–18 mm wingspan; fore wings more-or-less brown with a purplish sheen, and with irregular, yellowish-orange markings and silvery-white costal striae; hind wings dark brown (Fig. **657**). **Egg:** 0.7×0.6 mm; flat and oval, whitish to reddish. **Larva:** 8–11 mm long; brownish to salmon-pink, with brownish pinacula; head light brown; prothoracic and anal plates light greyish-brown. **Pupa:** 7–9 mm long; light brown; tip blunt.

657 Cherry bark tortrix moth, *Enarmonia formosana*.

LIFE HISTORY Adults appear from May or early June to September, the extended emergence period giving the impression of two generations. The moths are active in sunshine and often make repeated short flights to and from the branches or trunks of infested trees, but they are difficult to find when settled on the bark. Eggs are laid singly or in groups of two or three, usually on previously infested or otherwise injured parts of trees, such as frost-damaged or mechanically damaged bark, or adjacent to pruning wounds. They hatch in two or three weeks and the larvae then feed beneath the surface, excavating irregular, often deep galleries in the bark; the underlying cambium may also be damaged but tunnels do not extend into the wood. Larvae are usually fully grown by the following spring or early summer, passing through five instars. Each then pupates in a silken cocoon formed in the larval feeding gallery. Pupae remain protruding from the bark after the adult moths have emerged.

DAMAGE A considerable quantity of gum can exude from infested parts of host trees and this, along with accumulations of light-brown frass and silken webbing forced out of cracks in the bark, may be one of the first indications of an attack. Cherry trees are usually invaded near the base of the trunks, infestations producing cracks, swellings and cankers; severely damaged trees may be killed.

CONTROL Loose bark should be scraped away during March, while trees are still dormant, and the trunk and main branches coated liberally with creosote or treated with a proprietary winter wash.

19. Family PYRALIDAE

A variable family, the adults having moderately long, narrow bodies, prominent palps, relatively narrow fore wings and broad hind wings. The larvae usually have few body hairs and are often extremely active when disturbed.

Elophila nymphaeata (Linnaeus)
Brown china-mark moth

A generally common pest of various aquatic plants, including bur-reed (*Sparganium*), frog-bit (*Hydrocharis morus-ranae*), pondweed (*Potamogeton*) and ornamental water-lilies (Nympheaceae). Widespread in Europe.

DESCRIPTION **Adult:** 20–22 mm wingspan; wings white, marked irregularly with brown (Fig. **658**). **Larva:** 20–25 mm long; creamish, with a darker line down the back; head and prothoracic plate mainly brown.

LIFE HISTORY Adults occur from June to August, depositing eggs in batches on the underside or at the edges of leaves of host plants. Eggs hatch in about two weeks, and the young larvae then burrow into the leaves to begin feeding. They reappear about three days later, each constructing a small, flat, oval case out of leaf fragments. Larvae continue to feed on the surface of leaves throughout July and August, replacing their case with a larger one as necessary. In early September the larvae move to the edge of the pond or stream, where they hibernate until the following spring. Larval cases are often common on water-lily leaves and, during larval development, may also occur on submerged parts of host plants, the larvae surviving on air trapped within the confines of their habitations; when larvae are not feeding, the cases float freely on the surface of the water. Fully fed larvae pupate in the early summer within silken cocoons, into

658 Brown china-mark moth, *Elophila nymphaeata.*

which fragments of leaf are incorporated, spinning-up on a convenient leaf or stem a few centimetres above the water level.

DAMAGE Infested leaves are extensively holed and often become ragged, damaged tissue rotting and forming an unsightly mess.

CONTROL On a small scale, occupied larval cases can be picked off and destroyed; use of insecticides cannot be recommended if there is any possibility of affecting fish and other aquatic organisms.

659 Larva of *Udea prunalis*.

660 Adult of *Udea prunalis*.

Udea prunalis (Denis & Schiffermüller)
syn. *nivealis* (Fabricius)

A generally common species, associated mainly with herbaceous plants but also attacking the foliage of young ornamental trees and shrubs. The larvae feed within spun leaves; they occur briefly during the autumn and then hibernate, completing their development in the following spring. Individuals (up to 25 mm long) are greenish and shiny, with white subdorsal lines and a pale head (Fig. **659**). Adults (22–26 mm wingspan) are mainly brownish-grey (Fig. **660**); they appear in June and July.

661 Larva of *Palpita unionalis*.

Palpita unionalis (Hübner)

A migratory, Asiatic, African and southern European species, the very active, mainly greenish to yellowish-green larvae (Fig. **661**) feeding on the shoots of various members of the Oleaceae. Damage to the new shoots is often extensive and is sometimes noted in northern Europe on oleaceous plants such as jasmine (*Jasminum*); infestations are sometimes also found on decorative, container-grown olive (*Olea*) trees imported into northern Europe from Italy and other countries where the pest is endemic. The adults (28–30 mm wingspan) are mainly white, with a brown leading edge to each fore wing (Fig. **662**). In the British Isles this species occurs as a rare, non-resident migrant and is not regarded as a pest.

662 Adult of *Palpita unionalis*.

Numonia suavella (Zincken)
Porphyry knothorn moth

A local but widely distributed insect, associated mainly with wild blackthorn (*Prunus spinosa*) and hawthorn (*Crataegus*) but occasionally troublesome on ornamental *Cotoneaster*. The purplish-red, greyish- to whitish-marked adults (23–25 mm wingspan) occur in July but, although being attracted to light, are rarely seen. The larvae feed during the spring within distinctive whitish, silken galleries spun on the shoots beneath the leaves; dead leaf fragments and particles of frass accumulate on these larval shelters, making them even more conspicuous. The larvae are dark chestnut-brown, with a brown head, black prothoracic and anal plates, and a pair of distinctive dark spots on the second thoracic and the eighth abdominal segments. Pupation takes place in June, in a greyish-white cocoon spun either within or alongside the larval habitation.

663 Adult of *Phycita roborella*.

Phycita roborella (Denis & Schiffermüller)
syn. *spissicella* (Fabricius)

A locally common pest of oak (*Quercus*), and sometimes troublesome on young amenity trees and nursery stock. Widespread in Europe.

DESCRIPTION **Adult:** 20–27 mm wingspan; fore wings greyish-pink, suffused with purplish-red and brownish-black; hind wings pale brownish-grey, with a darker border (Fig. **663**). **Larva:** 12–15 mm long; dull greyish-green, with darker longitudinal lines; pinacula small and blackish; head pale brown; prothoracic plate pale greenish-yellow, spotted with black; first thoracic segment with a pair of dark, whitish-bordered spots (Fig. **664**).

LIFE HISTORY Adults occur in July and August, depositing eggs on oak trees. The larvae feed on the foliage from May onwards, individuals or groups

664 Larva of *Phycita roborella*.

665 Larval habitation of *Phycita roborella* on *Quercus*.

of individuals inhabiting distinctive clusters of webbed leaves (Fig. 665); these are very noticeable on the trees from late May to July. If disturbed, larvae are very active, rapidly wriggling backwards out of their shelters and falling to the ground. Fully grown larvae pupate in cocoons which they form in the soil, adults emerging shortly afterwards.

DAMAGE The growth of infested shoots is distorted, and some shoots may be completely defoliated (Fig. 666); infestations on mature trees are of no consequence but damage may be of considerable significance on young ones.

CONTROL On a small scale, occupied webs can be removed from infested trees and destroyed, but care is necessary to prevent larvae escaping as their habitations are disturbed.

666 *Phycita roborella* damage to shoots of *Quercus*.

20. Family **PIERIDAE**

A family of mainly white-winged butterflies. The larvae have setae on the head arising from raised tubercules; crochets on the abdominal prolegs are of two sizes (biordinal).

Pieris brassicae (Linnaeus)
Large white butterfly
Although of importance mainly as a pest of cruciferous vegetable crops, eggs of this often abundant species are sometimes deposited on related ornamentals, especially mignonette (*Reseda odorata*); they are also laid on nasturtium (*Tropaeolum*). The larvae are gregarious, feeding in large groups and causing extensive damage to the leaves, especially in the late summer. Individuals (up to 40 mm long) are pale green to yellow, variably marked with black (Fig. **667**). The mainly white, black-marked adults (55–65 mm wingspan) occur in the spring but are more numerous in the summer.

667 Larva of large white butterfly, *Pieris brassicae*.

668 Larva of small white butterfly, *Pieris rapae*.

669 Female small white butterfly, *Pieris rapae*.

Pieris rapae (Linnaeus)
Small white butterfly
This species also, occasionally, breeds on cruciferous ornamentals and nasturtium (*Tropaeolum*) but the larvae are usually present in only small numbers and, unlike those of the previous species, cause insignificant damage. Fully grown specimens are *c.* 25 mm long, green, finely specked with black, with a yellow line along the back and yellow markings along the sides (Fig. **668**). The mainly white to yellowish-white adults (45 mm wingspan) (Fig. **669**) appear in late April and May, with a second generation emerging in the summer.

670 Larva of December moth, *Poecilocampa populi*.

21. Family **LASIOCAMPIDAE**

Medium-sized to very large moths with bipectinated antennae (especially noticeable in males); males usually much smaller than females of the same species. The larvae are very hairy on both head and body; crochets on the abdominal prolegs are of two sizes (biordinal).

Poecilocampa populi (Linnaeus)
December moth
Widely distributed and locally common, the larvae feeding from late April to June on various broad-leaved trees, including birch (*Betula*), crab apple (*Malus*), hawthorn (*Crataegus*), lime (*Tilia*), oak (*Quercus*), ornamental cherry (*Prunus*) and poplar (*Populus*). The larvae are sometimes present in small numbers on ornamental plants but damage

671 Male December moth, *Poecilocampa populi*.

caused is not of significance. Fully grown larvae are 45 mm long and distinctly downy; they are mainly greyish and black above, with a red crossline just behind the head, but ochreous and black below (Fig. **670**). The moths (35–45 mm wingspan) are mainly brown and rather hairy, the wings marked with grey and ochreous; males have strongly bipectinated antennae (Fig. **671**). Adults fly during the winter, being most numerous in November and December.

Eriogaster lanestris (Linnaeus)
Small eggar moth

In parts of mainland Europe a minor pest of amenity trees such as birch (*Betula*) and lime (*Tilia*); also occurs on certain other hosts, including blackthorn (*Prunus spinosa*), hawthorn (*Crataegus*), oak (*Quercus*) and sallow (*Salix*). Widely distributed in central and northern Europe; in the British Isles, restricted mainly to blackthorn and hawthorn in a few parts of England and Ireland, and not regarded as a pest.

DESCRIPTION **Adult female:** 42 mm wingspan; wings thinly scaled, greyish-brown to pale reddish-brown, with a pale wavy crossline and, on each fore wing, a pale sub-central spot; abdomen with a greyish anal hair tuft (Fig. **672**). **Adult male:** 32 mm wingspan; similar to female but darker and without the anal hair tuft; antennae strongly bipectinated. **Egg:** olive-green, oval. **Larva:** 45–50 mm long; black or greyish-black, with reddish to whitish hairs and a series of brown, yellowish-edged patches along the back; head black (Fig. **673**); early instars are darker and recently moulted final-instar larvae brightly coloured with a gingery-brown head.

LIFE HISTORY Adults appear in February and March, females depositing batches of eggs on the twigs of host plants. The eggs are then covered with hairs from the anal tuft. Larvae feed from May to June or July, living gregariously and constructing dense, silken webs (Fig. **674**). Pupation occurs in large, yellowish-white to reddish-brown cocoons, usually formed in the ground. Most adults emerge in the following year but some individuals remain in the pupal stage through two or more winters.

DAMAGE Larvae cause considerable defoliation and the webs are disfiguring, but infestations rarely affect plant growth.

CONTROL Webs on young ornamentals should be cut out and transferred to a suitable wild hedge to complete their development.

672 Female small eggar moth, *Eriogaster lanestris*.

673 Final-instar larva of small eggar moth, *Eriogaster lanestris*.

674 Larval tent of small eggar moth on *Crataegus*.

675 Male lackey moth, *Malacosoma neustria*.

676 Larva of lackey moth, *Malacosoma neustria*.

Malacosoma neustria (Linnaeus)
Lackey moth

An often common pest of broad-leaved trees and shrubs, including alder (*Alnus*), birch (*Betula*), *Cotoneaster*, crab apple (*Malus*), elm (*Ulmus*), firethorn (*Pyracantha*), hawthorn (*Crataegus*), lilac (*Syringa vulgaris*), ornamental cherry (*Prunus*), rose (*Rosa*), sallow and willow (*Salix*); sometimes of importance in nurseries, parks and gardens. Eurasiatic. Present throughout much of Europe, except for the extreme north.

DESCRIPTION **Adult:** 30–40 mm wingspan; wings and body pale ochreous to dark brown; each fore wing with two crosslines, often (especially in female) enclosing a darker band (Fig. **675**). **Egg:** 0.5 mm across; cylindrical, laid in a large batch. **Larva:** 40–50 mm long; greyish-blue with a white dorsal stripe, and with orange-red, black-edged stripes running along the back and sides; body clothed in reddish-brown hairs; head blue, with two black spots (Fig. **676**). **Pupa:** 18 mm long; brownish-black and hairy.

LIFE HISTORY Moths occur from late July to September, each female depositing about 100–200 eggs in a characteristic band (6–14 mm wide) around a twig or spur of a host plant (Fig. **677**). Each egg mass, protected by a clear varnish-like coating secreted by the egg-laying female, remains *in situ* throughout the winter. The eggs hatch in the following spring, usually in late April and May. The young larvae are blackish but soon become more brightly coloured, feeding gregariously and forming a communal web or 'tent' (Fig. **678**).

677 Egg band of lackey moth, *Malacosoma neustria*.

678 Larval tent of lackey moth on *Prunus*.

These larval tents are very conspicuous and may exceed 30 cm in length, gradually being extended as the larvae and their feeding areas grow. Larvae are fully fed by late June or early July. They then pupate in white or yellowish, double-walled cocoons spun between leaves, in bark fissures or amongst herbage on the ground, adults emerging about three weeks later.

DAMAGE Defoliation is often severe, with infested branches or even whole trees stripped of leaves and covered in webbing. Growth, especially of young trees, may be seriously affected.

CONTROL Egg bands should be collected during the winter and destroyed; also, in early May, tents with young caterpillars should be cut out and burnt. If required, apply an insecticide in the spring.

22. Family **THYATIRIDAE**

Achlya flavicornis (Linnaeus)
Yellow-horned moth

A generally common species, associated with birch (*Betula*); found occasionally on ornamental and amenity trees, especially in the vicinity of birch woods. The larvae feed during June and July, sheltering during the daytime in a folded leaf. Young specimens are blackish-olive, marked with small, pale pinacula, and are superficially tortrix-like (Fig. **679**). Older individuals (up to 33 mm long) are whitish to pale greenish-white, more-or-less suffused with dark olive-green, and marked with black and white spots; the head is mainly yellowish-brown to reddish-brown (Fig. **680**). When fully grown, they enter the soil to pupate in flimsy cocoons. Adults are mainly greenish-grey, with dark grey or blackish markings (Fig. **681**). They occur in March and April.

679 Young larva of yellow-horned moth, *Achlya flavicornis*.

680 Larva of yellow-horned moth, *Achlya flavicornis*.

681 Male yellow-horned moth, *Achlya flavicornis*.

23. Family GEOMETRIDAE

A very large family of mainly slender-bodied moths with relatively large, usually broad, wings; most species are rather weak fliers. The larvae, commonly called 'looper caterpillars' or 'loopers', usually have just two functional pairs of abdominal prolegs (those on the third, fourth and fifth abdominal segments being rudimentary or absent), and they progress with a characteristic looping gait; some are brightly coloured but many are cryptic and, when at rest, bear a close resemblance to twigs.

682 Male March moth, *Alsophila aescularia*.

Alsophila aescularia (Denis & Schiffermüller)
March moth

An often common pest of ornamental trees and shrubs, including beech (*Fagus sylvatica*), birch (*Betula*), crab apple (*Malus*), hawthorn (*Crataegus*), hornbeam (*Carpinus betulus*), lilac (*Syringa vulgaris*), oak (*Quercus*), ornamental cherry (*Prunus*), privet (*Ligustrum vulgare*) and rose (*Rosa*). Eurasiatic. Present throughout most of Europe.

DESCRIPTION **Adult female:** wingless; body 8 mm long, shiny greyish-brown with a large anal tuft of hair. **Adult male:** 25–30 mm wingspan; fore wings pale grey to brownish-grey, with lighter crosslines; hind wings pale grey (Fig. **682**). **Egg:** dark brown, laid in a large batch encircling a twig. **Larva:** 25 mm long; pale green with a darker green dorsal line and yellowish lines along the sides, including one below the spiracles; body relatively narrow, with a vestigial pair of prolegs on the fifth abdominal segment (Fig. **683**) (cf. *Operophtera brumata*, pp. 277–278). **Pupa:** 10 mm long; brown and stumpy; cremaster with two curved, divergent spines.

683 Larva of March moth, *Alsophila aescularia*.

LIFE HISTORY Adults occur from mid-February to mid-April. The males, which are readily attracted to light, are commonly found at rest during the daytime on walls, fences and other surfaces; the flightless females, however, are rarely seen. Eggs, which are deposited in compact bands around the twigs of host plants, hatch in April. The larvae then feed on the foliage of various trees and shrubs, and on the blossoms of early-flowering species. The larvae often occur in company with those of *Operophtera brumata*; the latter, however, are usually more numerous. When fully grown, in late May or June, larvae enter the soil and each pupates in a silken cocoon.

DAMAGE Larvae may injure unopened buds but they cause most harm to the young foliage, often contributing to significant spring defoliation. They also damage the blossoms of flowering trees and shrubs, biting holes into the petals, and destroying the stamens and other sexual parts.

CONTROL Eggs bands discovered on the shoots should be destroyed. If required, apply an insecticide in the spring, preferably before the flowering stage.

Hemithea aestivaria (Hübner)
Common emerald moth

A widely distributed species, the larvae feeding during the autumn on low-growing plants and then hibernating. In the following spring, the larvae attack the foliage· of birch (*Betula*), hawthorn (*Crataegus*), lime (*Tilia*), oak (*Quercus*), rose (*Rosa*), sallow (*Salix*) and various other broad-leaved trees, sometimes causing minor damage to nursery plants and garden ornamentals. Fully grown larvae are *c.* 22 mm long, dull green marked with black and reddish-brown; the head and first thoracic segment are distinctly notched, and the body surface is noticeably roughened (Fig. **684**). The green, angular-winged adults (24–28 mm wingspan) (Fig. **685**) occur in late June and July.

684 Larva of common emerald moth, *Hemithea aestivaria.*

685 Common emerald moth, *Hemithea aestivaria.*

Xanthorhoe fluctuata (Linnaeus)
Garden carpet moth

A generally common species, the larvae feeding on the foliage of various Cruciferae from June onwards. They are sometimes found in small numbers on wallflower (*Cheiranthus cheiri*) but damage caused to cultivated plants is of no significance. Individuals (up to 32 mm long) are mainly yellowish-green to greyish-brown, with the body tapering gradually towards the small, pointed, black-marked head. Adults (26–29 mm wingspan) are mainly whitish to greyish, with darker markings on the fore wings. They occur most abundantly in May and June, and in August and September, resting openly during the daytime on walls, fences and other surfaces (Fig. **686**).

686 Garden carpet moth, *Xanthorhoe fluctuata.*

Chloroclysta truncata (Hufnagel)
Common marbled carpet moth

Minor infestations of this generally common and extremely varied species occur occasionally on herbaceous ornamentals such as *Geranium*, the larvae causing slight damage to the foliage; the larvae will also feed on seedling trees, including maple (*Acer*). Although most often reported on outdoor plants, minor attacks sometimes occur in glasshouses. The mainly black to brownish adults (32–35 mm wingspan) (Fig. **687**) occur in May and June, and in the autumn. The characteristic larvae (up to 35 mm long) are slender and greenish, with a darker dorsal line and a pair of yellowish subdorsal lines along the back; there is sometimes a complete or interrupted reddish stripe along the somewhat warty sides; the last body segment bears a pair of pointed projections (Fig. **688**). The larvae feed from September to April, hibernating during the winter months (but continuing to feed if in heated glasshouses); larvae of the second generation occur from late June to late July or early August, fully grown individuals pupating in withered leaves on the host plant or amongst debris on the ground.

687 Common marbled carpet moth, *Chloroclysta truncata*.

688 Larva of common marbled carpet moth, *Chloroclysta truncata*.

Hydriomena furcata (Thunberg)
syn. *sordidata* (Fabricius)
July highflier moth

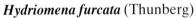

Widespread and generally common in and around open woodlands, and in association with hedges, the larvae occurring on hazel (*Corylus avellana*), poplar (*Populus*), sallow and willow (*Salix*). The larvae (up to 25 mm long) are stout-bodied, brownish to black, with partly whitish intersegmental rings and whitish lines along the back and sides (Fig. **689**). They feed between spun leaves in May and June, and are sometimes noted on ornamental or nursery trees and shrubs (especially hazel, sallow and willow); attacks have also occurred on conifers such as Sitka spruce (*Picea sitchensis*). The rather drab, greyish-green to brownish-yellow, round-winged adults (28–30 mm wingspan) (Fig. **690**) occur in July and August.

689 Larva of July highflier moth, *Hydriomena furcata*.

Epirrita dilutata (Denis & Schiffermüller)
November moth

A generally common woodland species and some-times a minor pest of trees and shrubs, including birch (*Betula*), crab apple (*Malus*), elm (*Ulmus*), hawthorn (*Crataegus*), maple (*Acer*) and oak (*Quercus*). The larvae occur in small numbers on cultivated plants, individuals grazing on the foliage from April to June, and contributing to damage caused by various other spring-feeding species. Fully grown larvae are 30 mm long and rather plump, bright velvety-green, greenish-white below, with small, pale pinacula, reddish-brown spiracles, a creamish subspiracular line and faint whitish lines down the back (Fig. **691**); the body is sometimes marked with purplish-red, the latter often forming a series of diamond-shaped markings along the back. Adults (32–38 mm wingspan) are dull greyish to black, with wavy crosslines (Fig. **692**). They occur mainly in October and November.

690 July highflier moth, *Hydriomena furcata*.

691 Larva of November moth, *Epirrita dilutata.*

Operophtera brumata (Linnaeus)
Winter moth

A generally common and often destructive pest of ornamental trees and shrubs, including *Cotoneaster*, crab apple (*Malus*), dogwood (*Cornus*), elm (*Ulmus*), hawthorn (*Crataegus*), hazel (*Corylus avellana*), hornbeam (*Carpinus betulus*), lilac (*Syringa vulgaris*), lime (*Tilia*), ornamental cherry (*Prunus*), ornamental pear (*Pyrus*), *Rhododendron*, rose (*Rosa*), spruce (*Picea*), sycamore (*Acer pseudoplatanus*) and willow (*Salix*); also an important pest of deciduous forest trees and fruit trees. Widely distributed in central and northern Europe.

692 November moth, *Epirrita dilutata.*

DESCRIPTION **Adult female:** wings reduced to stubs; body 5–6mm long, dark brown, mottled with greyish-yellow (Fig. **693**). **Adult male:** 22–28mm wingspan; fore wings pale greyish-brown with darker wavy crosslines; hind wings pale greyish (Fig. **694**). **Egg:** 0.5 × 0.4mm; oval, with a pitted surface; pale yellowish-green when newly laid but soon becoming orange-red. **Larva:** 25mm long; rather plump, pale green with a dark green dorsal stripe and several whitish or creamish-yellow stripes along the back and sides, including a pale yellow line passing through the spiracles (cf. *Alsophila aescularia*, p.000) (Fig. **695**). **Pupa:** 7–8mm long; brown and stumpy; cremaster bearing a pair of laterally directed spines.

LIFE HISTORY Adults occur from October to January but are most abundant in November and December. The males are active at night and are strongly attracted to light. They often rest openly on walls and fences during the daytime but the spider-like females secrete themselves on tree trunks and are less often seen. Mating takes place at night, copulating pairs occurring on the trees with the males characteristically standing head downwards, with the wings held outwards like a settled butterfly. Eggs, about 100–200 per female, are deposited singly in crevices in the bark. They hatch in late March or April and the newly emerged larvae then attack the buds, blossoms and expanding leaves. At this stage, the minute larvae are also blown about in the wind, each on a silken thread; they are often carried into previously uninfested nurseries and gardens from adjacent woodland trees and hedges. Feeding continues until late May or early June, the rather sluggish larvae spinning two leaves loosely together with silk, or feeding within the blossom trusses, sheltered by the overlying petals or calyxes. Fully grown larvae drop to the ground and enter the soil to pupate in flimsy cocoons about 8–10cm below the surface.

DAMAGE Larvae cause considerable defoliation, often completely stripping the leaves from the branches of heavily infested host plants. Attacks on the buds and blossoms of trees and shrubs are also of importance and, in severe cases, complete trusses may be destroyed.

CONTROL Proprietary grease bands, placed round the trunks of susceptible trees in October, can be used to prevent egg-laying females ascending into the branches. If required, apply an insecticide in the spring, ideally before the flowering stage.

693 Female winter moth, *Operophtera brumata*.

694 Male winter moth, *Operophtera brumata*.

695 Larva of winter moth, *Operophtera brumata*.

Operophtera fagata (Scharfenberg)

syn. *boreata* (Hübner)

Northern winter moth

Adults of this widely distributed species occur in October and November. They are similar in appearance and habits to those of *Operophtera brumata* (p. 278) but the females (Fig. **696**) have slightly longer wings and the males are lighter in colour. The larvae (up to 21 mm long) are mainly green with greyish-white lines along the back and sides, black spiracles and a black head (Fig. **697**). They feed on the foliage of host plants in May and June and, if numerous, can cause noticeable defoliation. Attacks are most frequently established on birch (*Betula*) and beech (*Fagus sylvatica*) but may also develop on other broad leaved trees and shrubs; minor infestations are sometimes noted on ornamental trees and nursery stock.

696 Female northern winter moth, *Operophtera fagata*.

Abraxas grossulariata (Linnaeus)

Magpie moth

The characteristic creamish-white and black-spotted larvae of this generally distributed species are sometimes damaging to the foliage of ornamental trees and shrubs, including almond (*Prunus dulcis*), cherry (*Prunus*), crab apple (*Malus*), flowering currant (*Ribes sanguineum*), hawthorn (*Crataegus*), hazel (*Corylus avellana*) and spindle (*Euonymus*). The mainly black, white and yellow adults (35–40 mm wingspan) occur in July and August; larvae feed from August onwards, individuals hibernating during the winter months and completing their development in the following May or June. Although capable of causing noticeable defoliation, numbers of larvae on ornamental plants are usually small and damage is rarely, if ever, of significance.

697 Larva of northern winter moth, *Operophtera fagata*.

Lomaspilis marginata (Linnaeus)

Clouded border moth

A generally distributed and often common species, associated with aspen (*Populus tremula*), black poplar (*Populus nigra*), hazel (*Corylus avellana*), sallow and willow (*Salix*). The larvae feed on the foliage from June onwards. They often occur on cultivated plants but damage caused is unimportant. Fully fed individuals are 18–20 mm long, and yellowish-green, with paired dark green lines along

698 Larva of clouded border moth, *Lomaspilis marginata*.

699 Clouded border moth, *Lomaspilis marginata*.

the back and a purplish-brown blotch on the last abdominal segment; the head is green, marked with purplish-brown (Fig. **698**). Adults (22–24 mm wingspan) are mainly white to yellowish-white, with an irregular greyish-black border and, sometimes, a partial or complete median band on each wing (Fig. **699**). The moths are most numerous in May and June; in parts of continental Europe, there are two generations annually.

700 Larva of brimstone moth, *Opisthograptis luteolata*.

Opisthograptis luteolata (Linnaeus)
Brimstone moth

A generally common species, the larvae sometimes feeding on rosaceous ornamental trees and shrubs, including crab apple (*Malus*), hawthorn (*Crataegus*) and ornamental cherry (*Prunus*). They cause damage to the foliage during the spring, summer and autumn but are usually present only in small numbers. Individuals are 25 mm long, stout-bodied, green to brownish and twig-like, with four pairs of functional abdominal prolegs (Fig. **700**). The mainly yellow, orange-marked adults (32–35 mm wingspan) (Fig. **701**) occur from April to August but are most numerous in May and June.

701 Brimstone moth, *Opisthograptis luteolata*.

Apeira syringaria (Linnaeus)
Lilac beauty moth

Although not a significant pest, the unusual larvae of this species are sometimes noticed on cultivated elder (*Sambucus*), honeysuckle (*Lonicera*), lilac (*Syringa vulgaris*) and privet (*Ligustrum vulgare*). Individuals, which are most often found in May, are yellowish-brown, marked with red and purplish-red, with a slightly hairy dorsal surface bearing two small projections on the second and third abdominal segments and a distinctive, forked projection on the fourth abdominal segment. They feed briefly in the autumn before hibernating, completing their development in the following spring. Fully grown larvae then pupate on the host plant, usually in early June, each in a pendulous, silken cocoon to which the cast skin of the larva is also attached. The adults (35–40 mm wingspan) are tawny-yellow, marked with purplish-white and red (Fig. **702**); they appear in June and July, often with a partial second generation in the autumn.

702 Female lilac beauty moth, *Apeira syringaria*.

703 Larva of early thorn moth, *Selenia dentaria*.

Selenia dentaria (Fabricius)
syn. bilunaria (Esper)
Early thorn moth

A generally abundant species, associated with various trees and shrubs including alder (*Alnus*), birch (*Betula*), crab apple (*Malus*), ornamental cherry (*Prunus*) and sallow (*Salix*). The orange-brown or reddish-brown, twig-like larvae (35–40 mm long when fully grown) (Fig. **703**) often feed in small numbers on nursery stock and ornamentals; they occur from May to June and, less commonly, in August and September. The butterfly-like, whitish-yellow to yellowish-orange adults (32–40 mm wingspan) (Fig. **704**) appear in April and early May, with a small second generation in July and August.

704 Early thorn moth, *Selenia dentaria*.

Odontopera bidentata (Clerck)
Scalloped hazel moth

The larvae of this polyphagous, widely distributed species occur from June to October on various trees and shrubs. They often attack cultivated plants, including beech (*Fagus sylvatica*), birch (*Betula*), hawthorn (*Crataegus*), oak (*Quercus*), willow (*Salix*) and conifers such as fir (*Abies*) and larch (*Larix*) but are usually present only in small numbers. Individuals (50 mm long) are dark green to greyish or purplish, with yellowish and pale brown lozenge-shaped marks down the back. When fully grown, they pupate in the soil in reddish-brown pupae, the mainly pale greyish-brown adults (38–42 mm wingspan) (Fig. **705**) appearing in the following April, May and June.

705 Scalloped hazel moth, *Odontopera bidentata*.

706 Larva of scalloped oak moth, *Crocallis elinguaria*.

Crocallis elinguaria (Linnaeus)
Scalloped oak moth

A polyphagous and generally distributed species, the larvae feeding from April to June on the leaves of various deciduous trees and shrubs, including blackthorn (*Prunus spinosa*), crab apple (*Malus*), honeysuckle (*Lonicera*) and ornamental cherry (*Prunus*); attacks sometimes occur on such plants in parks and gardens and also occur, occasionally, on nursery stock. The larvae are capable of causing noticeable defoliation, but are usually present only in small numbers so that damage caused is of little or no significance. The twig-like larvae are up to 45 mm long, greyish-yellow to greyish-black, tinged with purple, with a dark diamond-like pattern along the back and a slight, blackish-edged elevation on the eighth abdominal segment (Fig. **706**). Fully fed individuals pupate in the soil, the mainly whitish-yellow adults (32–35 mm wingspan) (Fig. **707**) occurring in July and August.

707 Scalloped oak moth, *Crocallis elinguaria*.

708 Larva of swallow-tailed moth, *Ourapteryx sambucaria.*

709 Swallow-tailed moth, *Ourapteryx sambucaria.*

Ourapteryx sambucaria (Linnaeus)
Swallow-tailed moth

A generally common species, the elongate (up to 60mm long), greyish-brown, purplish-marked, twig-like larvae (Fig. **708**) feeding on the foliage of various trees and shrubs, including elder (*Sambucus*), hawthorn (*Crataegus*), ivy (*Hedera helix*) and *Prunus*. Although sometimes infesting ornamentals, the larvae are rarely numerous but older individuals feeding in the spring can attract attention because of their size. The mainly pale yellow, butterfly-like adults (55–60mm wingspan) (Fig. **709**) appear in July.

710 Larva of feathered thorn moth, *Colotois pennaria.*

Colotois pennaria (Linnaeus)
Feathered thorn moth

Adults of this common woodland insect occur in October and November, the females depositing oblong batches of 100–200 olive-green, smooth-shelled eggs along the shoots of various trees and shrubs, including ornamental trees and nursery stock. Suitable hosts include birch (*Betula*), crab apple (*Malus*), hawthorn (*Crataegus*), hornbeam (*Carpinus betulus*), larch (*Larix*), oak (*Quercus*), ornamental cherry (*Prunus*), poplar (*Populus*) and sallow (*Salix*). Eggs hatch in the following spring, the larvae then feeding from April to June and eventually pupating in the soil. Fully grown larvae are 40–45mm long, stout-bodied and purplish-grey to slate-grey, with faint, yellowish diamond-shaped marks down the back and similarly coloured spots along the sides; the ninth abdominal segment bears two red-tipped projections (Fig. **710**). Adults (40 mm wingspan) are mainly pale orange to pale yellowish; males have strongly bipectinated antennae (Fig. **711**).

711 Male feathered thorn moth, *Colotois pennaria.*

712

712 Larva of pale brindled beauty moth, *Apocheima pilosaria*.

713 Male pale brindled beauty moth, *Apocheima pilosaria*.

Apocheima pilosaria (Denis & Schiffermüller)
syn. *pedaria* (Fabricius)
Pale brindled beauty moth
A generally distributed species, the greyish-brown to reddish-brown, twig-like (up to 40mm long) larvae (Fig. **712**) feeding in the spring on the foliage of various trees and shrubs, including birch (*Betula*), crab apple (*Malus*), elm (*Ulmus*), hawthorn (*Crataegus*), hornbeam (*Carpinus betulus*), larch (*Larix*), lime (*Tilia*), oak (*Quercus*), ornamental cherry (*Prunus*), poplar (*Populus*), rose (*Rosa*) and sallow (*Salix*). They sometimes attack such hosts in gardens, parks and nurseries but damage caused is slight. Adults occur mainly from January to March. Males (40–42mm wingspan) are grey, tinged with greenish or brownish and suffused with dark grey or brown (Fig. **713**); females (12mm long) are stout-bodied and virtually wingless (Fig. **714**).

714 Female pale brindled beauty moth, *Apocheima pilosaria*.

Lycia hirtaria (Clerck)
Brindled beauty moth
A widely distributed species, the larvae attacking the foliage of crab apple (*Malus*), elm (*Ulmus*), lime (*Tilia*), willow (*Salix*) and many other broad-leaved trees and shrubs, including ornamentals and nursery stock. Individuals are large (up to 55mm long), stout-bodied, purplish-grey to reddish-brown (the latter colour also forming wavy lines down the back), marked with dark speckles; there is also a yellow crossline just behind the head, and yellow spots on several of the abdominal segments (Fig. **715**). They feed from May to July, and then pupate

715 Larva of brindled beauty moth, *Lycia hirtaria*.

in the soil, the rather hairy, greyish to blackish, ochreous-marked adults (40–45 mm wingspan) (Fig. **716**) appearing in March and April.

Biston betularia (Linnaeus)
Peppered moth

Generally common on various trees and shrubs, including beech (*Fagus sylvatica*), birch (*Betula*), crab apple (*Malus*), elm (*Ulmus*), ornamental cherry (*Prunus*), larch (*Larix*) and rose (*Rosa*), and evergreens such as *Rhododendron* and spruce (*Abies*); larvae may also occur on herbaceous plants such as *Chrysanthemum* and pot marigold (*Calendula*). Eurasiatic. Widely distributed in Europe.

DESCRIPTION **Adult:** 42–55 mm wingspan; body and wings white, peppered with black (Fig. **717**); entirely black (ab. *carbonaria*) and intermediate (ab. *insularia*) forms also occur; male antennae strongly bipectinated. **Egg:** 0.7 × 0.5 mm; whitish-green. **Larva:** 50 mm long; brown or green, with pinkish markings and reddish spiracles; body stick-like, with a pair of dark purplish prominences on the fifth abdominal segment; head purplish-brown with a distinct central cleft (Fig. **718**). **Pupa:** 20–22 mm long; blackish-brown, terminating in a spike.

LIFE HISTORY Moths fly in May, June and July, eventually depositing eggs on various host plants. Larvae occur from July to September or October; they feed on the foliage but usually rest during the daytime, mimicking a shoot or a broken twig, with the body held straight out at an angle of about 45 degrees. Fully grown larvae enter the soil to pupate, adults emerging in the following year.

DAMAGE Larvae can cause noticeable defoliation, especially if present on herbaceous plants. However, damage to trees and shrubs is usually unimportant since the bulk of feeding occurs relatively late in the season.

CONTROL Larvae found on herbaceous plants should be picked off and destroyed; if required, apply a contact insecticide.

Agriopis aurantiaria (Hübner)
Scarce umber moth

This widespread and locally common species often attacks ornamental trees and shrubs, including beech (*Fagus sylvatica*), birch (*Betula*), crab apple (*Malus*), hornbeam (*Carpinus betulus*) and oak (*Quercus*), the larvae causing minor damage to the

716 Male brindled beauty moth, *Lycia hirtaria*.

717 Male peppered moth, *Biston betularia*.

718 Larva of peppered moth, *Biston betularia*.

719 Larva of scarce umber moth, *Agriopis aurantiaria*.

720 Larva of dotted border moth, *Agriopis marginaria*.

foliage. Larvae are 30–35 mm long, greyish to yellowish or brownish, marked with purplish lines and blackish patches (Fig. **719**); they occur from April to late May or early June. Adults occur mainly in October and November. Males (35–40 mm wingspan) have pale golden-yellow fore wings, marked with purplish speckles and crosslines. In females the wings are reduced to greyish stubs (2–4 mm long); the body (*c.* 8 mm long) is yellowish-brown, mottled with black, and marked dorsally with numerous, yellow scales.

721 Male dotted border moth, *Agriopis marginaria*.

Agriopis marginaria (Fabricius)
Dotted border moth

A generally common species, the larvae frequently occurring on ornamental trees and shrubs such as alder (*Alnus*), birch (*Betula*), crab apple (*Malus*), hornbeam (*Carpinus betulus*), oak (*Quercus*) and sallow (*Salix*). They feed on the foliage from April to late May or early June and then pupate in the soil, adults appearing in the following March or April. Larvae are 30 mm long, rather slender, olive-green to brownish, with pale patches along the sides and, often, a series of blackish x-shaped markings along the back (Fig. **720**). Adult males (30–35 mm wingspan) are mainly yellowish-brown, the fore wings marked with dark crosslines and a characteristic row of black dots along the outer margin (Fig. **721**). Females (7–10 mm long) are greyish-brown, mottled with black; the wings are reduced to short stubs, the hind wings being the longer pair.

722 Female mottled umber moth, *Erannis defoliaria*.

Erannis defoliaria (Clerck)
Mottled umber moth

A generally common pest of ornamental trees and shrubs, including birch (*Betula*), cherry (*Prunus*), elm (*Ulmus*), hazel (*Corylus avellana*), honeysuckle (*Lonicera*), hornbeam (*Carpinus betulus*), lime (*Tilia*), oak (*Quercus*), poplar (*Populus*) and rose (*Rosa*); also a pest of deciduous forest trees and fruit trees. Widely distributed in Europe; also present in Canada.

DESCRIPTION **Adult female:** wingless; body 10–15 mm long, mottled with black, yellow and, sometimes, white scales (Fig. **722**). **Adult male:** 35–38 mm wingspan; fore wings pale yellow to reddish-brown, more-or-less finely peppered with black, and often variably decorated with dark cross-markings (Fig. **723**). **Egg:** 0.9×0.5 mm; oval, yellowish to greyish; shell almost smooth. **Larva:** 35 mm long; reddish-brown with yellow or creamish-white areas on the sides of the first to seventh abdominal segments (Fig. **724**). **Pupa:** 12–14 mm long; dark yellowish-brown; cremaster with a short bifid tip.

LIFE HISTORY Adults occur mainly from mid-October to mid-January. The colourful males are sometimes noticed at rest during the daytime on trees, fences and walls but the wingless females are rarely seen. Eggs, which are laid in bark crevices, hatch in early April. The larvae then feed ravenously, usually resting fully exposed on the leaves or shoots; they are easily dislodged from the foodplant, remaining temporarily suspended by a silken thread with the head and thorax bent back at an angle to the abdomen. In June, when fully fed, the larvae pupate in the soil a few centimetres below the surface.

DAMAGE Larvae cause considerable defoliation, often stripping the leaves from the branches of host plants. Attacks on the buds and blossoms of trees and shrubs are also of importance; in severe cases complete trusses may be destroyed.

CONTROL Proprietary grease bands, placed round the trunks of susceptible trees in October, can be used to prevent egg-laying females ascending into the branches. If required, apply an insecticide in the spring, ideally before the flowering stage.

Menophra abruptaria (Thunberg)
Waved umber moth

The mainly greyish-brown, stick-like larvae (40 mm long) (Fig. **725**) of this widely distributed, locally

723 Male mottled umber moth, *Erannis defoliaria.*

724 Larva of mottled umber moth, *Erannis defoliaria.*

725 Larva of waved umber moth, *Menophra abruptaria.*

726 Male waved umber moth, *Menophra abruptaria.*

727 Larva of willow beauty moth, *Peribatodes rhomboidaria.*

common species feed from May to August on lilac (*Syringa vulgaris*), privet (*Ligustrum vulgare*) and certain other garden ornamentals such as jasmine (*Jasminum*) and rose (*Rosa*). They cause minor, unimportant damage to the foliage, each eventually pupating in a shallow depression formed by the larva on a branch of the host plant, hidden within a camouflaged cocoon incorporating masticated fragments of bark. Adults (35–40mm wingspan) are pale to whitish-ochreous, suffused with brown (Fig. **726**). They occur in April and May.

Peribatodes rhomboidaria (Denis & Schiffermüller)
Willow beauty moth

A generally common species, associated with various trees and shrubs. The larvae sometimes occur on ornamentals and nursery stock, including birch (*Betula*), *Clematis*, hawthorn (*Crataegus*), lilac (*Syringa vulgaris*), privet (*Ligustrum vulgare*), rose (*Rosa*) and yew (*Taxus baccata*), but cause only slight damage. They feed in the late summer and again in the following spring, each eventually pupating in a strong, silken cocoon formed on a twig or small branch of the foodplant. Individuals (up to 35mm long) are reddish-brown, mottled with ochreous, and with faint diamond-shaped markings along the back (Fig. **727**). The greyish to brownish-grey adults (40mm wingspan) (Fig. **728**) appear in July and August.

Deileptenia ribeata (Clerck)
syn. *abietaria* (Denis & Schiffermüller)
Satin carpet moth

A widely distributed species, associated with various trees but most often encountered on fir (*Abies*),

728 Male willow beauty moth, *Peribatodes rhomboidaria.*

729 Larva of satin carpet moth, *Deileptenia ribeata.*

730 Male satin carpet moth, *Deileptenia ribeata*.

731 Larva of mottled beauty moth, *Alcis repandata*.

larch (*Larix*), spruce (*Picea*) and yew (*Taxus bac-cata*). The greyish-brown larvae (Fig. **729**) sometimes cause slight damage to ornamentals but they are usually present in only small numbers. They feed from August onwards, hibernating during the winter and completing their development in the following June. Adults (40–48 mm wingspan) are pale yellowish-grey to black (Fig. **730**); they fly from late June to early August.

Alcis repandata (Linnaeus)
Mottled beauty moth

An often common but minor pest of trees and shrubs, including birch (*Betula*), elm (*Ulmus*), hawthorn (*Crataegus*), hazel (*Corylus avellana*) and sycamore (*Acer pseudoplatanus*). The larvae are elongate (up to 40 mm long) and mainly brown (Fig. **731**). They feed throughout the summer, completing their development in the following spring. Specimens are sometimes found on ornamental trees in parks, gardens and nurseries but they cause only slight damage. The greyish-white to dark yellowish-grey adults (40–44 mm wingspan) (Fig. **732**) occur from June to July. In some seasons, there may be a partial second generation.

732 Male mottled beauty moth, *Alcis repandata*.

Ectropis bistortata (Goeze)
Engrailed moth

A generally common, double-brooded species, the pale greenish-grey, shoot-like larvae (up to 30 mm long) (Fig. **733**) feeding on various trees and shrubs,

733 Larva of engrailed moth, *Ectropis bistortata*.

including beech (*Fagus sylvatica*), birch (*Betula*) and oak (*Quercus*). They occur from May onwards, sometimes causing minor damage to the foliage of ornamentals, but are not important pests. The pale yellowish-grey (38–40mm wingspan) adults (Fig. **734**) occur in the early spring and again in mid-summer.

734 Engrailed moth, *Ectropis bistortata*.

Bupalus piniaria (Linnaeus)
Bordered white moth

Generally common in the vicinity of coniferous woodlands, especially on Scots pine (*Pinus sylvestris*); also occurs on Douglas fir (*Pseudotsuga menziesii*), larch (*Larix*), Norway spruce (*Picea abies*) and silver fir (*Abies alba*). A notorious forestry pest; attacks on coniferous ornamentals and nursery stock are less frequent and usually of little or no importance. Widespread in Europe, except in the extreme north.

DESCRIPTION **Adult female:** 32–35mm wingspan; wings mainly orange or orange-yellow, suffused with brown; antennae thread-like (Fig. **735**). **Adult male:** 32–35mm wingspan; wings whitish-yellow to yellowish, bordered and variably suffused with brownish-black; antennae strongly bipectinated. **Egg:** 0.5mm long; light green. **Larva:** 25–30mm long; bright green, with white or yellowish-white, longitudinal stripes (Fig. **736**). **Pupa:** 10–15mm long; shiny brown.

LIFE HISTORY Larvae feed from June or July to October or November. They occur mainly on mature conifers but are sometimes found on small trees, especially those growing in the vicinity of more normal hosts. Fully grown larvae pupate in the soil at the base of host trees, adult moths appearing in the following May and June.

DAMAGE Larvae cause extensive defoliation, such depredations being especially severe on sandy-soil sites in low-rainfall areas but attacks on ornamentals are seldom, if ever, of significance.

Theria primaria Haworth
Early moth

A generally common species, the greenish to greenish-brown, white-striped larvae (*c.* 20mm long) (Fig. **737**) feeding on the leaves of hawthorn (*Crataegus*) throughout April and May. Various kinds of *Prunus*, especially blackthorn (*Prunus spinosa*), are also attacked; although associated

735 Female bordered white moth, *Bupalus piniaria*.

736 Larva of bordered white moth, *Bupalus piniaria*.

737 Larva of early moth, *Theria primaria*.

738 Female early moth, *Theria primaria*.

mainly with hedgerows, minor infestations are sometimes noted on nursery stock. The moths appear in January and February. Males (30–32 mm wingspan) are mainly pale greyish-brown; the virtually wingless females are grey, with their rather angular wing stubs each marked with a dark crossband (Fig. **738**).

Campaea margaritata (Linnaeus)
Light emerald moth

Larvae of this widely distributed woodland species feed mainly on birch (*Betula*), crab apple (*Malus*), elm (*Ulmus*), hornbeam (*Carpinus betulus*) and oak (*Quercus*); they also occur in small numbers on ornamental trees in gardens and nurseries. The delicate, pale green moths (38–45mm wingspan) (Fig. **739**) appear in June and July, depositing eggs in batches on the underside of leaves. The larvae feed from September onwards, lying stretched out flat against the twigs when at rest. Unlike most other species, they continue to feed during the winter, grazing the bark off the young shoots to expose the pale wood and also attacking the dormant buds. In the following spring, when about 20 mm long, they feed on the expanding buds and unfurled leaves, most individuals pupating by the end of May. Fully grown larvae are 30–35mm long, greyish to greenish-brown or purplish-brown, with orange, brown rimmed spiracles; they possess a characteristic skirt-like fringe of pale bristles along either side of the body and, unlike most geometrids, three pairs of functional prolegs (Fig. **740**); younger larvae are generally greyish, often with T-shaped markings along the back.

739 Light emerald moth, *Campaea margaritata*.

740 Larva of light emerald moth, *Campaea margaritata*.

24. Family SPHINGIDAE (hawk moths)

Large-bodied, strong-flying moths with elongate fore wings. The larvae are stout-bodied, with a characteristic horn arising from the eighth abdominal segment.

Sphinx ligustri Linnaeus
Privet hawk moth

A minor pest of privet (*Ligustrum vulgare*), including garden hedges; also associated with ash (*Fraxinus excelsior*), lilac (*Syringa vulgaris*) and, in continental Europe, certain other hosts such as elder (*Sambucus*) and *Viburnum*. Palearctic. Widely distributed in Europe.

DESCRIPTION **Adult:** 100–120 mm wingspan; fore wings mainly greyish-brown to blackish-brown, with black markings, and pinkish tinged basally; hind wings pale pinkish-white, with broad, blackish bands; thorax black, with whitish tegulae; abdomen yellowish-brown, with pink and black crossbands (Fig. **741**). **Larva:** 75–80 mm long; mainly green, with seven oblique white, purple-edged stripes on each side; caudal horn black above and yellow below.

LIFE HISTORY Moths occur in June and July, depositing eggs singly on the leaves and stems of host plants. The larvae feed on the foliage during July and August, often resting on the shoots during the daytime, several specimens sometimes occurring close together on the same plant. Fully grown individuals enter the soil to pupate, burrowing down well below the surface before excavating a suitable chamber. Although adults normally emerge in the following summer, the pupal stage sometimes extends over two winters.

DAMAGE Larvae rapidly defoliate shoots but, as the number of individuals present is usually small, damage caused is of no importance.

CONTROL If necessary, gently remove larvae by hand and transfer them to a suitable wild host so they may continue their development.

741 Privet hawk moth, *Sphinx ligustri.*

Mimas tiliae (Linnaeus)
Lime hawk moth

A widely distributed species, the larvae feeding mainly on lime (*Tilia*); especially in continental Europe, alder (*Alnus*), ash (*Fraxinus excelsior*), birch (*Betula*), elm (*Ulmus*), oak (*Quercus*) and walnut (*Juglans*) are also attacked. The larvae feed from June or July onwards; they often occur on garden and amenity trees but do not cause significant damage. Individuals (50–60 mm long when fully grown) are green, with seven oblique yellow stripes along each side and red spiracles; the caudal horn is blue above but red and yellow below (Fig. **742**). The larvae are most often noticed in August or September when they have completed feeding and are about to enter the soil to pupate, individuals at this stage becoming tinged with purple or purplish-brown. The adult moths (60–75 mm wingspan) are greyish-ochreous to pinkish-ochreous, suffused greenish-grey, the fore wings usually possessing a dark olive-green median fascia (Fig. **743**). They occur in May and June.

742 Larva of lime hawk moth, *Mimas tiliae*.

743 Lime hawk moth, *Mimas tiliae*.

744 Larva of eyed hawk moth, *Smerinthus ocellata*.

Smerinthus ocellata (Linnaeus)
Eyed hawk moth

Generally distributed and locally common, the larvae feeding on crab apple (*Malus*), sallow and willow (*Salix*) and often associated with nursery stock and young trees in gardens. They feed during July and August, rapidly defoliating the branches. Larvae (60–70 mm long when fully grown) are mainly green, with seven oblique white stripes along the sides and a bluish caudal horn (Fig. **744**). The adults (75–85 mm wingspan), which occur from May to July, have mainly greyish-brown to brown fore wings; the hind wings are brown, suffused with red, each with a large, eye-like mark (Fig. **745**).

Laothoe populi (Linnaeus)
Poplar hawk moth

Widespread and generally common on poplar (*Populus*), sallow and willow (*Salix*); in mainland Europe also associated with ash (*Fraxinus excelsior*), birch (*Betula*) and crab apple (*Malus*). The larvae feed on the foliage from July to September; they are often noticed on cultivated plants, especially at the final stages of larval development, but are usually present only in small numbers. Individuals

745 Eyed hawk moth, *Smerinthus ocellata*.

(70 mm long when fully grown) are green or bluish-green, densely speckled with yellow, with seven lateral pairs of oblique yellow stripes, the last extending into the yellowish-green, often red-tipped, caudal horn (Fig. 746). Adults (70–85 mm wingspan) are mainly greyish-brown, suffused with pinkish-brown, with a rusty-red patch on the base of each hind wing (Fig. 747). They fly in May and June, a partial second generation sometimes appearing in the autumn.

746 Larva of poplar hawk moth, *Laothoe populi*.

Deilephila elpenor (Linnaeus)
Elephant hawk moth

Generally common on wild bedstraw (*Galium*) and willowherb (*Epilobium*) but also a minor pest of cultivated plants such as busy lizzie (*Impatiens sultani*) and *Fuchsia*, both out-of-doors and in glasshouses. Palearctic. Present throughout Europe, except for the extreme north.

DESCRIPTION **Adult:** 60–70 mm wingspan; body olive-brown, marked with pink; fore wings olive-brown with a pinkish-grey subterminal line and costa; hind wings bright pink, basally black (Fig. 748). **Larva:** 75–85 mm long; brown or green, the abdominal segments speckled with black; second and third abdominal segments distinctly swollen and each marked with a pair of lilac-centred, black, eye-like patches; head and thoracic segments retractible; caudal horn relatively short. **Pupa:** 40–45 mm long; brown, speckled with darker brown.

747 Poplar hawk moth, *Laothoe populi*.

LIFE HISTORY Moths occur mainly in June, but sometimes also later in the year, depositing eggs singly on the leaves of host plants. Larvae feed during July and August; they may also occur in the autumn. Individuals often bask in sunshine, if disturbed immediately retracting the head and thoracic segments and dilating the anterior abdominal segments to display the eye-like markings. Fully grown larvae pupate in fragile, silken cocoons formed on or just below the surface of the ground.

DAMAGE Larvae cause considerable defoliation, especially in their later instars, but are usually present in only small numbers.

748 Elephant hawk moth, *Deilephila elpenor*.

CONTROL As for *Sphinx ligustri* (p. 292). Application of an insecticide is rarely, if ever, justified.

25. Family NOTODONTIDAE

Medium-sized to large, plump-bodied, often downy moths. The larvae are variable in appearance, often with the anal prolegs modified into a pair of raised filaments; crochets on the abdominal prolegs are of two sizes (biordinal).

749 Buff-tip moth, *Phalera bucephala*.

Phalera bucephala (Linnaeus)
Buff-tip moth

A generally common pest of broad-leaved trees and shrubs, including beech (*Fagus sylvatica*), birch (*Betula*), elm (*Ulmus*), hazel (*Corylus avellana*), hornbeam (*Carpinus betulus*), lime (*Tilia*), oak (*Quercus*), ornamental cherry (*Prunus*), rose (*Rosa*), sallow (*Salix*), sweet chestnut (*Castanea sativa*) and *Viburnum*; often present on amenity trees, ornamental and nursery stock. Widely distributed in Europe.

DESCRIPTION **Adult:** 50–60 mm wingspan; fore wings ash-grey to silver-grey, with dark brown and reddish-brown markings and a large, pale yellow apical blotch; hind wings pale yellow to whitish (Fig. 749). **Egg:** 1 mm across; hemispherical; pale bluish-white above, with a dark central spot; green below. **Larva:** 50–55 mm long; downy and yellow, with several incomplete, black, longitudinal lines and orange intersegmental crosslines; head black (Fig. 750). **Pupa:** 25–28 mm long; dark purplish-brown; cremaster with two pairs of short spines.

750 Larva of buff-tip moth, *Phalera bucephala*.

LIFE HISTORY Moths fly from late May to July or early August, depositing eggs in batches of about 50 on the underside of leaves. Eggs hatch in about two weeks. The larvae feed at first on the underside of leaves, grazing away the lower epidermis; later they devour complete leaves, feeding gregariously (Fig. 751) until the final stages of their development. They are fully grown by the autumn, and then enter the soil to pupate in earthen chambers where they overwinter.

DAMAGE The damaged portion of leaves grazed from below by the youngest larvae turns brown, visible from above. Older larvae cause extensive damage, rapidly defoliating the shoots and branches; attacks on young trees and shrubs are of particular importance.

751 Group of young buff-tip larvae on *Betula*.

CONTROL Leaves with egg batches or young larvae should be picked off and destroyed; if necessary, apply a contact insecticide before damage becomes extensive.

Cerura vinula (Linnaeus)
Puss moth

A generally common pest of aspen (*Populus tremula*), poplar (*Populus*), sallow and willow (*Salix*), and often associated with trees in gardens, parks and nurseries; in mainland Europe also found occasionally on birch (*Betula*). Eurasiatic. Widely distributed in Europe.

DESCRIPTION **Adult:** 65–80 mm wingspan; body fluffy, the thorax yellowish-grey with black spots and the abdomen greyish-white with black crossbars; wings greyish-white, marked with greyish-black, the veins yellowish, edged with black (Fig. **752**). **Egg:** 1.8 mm across; hemispherical and reddish-purple. **Larva:** 65–70 mm long; bright green with a broad, purplish, white-edged (sometimes yellowish-edged) dorsal stripe, widest and often reaching to the prolegs on the fourth abdominal segment; third thoracic segment humped; anal prolegs modified into a pair of fang-like appendages, each enclosing a red, extendible filament; head brownish, marked with purplish or black (Fig. **753**); young larva reddish-brown with a pair of broad, spinose projections on the first thoracic segment (Fig. **754**). **Pupa:** 25–30 mm long; black and stumpy.

LIFE HISTORY Adults occur from May to July. Eggs are laid singly, but more frequently in twos or threes, on the upper surface of aspen, poplar or willow leaves. They hatch in about ten days. The larvae occur from June or July onwards, often feeding in pairs. There are five larval instars, the distinctive thoracic projections of the first-instar larva persisting to the third instar (Fig. **755**) but then diminishing (fourth instar) and finally disappearing. The older larvae rest fully exposed on the shoots but, in spite of their size, are easily overlooked. The presence of such larvae is often discovered only following a close examination of branches immediately above large pellets of frass which have accumulated on the ground beneath infested trees. Larvae complete their development in August or early September. They then wander away from the branches onto tree trunks or nearby fence posts where they construct tough, brown cocoons of silk and masticated wood. Individuals then pupate, adults emerging in the following summer.

DAMAGE Larvae cause considerable defoliation but, except on small trees, damage is unimportant.

CONTROL Hand-picking of larvae should be adequate, specimens ideally being transferred to a suitable wild host and allowed to continue their development.

752 Puss moth, *Cerura vinula*.

753 Larva of puss moth, *Cerura vinula*.

754 First-instar larva of puss moth, *Cerura vinula*.

Furcula furcula (Clerck)
Sallow kitten moth

Relatively common on sallow (*Salix*) but of minor importance; sometimes also found on aspen (*Populus tremula*), poplar (*Populus*) and, in continental Europe, birch (*Betula*). Eurasiatic; also present in North America. Widely distributed in Europe.

DESCRIPTION **Adult:** 34–38 mm wingspan; fore wings pale greyish-white marked with pale greyish, the markings partly edged with black and orange-yellow, and with blackish dots along the outer margin between the veins; hind wings whitish to greyish-white, with blackish dots around the margin between the veins (Fig. **756**). **Larva:** 35 mm long; pale green flecked with yellow, and with a broad, greyish-pink to purplish, yellow-edged dorsal stripe widening to become saddle-like on the fourth abdominal segment; anal prolegs modified into eversible filamental appendages; head purplish-brown (Fig. **757**).

LIFE HISTORY Adults emerge in late May and early June, depositing eggs singly or in twos or threes on the upper surface of sallow or willow leaves. Larvae feed from June onwards. When fully grown, each forms a tough, relatively flat cocoon in a crevice or a hollow in the bark of the foodplant. The cocoons are constructed of silk and masticated wood, and are difficult to distinguish from the surrounding plant tissue, remaining intact until adults emerge in the following spring. In favourable districts a second generation of adults emerges in August, larvae then feeding in the late summer and autumn.

DAMAGE Larvae cause slight defoliation but infestations are of no significance.

755 Third-instar larva of puss moth, *Cerura vinula*.

756 Sallow kitten moth, *Furcula furcula*.

Drymonia ruficornis (Hufnagel)
Lunar marbled brown moth

Larvae of this widely distributed species feed on the leaves of oak (*Quercus*) from June to August; they are found occasionally on specimen trees in parks and gardens, especially in southerly areas, but cause only slight damage. Fully fed individuals

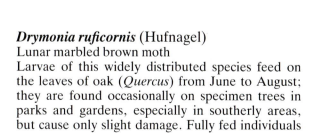

757 Larva of sallow kitten moth, *Furcula furcula*.

758 Larva of lunar marbled brown moth, *Drymonia ruficornis*.

759 Male lunar marbled brown moth, *Drymonia ruficornis*.

(*c.* 40mm long) are bluish-green to whitish-green, marked with two thin, yellow lines down the back and a wider spiracular line along each side (Fig. **758**). Adults (37–40mm wingspan) are whitish, marked with grey and greyish-black (Fig. **759**). They are most numerous in May.

26. Family **DILOBIDAE**

A small group of medium-sized moths. The larvae have dorsal setae arising from raised verrucae.

760 Larva of figure of eight moth, *Diloba caeruleocephala*.

Diloba caeruleocephala (Linnaeus)
Figure of eight moth
Widely distributed and common, the larvae occasionally infesting ornamental trees and shrubs such as *Cotoneaster*, crab apple (*Malus*), hawthorn (*Crataegus*) and *Sorbus*. They feed from spring to June or early July, causing minor damage to the leaves. Although several individuals may occur on one and the same host, damage is rarely of significance. Larvae (35mm long when fully grown) are greyish-blue with an incomplete yellow dorsal stripe and a yellow stripe along each side, and prominent black verrucae, each bearing a short, black spine (Fig. **760**). The adults (30–35mm wingspan) have brownish-grey fore wings, each with a large, pale greenish-yellow, irregular 'figure of eight' mark (Fig. **761**); they occur in October and November, depositing eggs in small groups on the trunks and branches of host plants.

761 Figure of eight moth, *Diloba caeruleocephala*.

27. Family LYMANTRIIDAE

Medium-sized moths with hairy bodies, the males with strongly bipectinated antennae. The larvae are hairy, with a pair of eversible dorsal glands on the abdomen or with brush-like tufts of hairs arising from the first to fourth abdominal segments.

Orgyia antiqua (Linnaeus)
Vapourer moth

An often abundant pest of trees and shrubs in nurseries, parks and gardens, feeding indiscriminately on various ornamentals; most important damage is likely on younger plants, especially buddleia (*Buddleja*), *Camellia*, *Ceanothus*, crab apple (*Malus*), firethorn (*Pyracantha*), heather (*Erica*), ornamental cherry (*Prunus*), *Rhododendron* and rose (*Rosa*); infestations also occur on conifers, including Douglas fir (*Pseudotsuga menziesii*), fir (*Abies*), larch (*Larix*) and pine (*Pinus*). Holarctic. Widespread in Europe.

DESCRIPTION **Adult female:** 10–15 mm long and virtually wingless; body dark yellowish-grey, fat and sack-like (Fig. **762**). **Adult male:** 25–33 mm wingspan; wings ochreous-brown or chestnut-brown; fore wings with darker markings and with a large white spot near the hind angle (Fig. **763**). **Egg:** 0.9 mm across; rounded, brownish-grey to reddish-grey, with a central spot and a dark rim-like band (Fig. **762**). **Larva:** 25–35 mm long; greyish or violet, with various red, black and yellow markings; body very hairy, including four brush-like tufts of yellow or greyish hairs on the back, long blackish pencil-like tufts near the head and similar brownish tufts near the tail (Fig. **764**). **Pupa:** shiny brownish-black and rather hairy.

LIFE HISTORY Adults occur from July to September or October, the males flying in sunshine in search of newly emerged, unmated females. The latter are sluggish and, on emerging from the pupa, each remains alongside the pupal cocoon upon which, after mating, about 100–300 eggs are laid in a conspicuous batch. The eggs hatch intermittently in the following spring. The larvae then wander away to feed on the foliage of the same or nearby host plants. They sometimes rest together on flowers or foliage whilst undergoing the change from one instar to the next. Fully grown individuals eventually pupate in silken cocoons, incorporating body hairs, spun on a twig, branch or trunk of the host plant or on a suitable support such as a nearby fence or wall, adults emerging shortly afterwards. In favourable conditions there is a second generation in the autumn.

762 Female vapourer moth, *Orgyia antiqua*, depositing eggs on pupal cocoon.

763 Male vapourer moth, *Orgyia antiqua*.

764 Larvae of vapourer moth, *Orgyia antiqua*.

DAMAGE Larvae can cause noticeable defoliation and will also damage buds and flowers but infestations are rarely significant.

CONTROL Egg batches discovered during the winter should be destroyed. If necessary, apply an insecticide to kill the young larvae.

Calliteara pudibunda (Linnaeus)
Pale tussock moth

The colourful, mainly green or yellow, very hairy larvae (up to 50 mm long) (Fig. **765**) of this widely distributed species are associated mainly with hop (*Humulus lupulus*) but will also infest ornamental trees and shrubs, including beech (*Fagus sylvatica*), birch (*Betula*), cherry (*Prunus*), elm (*Ulmus*), hazel (*Corylus avellana*), hornbeam (*Carpinus betulus*), oak (*Quercus*), poplar (*Populus*), pussy willow (*Salix caprea*) and walnut (*Juglans*). They feed from July to October but are of no importance. The mainly grey adults (45–55 mm wingspan) (Fig. **766**) occur in May and June.

Euproctis chrysorrhoea (Linnaeus)
syn. *phaeorrhoeus* (Haworth)
Brown-tail moth

A locally important pest of blackthorn (*Prunus spinosa*), hawthorn (*Crataegus monogyna*) and sea-buckthorn (*Hippophaë rhamnoides*); in some years attacks also spread to trees and shrubs such as ash (*Fraxinus excelsior*), crab apple (*Malus*), dogwood (*Cornus*), elder (*Sambucus*), false acacia (*Robinia pseudoacacia*), lilac (*Syringa vulgaris*), ornamental cherry (*Prunus*), ornamental pear (*Pyrus*), privet (*Ligustrum vulgare*), sallow and willow (*Salix*). Palearctic; also introduced into North America. Widely distributed in central and southern Europe but more restricted in northern Europe.

DESCRIPTION **Adult:** 30–38 mm wingspan; wings white, but fore wings of male sometimes with a few black dots; head and thorax white and fluffy; abdomen dark brown, with a large anal tuft of hair. **Larva:** 35–40 mm long; blackish-grey with tufts of gingery-brown hairs arising from brownish verrucae, two rows of bright red marks down the back, a series of downy white patches towards each side and bright, orange-red glands on the sixth and seventh abdominal segments; head black (Fig. **767**). **Pupa:** 15–18 mm long; brownish-black and hairy.

LIFE HISTORY Adults occur in July and August, eggs being laid in elongate batches on leaves or stems and then covered with brown hairs from the female's anal tuft. The eggs hatch from mid-August

765 Larva of pale tussock moth, *Calliteara pudibunda*.

766 Male pale tussock moth, *Calliteara pudibunda*.

767 Larva of brown-tail moth, *Euproctis chrysorrhoea*.

to early September, the larvae then constructing a strong, silken, communal retreat in which they shelter during inclement weather. Larvae feed on the foliage in decreasing numbers until the end of October and then hibernate. Activity is resumed in the following April, the larvae appearing on the outside of the webbing in increasing numbers, to bask in the spring sunshine, but little or no feeding occurs until May. Young foliage is then devoured ravenously. As the larvae grow, they wander further and further from their communal tent, spinning additional, less substantial webs and establishing trails of silk along the branches. In the later stages of the development, when about 25 mm long, they often become solitary and may wander away to feed elsewhere. Individuals are fully grown by late June. They then spin silken cocoons between the leaves, either singly or in groups, and pupate, adults emerging about two weeks later.

DAMAGE The larvae are voracious feeders, especially in their later instars, and often cause considerable defoliation of roadside hedges; their urticating hairs can also constitute a public nuisance, sometimes requiring local authorities to conduct eradication campaigns. On ornamentals, feeding damage may be of considerable significance; the larval tents are also disfiguring.

CONTROL Where necessary, larval tents should be cut out during the winter months and burnt. Alternatively, apply an insecticide in about mid-September, when all eggs have hatched but before the protective tents become too dense; spraying in the spring (in about mid-May) will give some control but is less effective.

768 Female yellow-tail moth, *Euproctis similis*.

769 Larva of yellow-tail moth, *Euproctis similis*.

Euproctis similis (Fuessly)
syn. *chrysorrhoea* sensu Haworth
Yellow-tail moth
Locally common on various trees and shrubs, including beech (*Fagus sylvatica*), birch (*Betula*), crab apple (*Malus*), hawthorn (*Crataegus*), oak (*Quercus*), ornamental cherry (*Prunus*), rose (*Rosa*), sallow (*Salix*) and *Viburnum*. Minor infestations often occur on garden trees and nursery stock. Eurasiatic. Widely distributed in Europe.

DESCRIPTION **Adult female:** 40–45 mm wingspan; wings white; body with a bulbous, yellow tip (Fig. **768**). **Adult male:** 30–40 mm wingspan; wings white, with black markings on the trailing edge of each fore wing; antennae strongly bipectinated; body with a conspicuous yellow anal hair tuft. **Larva:** 35 mm long; velvet-black, prominently marked with white and with an interrupted bright red, black-centred dorsal stripe; sixth and seventh abdominal segments each with an eversible orange dorsal gland; body hairs long and whitish; head black (Fig. **769**). **Pupa:** 15–16 mm long; dark brown, plump and slightly hairy.

LIFE HISTORY Adults occur in late July and August. Eggs are laid in batches on the twigs of host plants and then coated with hairs from the female's anal tuft. They hatch 7–10 days later. The larvae feed gregariously for a short while and then spin small (about 6 mm long), roughly oval, greyish cocoons under flakes of bark or in other sheltered positions. Individuals hibernate within these cocoons, becoming active again in the following spring. The larvae continue to feed on the foliage until about June. They then pupate in oval, greyish-brown silken cocoons incorporating numerous body hairs, adult moths emerging a few weeks later.

DAMAGE Larvae cause slight defoliation but numbers present on ornamentals are usually small.

Leucoma salicis (Linnaeus)
White satin moth

A local pest of poplar (*Populus*) and sallow (*Salix*); in mainland Europe other hosts include birch (*Betula*), crab apple (*Malus*) and snowy mespilus (*Amelanchier laevis*). Eurasiatic; introduced into North America. Widely distributed in Europe.

DESCRIPTION **Adult:** 45–55 mm wingspan; wings satin-white to creamish-white and thinly scaled; male distinctly smaller and with strongly bipectinated antennae (Fig. **770**). **Egg:** 0.65–0.85 mm diameter; light green. **Larva:** 40–45 mm long; hairy and mainly reddish-brown, with a bright red and distinctive white patch on each segment, and a blue subspiracular line; pinacula orange-red to orange; head blackish-grey to black.

LIFE HISTORY Moths appear in July or August. Eggs are laid in batches on the twigs of host plants and then coated with a white secretion incorporating hairs from the female's abdomen. Larvae emerge from late August onwards, feeding for a short time and then spinning small webs in which to hibernate. Activity is resumed in the following spring, usually in April. The larvae feed ravenously, often in groups, completing their development in late June or early July. They then pupate in silken cocoons spun in crevices in the bark of host trees, adults emerging a few weeks later.

DAMAGE Larvae destroy the foliage and, if numerous, can cause severe defoliation, which affects the growth and appearance of host trees.

CONTROL If practical, pick off and destroy the larvae or apply an insecticide.

Lymantria dispar (Linnaeus)
Gypsy moth

An important and often destructive pest of trees and shrubs, especially oak (*Quercus*), but also beech (*Fagus sylvatica*), hazel (*Corylus avellana*), hornbeam (*Carpinus betulus*) and many others. Palearctic. Widely distributed in mainland Europe, from mid-Sweden to the Mediterranean; extinct in Britain. An introduced pest in North America.

DESCRIPTION **Adult female:** 50–70 mm wingspan; wings mainly brownish-white, with irregular greyish to blackish cross-markings. **Adult male:** 35–50 mm wingspan: wings greyish-brown with blackish

770 Male white satin moth, *Leucoma salicis*.

markings; antennae strongly bipectinated. **Larva:** 40–75 mm long; grey to greyish-yellow, with darker markings; body with long, brown hairs arising from prominent verrucae; verrucae on the thoracic segments and first two abdominal segments blue, those on the third to eighth abdominal segments brownish-red.

LIFE HISTORY Adults occur from late July onwards, the flight period varying according to local conditions. The females do not fly but crawl about on the trunks and branches of host trees, sometimes also gliding to the ground from the tops of tall trees. Adult males are especially active at mid-day, flying about rapidly in their search for newly emerged females. After mating, each female deposits a large batch of eggs on the trunk or under a branch of a tree, the eggs being camouflaged by a spongy coating incorporating numerous body hairs. The eggs hatch in the following April or May, the young larvae eventually wandering away to feed on the bursting buds and leaves. Larvae develop through six instars, becoming fully grown in about 2–3 months. They then pupate on the host plant or elsewhere, adults emerging 2–3 weeks later.

DAMAGE The larvae are capable of causing considerable defoliation, infestations being especially severe on small trees and on those growing in light, sunny positions.

CONTROL In vulnerable areas, the likelihood of significant infestations developing on nursery trees will be lessened by avoiding oak and by introducing mixed plantings. Also, young trees should be planted as close together as possible, thereby reducing light intensity. Application of insecticides is rarely practical.

Lymantria monacha (Linnaeus)
Black arches moth
Polyphagous on broad-leaved trees and shrubs in mainland Europe, but associated mainly with conifers such as fir (*Abies*), pine (*Pinus*) and spruce (*Picea*) and often a serious forestry pest. Eurasiatic. Widespread in mainland Europe; in Britain restricted to southern England and parts of Wales, and associated mainly with oak (*Quercus*).

DESCRIPTION **Adult female:** 45–50 mm wingspan; body and fore wings mainly white, irregularly marked with black, the abdomen tinged with pink; hind wings mainly pale greyish-brown. **Adult male:** 37–42 mm wingspan; similarly coloured to female but antennae strongly bipectinated (Fig. **771**). **Larva:** 35 mm long; dark grey with an irregular brownish-black, black-edged dorsal stripe interrupted by a whitish, black-centred, mark on the third thoracic segment and by whitish, sometimes red-centred, patches on the fourth to sixth abdominal segments; pale hair tufts arise from pinacula along the dorsal and spiracular lines.

LIFE HISTORY Eggs are laid during the summer on the bark of host trees, either singly or in pairs, hatching in the following spring. The larvae feed on the foliage until June or early July. They then pupate in silken cocoons spun in bark crevices, adults emerging in late July or August.

DAMAGE Leaf damage is usually insignificant but severe infestations can cause considerable defoliation and affect the vigour of host trees.

CONTROL Populations are usually kept in check by naturally occurring parasites but, if necessary on young trees, apply an insecticide when damage is first seen.

771

771 Male dark arches moth, *Lymantria monacha*.

772

772 Garden tiger moth, *Arctia caja*.

28. Family **ARCTIIDAE** (ermine and tiger moths)

Medium-sized to large, often brightly coloured moths. The larvae are very hairy but the head is virtually hairless; the body hairs arise in tufts from distinctive verrucae.

Arctia caja (Linnaeus)
Garden tiger moth
A generally common but minor pest of herbaceous plants, especially in weedy situations; also sometimes associated with young trees and shrubs, including alder (*Alnus*), birch (*Betula*), hawthorn (*Crataegus*), hazel (*Corylus avellana*), heather (*Erica*), honeysuckle (*Lonicera*), lime (*Tilia*), oak (*Quercus*), ornamental cherry (*Prunus*), poplar (*Populus*), rowan (*Sorbus aucuparia*), sallow and willow (*Salix*). North Palearctic. Widespread in Europe.

DESCRIPTION **Adult:** 60–75 mm wingspan; fore wings chocolate-brown, irregularly marked with creamish-white; hind wings orange-red, with several blue blotches (Fig. **772**). **Egg:** 0.8 mm across; hemispherical, glossy, yellowish to green. **Larva:** 50–60 mm long; mainly blackish with a thick coat of long, gingery, often pale-tipped hairs; head black (Fig. 773). **Pupa:** 22–28 mm long; black to brownish-black; cremaster with several distorted bristles.

LIFE HISTORY Adults fly in July and August. Eggs are deposited in large batches on the underside of leaves of herbaceous plants, especially weeds such as dandelion (*Taraxacum*), dock (*Rumex*) and plantain (*Plantago*); they hatch from August onwards. The larvae (commonly known as 'woolly bears') feed during August and September and then hibernate, resuming activity in the following spring. They are then often found sunning themselves on the foliage of low plants or on nearby fences and walls. Larvae are fully grown by mid- or late June. Individuals pupate in yellowish cocoons spun amongst debris on the ground.

DAMAGE Foliage damage is indiscriminate and usually of no significance; however, on rare occasions larvae may be locally abundant and cause extensive defoliation.

CONTROL If necessary, apply an insecticide when larvae are first seen; infestations are unlikely to develop in weed-free sites.

Hyphantria cunea (Drury)

larva = fall webworm

An important North American pest, first noted in Europe in 1940 in Hungary; since reported in certain other countries, including Austria, France, Italy and Yugoslavia. The larvae attack various broad leaved trees and shrubs, including ash (*Fraxinus excelsior*), European hop-hornbeam (*Ostrya carpinifolia*), horse chestnut (*Aesculus hippocastanum*), lime (*Tilia*), rose (*Rosa*), *Sorbus*, southern nettle-tree (*Celtis australis*) and tree of heaven (*Ailanthus altissima*), but are especially common on maple (*Acer*), notably the so-called box elder (*Acer negundo*), and mulberry (*Morus*).

DESCRIPTION **Adult:** 26–30 mm wingspan; mainly white (Fig. 774), the fore wings sometimes flecked with black; male with noticeably bipectinated antennae (Fig. 775). **Egg:** light green, laid in a large batch (Fig. 776). **Larva:** 30–35 mm long; varying from yellow or yellowish-green to brown, with tufts of whitish hairs arising from black verrucae; spira-

773 Larva of garden tiger moth, *Arctia caja*.

774 Adult female of *Hyphantria cunea*.

775 Adult male of *Hyphantria cunea*.

cules white, ringed with black; head shiny black (Fig. **777**). **Pupa:** 10–12 mm long; shiny blackish-brown; cremaster with twelve hooks.

LIFE HISTORY Adults of the first generation occur in the spring from April onwards, the egg-laying females often showing a particular preference for maple and each depositing several hundred eggs on the spurs or on the underside of the expanded leaves; the egg batches are then partly covered with whitish hairs. The larvae are gregarious. They feed ravenously on the foliage from May to July, sheltering during the day within a large but flimsy (cf. *Euproctis chrysorrhoea*, p. 300) communal web (Fig. **778**). When fully fed, the larvae wander away to pupate on the food plant, each in a slight, greyish-brown cocoon. Moths of a second generation appear in July and August, eventually giving rise to larvae which complete their development in the autumn. These second-generation larvae usually pupate within bark crevices or amongst dead leaves, and then overwinter.

DAMAGE Host trees are clothed in webbing and extensively defoliated. Attacks are often severe on hedgerow, shade or ornamental trees in urban and suburban areas.

CONTROL Elimination of this pest is often impractical but, on a small scale, destruction of occupied webs or application of insecticides to kill the young larvae can be successful. Suspected infestations in countries where this insect is not established should be reported to Plant Health authorities.

776

776 Egg batch of *Hyphantria cunea*.

777

777 Fall webworm, *Hyphantria cunea*.

778

778 Web of fall webworms, *Hyphantria cunea*.

779 Larva of white ermine moth, *Spilosoma lubricipeda*.

780 Young larva of white ermine moth, *Spilosoma lubricipeda*.

Spilosoma lubricipeda (Linnaeus)
White ermine moth

Larvae of this widespread and generally common species feed from late July to September on various low-growing plants. They sometimes occur on cultivated plants in herbaceous borders, feeding indiscriminately on the foliage, but damage caused is of no importance. Minor infestations may also occur on ornamental shrubs such as elder (*Sambucus*), honeysuckle (*Lonicera*), lilac (*Syringa vulgaris*) and privet (*Ligustrum vulgare*). Larvae (35–40 mm long when fully grown) are dark brown with a red or orange dorsal stripe, and tufts of brown hairs arising from black verrucae (Fig. **779**); young specimens are dull greenish to yellowish-green, with especially prominent verrucae (Fig. **780**). Pupation occurs in silken cocoons, formed amongst debris on the ground, the mainly white or creamish-white, black-spotted adults (35–45 mm wingspan) (Fig. **781**) appearing in the following May or June.

781 Male white ermine moth, *Spilosoma lubricipeda*.

Spilosoma luteum (Hufnagel)
Buff ermine moth

Widespread and common, the greyish-brown, hairy larvae (Fig. **782**) attacking the foliage of various garden plants, including herbaceous ornamentals, but occurring mainly on weeds such as dandelion

782 Larva of buff ermine moth, *Spilosoma luteum*.

(*Taraxacum*) and dock (*Rumex*). They feed during the summer, pupating in the autumn; the pale yellow to ochreous, black-marked adults (36–42 mm wingspan) (Fig. **783**) appear in the following June.

783 Male of buff ermine moth, *Spilosoma luteum*.

Phragmatobia fuliginosa (Linnaeus)
Ruby tiger moth

This locally common species is associated mainly with plants such as dandelion (*Taraxacum*), dock (*Rumex*), golden rod (*Solidago virgaurea*) and plantain (*Plantago*) but is sometimes a minor pest in flower borders; slight damage may also be caused to glasshouse-grown ornamentals. Adults (30–35 mm wingspan) are mainly reddish-brown, the fore wings thinly scaled centrally, the hind wings pink, suffused with dark grey; the abdomen is red, marked with black (Fig. **784**). The moths occur mainly during May and June, eggs being laid in batches on the foliage of various herbaceous plants. Larvae feed from late May or June onwards, most completing their development in the autumn and then overwintering. Individuals reappear in the following spring; they do not feed but, instead, spin brown, silken cocoons amongst foliage or debris on the ground and then pupate, adults appearing a few weeks later. Some larvae develop more rapidly than normal, becoming fully grown by August and giving rise to a partial second generation of adults in September. The larvae (up to 30–35 mm long) are blackish with a red dorsal stripe, red spots subdorsally and yellowish, reddish or brownish hair tufts arising from large, grey verrucae (Fig. **785**).

784 Ruby tiger moth, *Phragmatobia fuliginosa*.

Callimorpha dominula (Linnaeus)
Scarlet tiger moth

This widely distributed and distinctly local species is polyphagous on herbaceous plants, but especially abundant on comfrey (*Symphytum officinale*), green alkanet (*Pentaglottis sempervirens*) and hounds-tongue (*Cynoglossum officinale*). In areas where colonies are established, older larvae sometimes feed in the spring on ornamental herbaceous plants and young trees, including ash (*Fraxinus excelsior*), blackthorn (*Prunus spinosa*), cherry (*Prunus*), elm (*Ulmus*), oak (*Quercus*), rowan (*Sorbus aucuparia*), sallow and willow (*Salix*). Although causing slight

785 Larva of ruby tiger moth, *Phragmatobia fuliginosa*.

damage to the leaves, this attractive insect is not of pest status and specimens found on cultivated plants should *not* be destroyed. The young larvae feed briefly in the late summer or early autumn, before hibernating, and complete their development in the following spring. Fully grown larvae are about 40 mm long and mainly black, marked prominently with white and bright yellow (Fig. **786**). The spectacular, bright red, yellow, white and black, day-flying adults (Fig. **787**) occur in June and July.

786 Larva of scarlet tiger moth, *Callimorpha dominula*.

29. Family NOCTUIDAE

A large family of mostly medium-sized, stout-bodied, dull-coloured moths, often with a kidney-shaped mark and a small circle on each fore wing (this pattern is sometimes highlighted). The larvae of most species have five pairs of abdominal prolegs but groups with a reduced number also occur; crochets are of one size (uniordinal), arranged in a half-circle.

787 Scarlet tiger moth, *Callimorpha dominula*.

Agrotis segetum (Denis & Schiffermüller)
Turnip moth

An important horticultural pest, sometimes affecting herbaceous ornamentals such as China aster (*Callisterephus chinensis*), *Chrysanthemum*, *Dahlia*, Michaelmas daisy (*Aster*), *Petunia*, *Phlox*, pot marigold (*Calendula*), primrose (*Primula*) and *Zinnia*; seedling trees and shrubs may also be attacked. Eurasiatic. Widespread in Europe.

DESCRIPTION **Adult:** 38–44 mm wingspan; fore wings whitish-brown with brownish-black or blackish markings; hind wings pearly-whitish; male antennae distinctly bipectinated (Fig. **788**). **Larva:** 35 mm long; plump-bodied, glossy greyish-brown with a yellowish or pinkish tinge and indistinct darkish lines along the back; pinacula black (Fig. **789**). **Pupa:** 18–20 mm long; light reddish-brown; cremaster with two divergent spines.

LIFE HISTORY Adults occur from late May to June or early July, depositing eggs on various weeds and cultivated plants. The eggs hatch in about 1–3 weeks, depending on temperature. Young larvae then browse on the foliage of host plants, this feeding habit persisting only for the first two instars. Older individuals enter the soil and become typical sluggish 'cutworms', attacking plants to feed on the roots, crowns and underground portions of the

788 Male turnip moth, *Agrotis segetum*.

stems. Cutworms are active at night, resting during the day in the soil close to the plant roots. Most larvae become fully fed in the autumn but they do not normally pupate until the following spring; under favourable conditions, however, some individuals may develop more rapidly and give rise to a small second generation of adults in the autumn.

DAMAGE Larvae graze or burrow into corms, crowns, roots and tubers, causing plants to wilt; also, stems of plants may be girdled or completely severed, typically at about soil level. Infestations are most severe on light soils and in hot, dry conditions (cf. slug damage, p. 424), and tend to be most significant on younger, unirrigated, slower-growing hosts. Attacks sometimes interfere with the establishment and growth of seedling trees.

CONTROL Infestations are unlikely to develop in weed-free, well-irrigated sites. If necessary, apply an insecticide to kill the young larvae before they adopt the typical cutworm habit.

789 Larva of turnip moth, *Agrotis segetum*.

790 Larva of heart and dart moth, *Agrotis exclamationis*.

Agrotis exclamationis (Linnaeus)
Heart and dart moth
Generally abundant and sometimes a minor pest of young trees and shrubs in nurseries. The larvae occur from July onwards, at first feeding briefly on the foliage of host plants but later, in common with related species such as *Agrotis segetum*, attacking the roots and adopting a typical 'cutworm' habit. Individuals (up to 38 mm long) are dull brownish to greenish-brown, with dark-edged longitudinal markings along the back and relatively large, black spiracles (Fig. **790**). They complete their development in the autumn but usually do not pupate until the following spring, adults appearing in June and July. Under favourable conditions, however, some larvae may pupate early to produce a partial second generation of moths in the autumn. Adults (38–40 mm wingspan) are whitish-brown to dark brown, each fore wing including a reniform and a dart-like mark (Fig. **791**); the hind wings are pale greyish-brown in females but whitish in males.

791 Heart and dart moth, *Agrotis exclamationis*.

Noctua pronuba (Linnaeus)
Large yellow underwing moth

A generally common pest of low-growing plants, including herbaceous ornamentals such as *Anemone*, carnation and pink (*Dianthus*), *Chrysanthemum*, *Dahlia*, pot marigold (*Calendula*) and primrose (*Primula*); attacks may also occur on young trees and shrubs. Although mainly an outdoor pest, infestations also occur on glasshouse-grown plants. Palearctic. Present throughout Europe.

DESCRIPTION **Adult:** 50–60 mm wingspan; fore wings yellowish-brown, greyish-brown to dark rusty-brown; hind wings yellow with a blackish-brown border (Fig. **792**). **Egg:** creamish-white to purplish-grey (Fig. **793**). **Larva:** 45–50 mm long; greyish-brown, dull yellowish or greenish, with three pale lines along the back, the outer pair bordered inwardly with short blackish bars on each abdominal segment; spiracles white with black rims; head relatively small, pale brown, marked with black (Fig. **794**). **Pupa:** 22 mm long; plump and reddish-brown; cremaster with two strong, divergent spines.

LIFE HISTORY Adults occur from mid-June to August. Although active mainly at night, the moths are readily disturbed during the daytime; they then career wildly through the air before resettling, the flash of colour from the hind wings making them especially obvious. Eggs are laid in large, neat batches, often at the leaf tips of monocotyledonous plants such as *Gladiolus* and *Iris*. They hatch in 2–3 weeks, earlier or later depending on temperature. Larvae occur from July onwards, feeding both above ground on leaves and flowers, and in the soil on roots and crowns. A few individuals feed up rapidly to produce a second generation of adults in the autumn; however, most do not complete their development until the following May. Pupation takes place in a subterranean, earthen cell but without forming a cocoon. Unlike *Agrotis segetum* (p. 308), larvae of this species are less affected by weather conditions and they frequently cause damage to crops in cool, wet seasons.

DAMAGE In addition to weakening or killing plants by attacking the roots and crowns, larvae are also directly harmful to the buds, foliage, flowers and stems, sometimes destroying complete shoots or inflorescences.

CONTROL Apply an insecticide as soon as damage is seen.

792 Large yellow underwing moth, *Noctua pronuba*.

793 Egg batches of large yellow underwing moth on leaves of *Iris*.

794 Larva of large yellow underwing moth, *Noctua pronuba*.

Naenia typica (Linnaeus)
Gothic moth

An occasionally troublesome pest of herbaceous plants, shrubs and young trees, including *Chrysanthemum*, crab apple (*Malus*), ornamental cherry (*Prunus*) and *Rhododendron*; infestations may also occur under glass, especially on *Chrysanthemum*, *Fuchsia*, *Geranium* and other pot plants. Eurasiatic. Widespread in Europe.

DESCRIPTION **Adult:** 36–46 mm wingspan; fore wings whitish-brown suffused with blackish-brown, each marked with pale stigmata and crosslines, the reniform stigma enclosing a pale line; hind wings brownish-grey (Fig. 795). **Larva:** 40–45 mm long; greyish-brown flecked with darker brown, the sides marked with pale oblique streaks and a pinkish, undulating spiracular line edged above by black; second and third thoracic segments with a pair of creamish-white spots; seventh and eighth abdominal segments each with distinctive, black, oblique markings; pinacula whitish; head pale brownish marked with brown (Fig. 796). **Pupa:** 15–18 mm long; dark chestnut-brown; cremaster with a pair of downwardly curved, converging spines.

LIFE HISTORY Adults occur in June and July, depositing eggs on the foliage of various plants. Eggs hatch in 10–14 days. Young larvae then feed gregariously from late July or August onwards. The larvae hibernate throughout the winter, reappearing in the following spring. They then feed singly, becoming fully grown in May and finally pupating in the soil in flimsy cocoons.

DAMAGE Larvae cause defoliation which, if extensive, may be of importance, especially on glasshouse-grown chrysanthemums; flowers are also destroyed.

CONTROL Apply an insecticide as soon as damage is seen.

Mamestra brassicae (Linnaeus)
Cabbage moth

A generally common and often important pest of herbaceous ornamentals; infestations also occur in glasshouses on carnation and pink (*Dianthus*), *Chrysanthemum*, Peruvian lily (*Alstroemeria*) and other plants. Young trees and shrubs, including birch (*Betula*), hawthorn (*Crataegus*), larch (*Larix*), oak (*Quercus*), ornamental cherry (*Prunus*) and sallow (*Salix*), are also attacked. Eurasiatic. Widely distributed in Europe.

DESCRIPTION **Adult:** 38–45 mm wingspan; fore wings greyish-brown to blackish-brown, with pale, often black-edged markings; hind wings brownish-grey (Fig. 797). **Egg:** 0.8 mm across; hemispherical

795 Gothic moth, *Naenia typica*.

796 Larva of gothic moth, *Naenia typica*.

797 Cabbage moth, *Mamestra brassicae*.

and distinctly ribbed; whitish with a dark central spot and girdle. **Larva:** 35–45 mm long; green, brownish-green, greenish-brown or blackish-brown; dorsal line black, the subdorsal lines comprising segmentally arranged blackish bars, the pair on the eighth abdominal segment meeting to form a saddle-like mark; spiracular line pale; spiracles white with black rims; a slight hump on the eighth abdominal segment; head pale brown (Fig. **798**); young larva pale green, marked with yellow intersegmentally (Fig. **799**). **Pupa:** 17–22 mm long; reddish-brown and finely punctured; cremaster bearing two, often pale, hooked spines.

LIFE HISTORY Although the moths may occur at any time of year, they are most frequent in June and July, and from the end of August to late September. Eggs are deposited in large batches on the leaves of various plants. They hatch in 8–15 days. Larvae then feed from late June or July onwards. The young larvae graze the leaves; older individuals feed more extensively, often damaging buds and flowers. Fully fed larvae pupate in flimsy subterranean cocoons. This species is basically univoltine but, especially under glass, breeding may be sufficiently rapid for the completion of two generations.

DAMAGE Infestations on glasshouse ornamentals are especially important, damage to buds, leaves and open blooms rendering plants unmarketable.

CONTROL On a small scale, glasshouse plants should be inspected regularly and egg batches or young larvae destroyed. Alternatively, apply an insecticide as soon as larvae are found and repeat as necessary.

Melanchra persicariae (Linnaeus)
Dot moth

Generally common in gardens and nurseries on ornamentals such as *Anemone*, *Dahlia* and lupin (*Lupinus*); minor infestations may also occur on trees and shrubs, including alder (*Alnus*), birch (*Betula*), crab apple (*Malus*), larch (*Larix*) and sallow (*Salix*). Holarctic. Widely distributed in Europe.

DESCRIPTION **Adult:** 38–48 mm wingspan; fore wings bluish-black, each with a prominent, white, kidney-shaped stigma; hind wings greyish-brown (Fig. **800**). **Larva:** 35–45 mm long; pale green or pale brown, with darker chevron-like markings on the back and sides, and a thin, pale, dorsal stripe; prothoracic plate with distinct dorsal and subdorsal lines; eighth abdominal segment with a bluntly pointed hump (Fig. **801**). **Pupa:** 22–24 mm long;

798 Larva of cabbage moth, *Mamestra brassicae.*

799 Young larvae of cabbage moth, *Mamestra brassicae.*

800 Dot moth, *Melanchra persicariae.*

801 Larva of dot moth, *Melanchra persicariae*.

802 Tomato moth, *Lacanobia oleracea*.

dark chestnut-brown; cremaster with two divergent, barb-tipped spines.

LIFE HISTORY Adults occur from June to August. Eggs are laid on leaves of various plants, hatching in about eight days. Larvae feed slowly from July onwards, usually completing their development in September or October. They then enter the soil to pupate in flimsy cocoons, moths emerging in the following summer.

DAMAGE Larvae devour large amounts of foliage and rapidly strip the leaves from host plants. However, injury is usually important only on young plants or where numbers of larvae are large.

CONTROL Apply an insecticide as soon as damage is seen.

Lacanobia oleracea (Linnaeus)
Tomato moth

An often common pest of herbaceous plants, including various ornamentals growing out-of-doors and in glasshouses; most significant damage occurs on carnation and pink (*Dianthus*) and *Chrysanthemum*. Eurasiatic. Widely distributed in Europe.

DESCRIPTION **Adult:** 35–45 mm wingspan; fore wings reddish-brown to purplish-brown, with a small, yellowish stigma and a whitish subterminal line; hind wings pale brownish-grey (Fig. **802**). **Egg:** greenish, hemispherical, slightly ribbed and reticulated. **Larva:** 40 mm long; green, yellowish-brown or brown, finely speckled with white and, less densely, with black; dorsal and subdorsal lines pale, the spiracular line broad and yellow or orange-yellow, darkly-edged above; spiracles white, ringed with black; head light brown (Figs. **803, 804**). **Pupa:**

803 Larva of tomato moth, *Lacanobia oleracea* — green form.

804 Larva of tomato moth, *Lacanobia oleracea* — brown form.

16–19mm long; dark brown to black, and coarsely punctured; cremaster with a pair of blunt-headed spines.

LIFE HISTORY Adults emerge out-of-doors from late May or early June onwards but the first individuals can appear from late January onwards in heated glasshouses. Eggs are deposited in large batches of 30–200 on the underside of leaves. They hatch in 1–2 weeks, larvae then grazing in groups on the lower surface of leaves. After the second instar, they disperse and tend to occur singly, each larva feeding voraciously and becoming fully grown a few weeks later. Larvae are most abundant from July to September. Pupation takes place in a flimsy cocoon, either in the soil or amongst debris or attached to a suitable surface. In favourable conditions there is a second generation of adults in the autumn; under glass there is commonly a second generation in the summer and a partial third in the autumn.

DAMAGE Although young larvae merely graze away the lower surface of leaves, older larvae bite through the complete lamina and often reduce the foliage to a skeleton of major veins. Leaves, stems and flowers are often damaged severely, especially in glasshouses, attacked plants being rendered unmarketable if not completely destroyed.

CONTROL Apply an insecticide as soon as damage is seen.

805 Clouded drab moth, *Orthosia incerta.*

806 Larva of clouded drab moth, *Orthosia incerta.*

Hadena bicruris (Hufnagel)
syn. *capsincola* (Denis & Schiffermüller)
Lychnis moth
A widely distributed but minor pest, the yellowish-brown, blackish-marked larvae (up to 35mm long) feed mainly on the seeds of red campion (*Silene dioica*) and sweet william (*Dianthus barbatus*) but sometimes also damage the buds and seedheads of cultivated carnation (*Dianthus caryophyllus*). Adults (30–40mm wingspan) are greyish-brown to blackish, the fore wings variegated with whitish and yellowish-white markings. They occur from early June onwards, the larvae feeding in June and July; in favourable situations larvae also occur from August to September.

Orthosia incerta (Hufnagel)
Clouded drab moth
A generally common pest of broad-leaved trees and shrubs, including ash (*Fraxinus excelsior*), beech (*Fagus sylvatica*), birch (*Betula*), crab apple (*Malus*), elm (*Ulmus*), hawthorn (*Crataegus*), hornbeam (*Carpinus betulus*), lime (*Tilia*), oak (*Quercus*), ornamental cherry (*Prunus*), poplar (*Populus*), rose (*Rosa*) and sallow (*Salix*); often troublesome on garden trees and nursery stock. Eurasiatic. Widespread throughout Europe.

DESCRIPTION **Adult:** 34–40 mm wingspan; fore wings light grey to reddish-brown or purplish-brown, with darker markings and a pale submarginal line partly edged with brown; hind wings greyish (Fig. **805**). **Egg:** 0.7 mm across; dirty creamish-white with a purplish band and micropyle. **Larva:** 35–40 mm long; blackish-green, bluish-green or pale green, dotted with white; a prominent white dorsal stripe down the back, a pair of often indistinct subdorsal lines and a white stripe along each side, the latter edged above with blackish-green and, often yellow; head pale bluish-green, brownish-green or yellowish-green (Fig. **806**). **Pupa:** 14 mm long; shiny, dark reddish-brown; cremaster with two spines.

LIFE HISTORY Adults occur from March to late May or early June, but are most numerous in April and early May, eggs being laid in groups in cracks and crevices in the bark of host plants. The eggs hatch in 10–14 days and the tiny, very mobile, larvae immediately invade the bursting buds or unfurling leaves. The larvae feed mainly at night, young individuals hiding during the day in the shelter of young blossom trusses and partially un-furled leaves. Larvae are fully grown in about six weeks, usually in June. They then enter the soil to pupate in flimsy silken cocoons formed a few centimetres below the surface. Adults emerge in the following spring.

DAMAGE The larvae are capable of causing exten-sive defoliation and also damage buds and flowers. On rose, the larvae will feed on the outer tissue and also burrow into the centre of the blooms.

CONTROL Apply an insecticide in the spring, as soon as damage is seen, but do not spray open blossoms.

807 Larva of Hebrew character moth, *Orthosia gothica*.

808 Hebrew character moth, *Orthosia gothica*.

Orthosia gothica (Linnaeus)
Hebrew character moth

The larvae of this widely distributed and often abundant species are polyphagous on trees and shrubs, including birch (*Betula*), broom (*Cytisus*), hawthorn (*Crataegus*), lime (*Tilia*), oak (*Quercus*), poplar (*Populus*), *Sorbus* and willow (*Salix*); they also attack herbaceous plants. Although occurring mainly on wild hosts, individuals (which are up to 35 mm long, green, striped with white or whitish-yellow) (Fig. **807**) sometimes cause minor damage to ornamentals, feeding from April to June. The pale purplish-brown to reddish-brown adults (32–34 mm wingspan) are readily distinguished from other species of the genus by the characteristic brownish-black mark on each fore wing (Fig. **808**). They occur in greatest numbers in March and April.

Orthosia gracilis (Denis & Schiffermüller)
Powdered quaker moth

Generally common in association with various trees, shrubs and herbaceous plants; sometimes a minor pest in gardens and nurseries. Adults (35–40 mm wingspan) have whitish-beige fore wings, each more-or-less tinged with greyish, and marked with a few

809 Powdered quaker moth, *Orthosia gracilis*.

810 Larva of powdered quaker moth, *Orthosia gracilis*.

black dots and a pale yellowish or pinkish submarginal line (Fig. 809). They occur in late April and May, rather later than other members of the genus. Eggs are laid in indiscrete batches on twigs or senescent foliage, hatching about ten days later. The larvae, which feed from May to July, cause slight defoliation and can also injure buds and flowers. Fully grown individuals are 40–45 mm long and yellowish-green to pinkish-green, with several white spots and three pale lines along the back, a yellow stripe (broadly edged above with greenish-black) along each side and a yellowish-brown head (Fig. 810).

811 Larva of blossom underwing moth, *Orthosia miniosa*.

Orthosia miniosa (Denis & Schiffermüller)
Blossom underwing moth

A widespread but local species, confined mainly to oak (*Quercus*) woodlands. The moths occur in March and April, eggs being laid in batches on the shoots of scrub-oaks. On hatching, the young larvae feed gregariously, clustering together within the shelter of slight webs spun close to the shoot tips. Such webs are sometimes noted during May on young amenity or nursery trees growing in the vicinity of oak woods. Later, the larvae disperse to feed individually; they may then infest various other trees and shrubs, including beech (*Fagus sylvatica*), birch (*Betula*) and *Sorbus*. Fully grown larvae (35–40 mm long) are mainly blue to bluish-grey, with three yellow lines along the back and prominent, black, dorsal spots (Fig. 811). They complete their development in June and then pupate, the mainly pinkish-grey to reddish-grey adults (32–35 mm wingspan) (Fig. 812) appearing in the following spring.

812 Blossom underwing moth, *Orthosia miniosa*.

813 Larva of twin-spotted quaker moth, *Orthosia munda*.

814 Twin-spotted quaker moth, *Orthosia munda*.

Orthosia munda (Denis & Schiffermüller)
Twin-spotted quaker moth

This widely distributed, mainly woodland species is associated with various deciduous trees and shrubs, including crab apple (*Malus*), elm (*Ulmus*), honeysuckle (*Lonicera*), oak (*Quercus*) and sallow (*Salix*). The larvae (40–45 mm long) are pale brown to dark brown, with brownish-yellow dorsal and subdorsal lines, and a pale yellowish-brown spiracular stripe (Fig. **813**). They feed from April to June, sometimes causing minor damage to the young foliage of nursery stock and garden trees. The reddish-ochreous to pale greyish-ochreous adults (38–40 mm wingspan) (Fig. **814**) occur mainly in March and April.

815 Common quaker moth, *Orthosia stabilis*.

Orthosia stabilis (Denis & Schiffermüller)
Common quaker moth

A generally common species, the larvae occasionally attacking trees and shrubs in gardens, parks and nurseries but more usually associated with forest and hedgerow trees such as beech (*Fagus sylvatica*), birch (*Betula*), elm (*Ulmus*) oak (*Quercus*) and sallow (*Salix*). The pale reddish-ochreous to pale greyish-ochreous adults (32–35 mm wingspan) (Fig. **815**) occur in March and April. The larvae feed on foliage from April to June but numbers on cultivated plants are usually small. Individuals are plump, 35–40 mm long, yellowish-green and finely dotted with yellow; there are three yellow stripes along the back, a broader one along each side, and a prominent yellow bar across the first thoracic and last abdominal segments; the head is bluish-green (Fig. **816**).

816 Larva of common quaker moth, *Orthosia stabilis*.

Cucullia verbasci (Linnaeus)
Mullein moth

Often common, especially on light soils, infesting figwort (*Scrophularia*) and mullein (*Verbascum*), and sometimes a minor pest of cultivated plants; attacks also occur on buddleia (*Buddleja*). Eurasiatic. Widely distributed in Europe.

DESCRIPTION **Adult:** 45–55 mm wingspan; fore wings lanceolate, mainly ochreous-brown to dark brown; hind wings dark brown, grading to ochreous-brown basally. **Larva:** 50–55 mm long; whitish to pale green or pale blue, each segment marked posteriorly with a yellow dorsal crossband and on the sides with black lines and spots; pinacula black; head yellow, spotted with black (Fig. **817**).

LIFE HISTORY Adults appear from mid-April to the end of May, depositing eggs singly on the underside of leaves of host plants. The larvae feed on the foliage in June and July, each eventually pupating in the soil in a strong, silken cocoon several centimetres below the surface. The period of pupation is protracted, often persisting through four or five winters. This probably accounts for the species being most frequent on light soils and rather scarce in heavy-clay districts.

DAMAGE Young larvae bite out small holes in the leaves but older individuals feed ravenously, often causing significant defoliation.

CONTROL Hand-picking of the conspicuous larvae can provide adequate protection.

817 Larva of mullein moth, *Cucullia verbasci.*

818 Larva of Blair's shoulder knot moth, *Lithophane leautieri hesperica.*

Lithophane leautieri hesperica Boursin
Blair's shoulder knot moth

A migratory, Mediterranean insect associated with common juniper (*Juniperus communis*) and Italian cypress (*Cupressus sempervirens*). In recent years this moth has greatly extended its range. It now occurs abundantly in many more northerly areas, including much of southern England where it was first reported in 1951 and has since become adapted to Leyland cypress (× *Cupressocyparis leylandii*) and Monterey cypress (*Cupressus macrocarpa*). The larvae (up to 35 mm long) are green with a pair of broad white subdorsal stripes, a similar pair of lateral stripes, white pinacula and the abdominal spiracles surrounded by purplish-red patches (Fig. **818**). They feed mainly on the young growth, developing from February or March onwards. Fully grown individuals enter the soil in July, each spinning a flimsy cocoon in which they pupate after an extended prepupal stage. The adults (40–45 mm

819 Blair's shoulder knot moth, *Lithophane leautieri hesperica.*

wingspan) are mainly grey, irregularly marked with black (Fig. **819**). In northerly areas they occur in October and November.

Xylocampa areola (Esper)
Early grey moth

Adults of this widely distributed, mainly grey, blackish-marked moth (30–40 mm wingspan) occur from March to early May, and are often found at rest on garden fences, posts and walls (Fig. **820**). They are active at night, when they are frequent visitors to sallow (*Salix*) catkins. Eggs are deposited singly on the stems of honeysuckle (*Lonicera*), minor infestations sometimes occurring on ornamental bushes. The larvae feed in May and June, older individuals remaining flattened against the stem of the foodplant during the daytime and moving onto the leaves to feed at night. Individuals (up to 45 mm long) are elongate, pale yellowish-brown with a brown dorsal stripe and a fine spiracular line; the head is greyish-brown marked with grey (Fig. **821**). When fully fed, they pupate in the soil within tough, white cocoons. There is just one generation annually.

Allophyes oxyacanthae (Linnaeus)
Green-brindled crescent moth

Larvae of this widely distributed and often common species feed on various members of the Rosaceae, especially blackthorn (*Prunus spinosa*) and hawthorn (*Crataegus*); they are found occasionally on ornamental plants, including *Cotoneaster* and crab apple (*Malus*). Eggs laid in the previous autumn hatch in the spring, the larvae at first attacking the leaf buds and, later, the expanded foliage. Larvae are rather plump (*c.* 45 mm when fully grown), greyish-brown to reddish-brown, marked with blackish and with small, black-edged, pinkish-orange markings; the eighth abdominal segment is slightly humped and bears two pairs of pale projections (Fig. **822**). Feeding is completed by late May or early June. Larvae then enter the soil and eventually pupate in strong, silken cocoons. The mainly brown,

820 Early grey moth, *Xylocampa areola*.

821 Larva of early grey moth, *Xylocampa areola*.

822 Larva of green-brindled crescent moth, *Allophyes oxyacanthae*.

partly greenish-dusted adults (38–40mm wingspan) (Fig. **823**) closely resemble tree bark; they fly in October and November.

823 Green-brindled crescent moth, *Allophyes oxyacanthae*.

Conistra vaccinii (Linnaeus)
Chestnut moth

A generally distributed species, the larvae feeding on the foliage and flowers of various trees, shrubs and herbaceous plants, including alder (*Alnus*), elm (*Ulmus*), lime (*Tilia*), maple (*Acer*), oak (*Quercus*), poplar (*Populus*), *Sorbus* and willow (*Salix*); the larvae sometimes cause minor damage in gardens and nurseries but are not important pests. Adults appear in September and October, but do not deposit eggs until after emerging from hibernation in the following spring. Eggs hatch within a couple of weeks, larvae feeding on the foliage and flowers of host plants from April onwards. Fully grown individuals are about 32mm long, stout-bodied and mainly greyish-brown or greyish-green, with whitish dorsal pinacula and indistinct dorsal and subdorsal lines. When feeding is completed, usually in July, larvae enter the soil and construct silken cocoons in which, after a lengthy period of aestivation, they eventually pupate, moths emerging shortly afterwards. Adults (30–35 mm wingspan) are broad-bodied with a distinctly flattened abdomen; the fore wings are usually glossy chestnut-red with a prominent black spot in the reniform stigma; brownish variegated forms also occur; the hind wings are pinkish-brown to greyish-brown (Fig. **824**).

824 Chestnut moth, *Conistra vaccinii*.

Acronicta psi (Linnaeus)
Grey dagger moth

A generally common but minor pest of trees and shrubs, including alder (*Alnus*), birch (*Betula*), *Cotoneaster*, crab apple (*Malus*), hawthorn (*Crataegus*), lime (*Tilia*), ornamental cherry (*Prunus*), ornamental pear (*Pyrus*), rose (*Rosa*) and *Sorbus*; larvae also feed on herbaceous plants. Eurasiatic. Widely distributed in Europe.

DESCRIPTION **Adult:** 38–42mm wingspan; fore wings pale grey with black markings; hind wings greyish (Fig. **825**). **Larva:** 35mm long; greyish-black with a broad, white stripe along either side and a broad yellow dorsal band, the latter bordered by a blue stripe (interrupted by red, black-edged spots); a prominent, black, pointed hump on the first abdominal segment and a small black hump on

825 Grey dagger moth, *Acronicta psi*.

the eighth; head black (Fig. **826**). **Pupa:** 15 mm long; brown and rather slender, tapering towards the tip; cremaster with several strong spines.

LIFE HISTORY Adults occur from late May onwards, depositing eggs on the foliage of host plants; they are univoltine in northerly areas but bivoltine elsewhere. Larvae occur from June to September or October. When fully fed, each pupates in a greyish-brown cocoon constructed beneath dead leaves on the foodplant or secreted in a crevice in the bark.

DAMAGE Defoliation is usually unimportant but can be significant on nursery plants.

CONTROL If necessary, infestations on nursery stock can be controlled by spraying with a contact insecticide.

826 Larva of grey dagger moth, *Acronicta psi.*

Acronicta aceris (Linnaeus)
Sycamore moth

The colourful larvae of this locally common species are associated mainly with horse chestnut (*Aesculus hippocastanum*); they often occur on such trees planted in urban areas but will also feed on other hosts, including *Laburnum*, lime (*Tilia*), maple (*Acer*), rose (*Rosa*) and sycamore (*Acer pseudoplatanus*). Adults (40–45 mm wingspan) are mainly whitish to pale grey, with blackish-marked fore wings and whitish hind wings (Fig. **827**). Larvae are 35 mm long when fully grown; the body, which bears long tufts of yellow or reddish hairs, is mainly yellowish-brown, with a distinct series of white, black-bordered marks along the back. Adults occur from June to July or early August, and the larvae from July to September.

827 Sycamore moth, *Acronicta aceris.*

Acronicta alni (Linnaeus)
Alder moth

A widely distributed but local species, the larvae feeding in July and August on the foliage of various trees and shrubs, including alder (*Alnus*), beech (*Fagus sylvatica*), birch (*Betula*), elm (*Ulmus*), lime (*Tilia*), maple (*Acer*), ornamental cherry (*Prunus*), poplar (*Populus*), rose (*Rosa*), sallow (*Salix*) and *Sorbus*. The larvae usually occur singly, often resting fully exposed on the upper surface of a leaf. They are sometimes noticed on garden trees or nursery stock, their striking and unusual appearance immediately attracting attention, but they cause little or no damage and are of no consequence. Older individuals (up to 35 mm long) are black and yellow, with distinctive spatulate body hairs (Fig. **828**). The mainly grey, brownish-grey and blackish-

828 Larva of alder moth, *Acronicta alni.*

829 Alder moth, *Acronicta alni.*

830 Young larva of poplar grey moth, *Acronicta megacephala.*

marked adults (32–40 mm wingspan) (Fig. **829**) appear in May and June.

Acronicta megacephala (Denis & Schiffermüller)
Poplar grey moth

This species occurs mainly on black poplar (*Populus nigra*) but will also attack other kinds of poplar and goat willow (*Salix caprea*), the larvae commonly occurring on ornamental trees in urban areas. Adults occur from May to July, eggs being deposited singly on the leaves of host plants. The larvae feed from July to September, typically resting on the leaves during the daytime with the head turned back alongside the abdomen (Fig. **830**). Fully grown larvae (*c*. 35 mm long) are black finely speckled with yellowish-grey and marked with orange or reddish spots; there is a large pale patch on the seventh abdominal segment; the body is also partly clothed in long, fine hairs which arise in tufts from lateral pairs of brownish warts. The moths (40 mm wingspan) are mainly grey (Fig. **831**).

831 Poplar grey moth, *Acronicta megacephala.*

Acronicter rumicis (Linnaeus)
Knotgrass moth

A generally distributed and common species, the larvae feeding on various plants from July onwards. The larvae occasionally damage the foliage of bedding plants and various other cultivated hosts but attacks are of minor importance. Individuals overwinter as pupae in subterranean cocoons, adults emerging in the following May or June; in favourable areas there is also an autumn generation. Fully grown larvae are about 35 mm long, brownish-grey to black, with red markings and two lateral series of white patches along the back and lateral tufts of

832 Larva of knotgrass moth, *Acronicta rumicis.*

pale brownish hairs; the subspiracular line is creamish-white, interrupted by raised red verrucae; the first and eighth abdominal segments are mainly black and slightly humped; the head is black (Fig. **832**). Adults (38 mm wingspan) are mainly blackish to blackish-grey; the fore wings are variably marked with grey and each has a more-or-less distinct paler subcostal spot; the hind wings are pale brown (Fig. **833**).

Acronicta tridens (Denis & Schiffermüller)
Dark dagger moth

Larvae of this generally distributed species occur on various trees and shrubs but are associated most frequently with rosaceous hosts. They feed from July onwards but damage is limited to the foliage and of little or no importance. Fully grown individuals (*c.* 40 mm long) are mainly black, marked with red and white along the back and sides; there is a black, peg-like hump on the first abdominal segment and a smaller but broader swelling on the eighth; the body hairs are long and mainly blackish (Fig. **834**). Adults are virtually identical in appearance to those of *Acronicta psi* (p. 320) (Fig. **835**); they occur in June and July with, in favourable conditions, a partial second generation in September.

Amphipyra pyramidea (Linnaeus)
Copper underwing moth

Locally common in parkland and woodland areas, the moths occurring in greater or lesser numbers from late July to September or October. The larvae feed from April to June on the foliage of various trees and shrubs, including birch (*Betula*), hornbeam (*Carpinus betulus*), oak (*Quercus*), ornamental cherry (*Prunus*), rose (*Rosa*) and sallow (*Salix*); they sometimes cause minor damage to such plants in gardens and nurseries but are not important pests. Fully grown individuals (up to 45 mm long) are rather plump, and green to whitish-green, dotted with white or whitish-yellow; there are three incomplete white lines along the back and one, partly bordered above with yellow, along the sides; the eighth abdominal segment is humped, with a

833 Knotgrass moth, *Acronicta rumicis*.

834 Larva of dark dagger moth, *Acronicta tridens*.

835 Dark dagger moth, *Acronicta tridens*.

distinct horn-like apex (Fig. **836**). Adults (45–55 mm wingspan) have brownish fore wings, marked with black and pale yellowish-grey; the hind wings are coppery red (Fig. **837**).

836 Larva of copper underwing moth, *Amphipyra pyramidea*.

Euplexia lucipara (Linnaeus)
Small angle-shades moth

A generally common species, the larvae feeding on herbaceous plants and shrubs, and sometimes also attacking cultivated ferns such as *Dryopteris*. When present on ferns, the larvae graze the fronds but damage caused is rarely significant. Fully grown larvae are 30–35 mm long, rather stout-bodied, purplish-brown or green, with a V-shaped chevron-like mark on each segment and a whitish subspiracular line, whitish pinacula, black spiracles and a pair of white spots on the slightly humped eighth abdominal segment; the head is small, glossy and pale greenish. The larvae feed from August to September, eventually pupating in silken and earthen cocoons formed in the soil. Most individuals overwinter as pupae but, in favourable seasons, there may be a partial second generation. The purplish to light reddish-brown adults (30–35 mm wingspan) (Fig. **838**) occur mainly in June and July.

837 Copper underwing moth, *Amphipyra pyramidea*.

Phlogophora meticulosa (Linnaeus)
Angle-shades moth

An often common pest of herbaceous plants, including *Chrysanthemum*, *Cineraria*, *Geranium*, *Pelargonium*, violet (*Viola*) and various ferns; attacks are especially damaging in glasshouses with artificial lighting. Infestations also occur on *Anemone*, *Dahlia*, *Fuchsia*, hollyhock (*Althaea rosea*), *Iris*, *Primula*, wallflower (*Cheiranthus cheiri*) and many other ornamental plants growing in outdoor flower beds and borders. Eurasiatic. Present throughout Europe.

DESCRIPTION **Adult:** 40–50 mm wingspan; fore wings mainly pale pinkish-brown, each with a darker base and a large inverted triangular olive-green median mark; hind wings whitish-brown (Fig. **839**). **Egg:** 0.8 mm across; hemispherical and strongly ribbed; pale yellow with darker mottling. **Larva:** 35–40 mm long; velvety yellowish-green or brownish, with a fine, white, interrupted dorsal line, bounded on each side by a series of faint V-shaped marks; spiracles usually white with black rims; body rather

838 Small angle-shades moth, *Euplexia lucipara*.

plump (Fig. **840**); early instars are leech-like, bright green and translucent, with indistinct dusky segmental markings. **Pupa:** 18mm long; reddish-brown and plump; cremaster with a pair of elongate spines.

LIFE HISTORY Adults occur from May to October, mainly as one generation in May and June and another in the autumn. The moths often rest on glasshouse walls and other structures, with the wings folded lengthwise to resemble a crumpled leaf. If disturbed during the daytime, they fly erratically for a short distance before resettling. Eggs are deposited on leaves, either singly or in small groups. At normal glasshouse temperatures they hatch within a few days. Larvae feed in greater or lesser numbers in most months of the year but are most abundant from July to September, their rate of development being greatly affected by both temperature and humidity. Pupation takes place in the soil within a slight silken cocoon.

DAMAGE Young larvae 'window' the leaves but older individuals bite right through the lamina, causing considerable defoliation; the larvae also attack the growing points, flower buds and blossom trusses.

CONTROL If practical, glasshouse plants should be inspected regularly and egg batches or young larvae destroyed. If required, apply an insecticide, and repeat as necessary.

839 Angle-shades moth, *Phlogophora meticulosa*.

840 Larva of angle-shades moth, *Phlogophora meticulosa*.

Cosmia trapezina (Linnaeus)
Dun-bar moth

A generally common but minor pest of trees and shrubs, including ash (*Fraxinus excelsior*), beech (*Fagus sylvatica*), birch (*Betula*), common buckthorn (*Rhamnus cathartica*), crab apple (*Malus*), elm (*Ulmus*), hawthorn (*Crataegus*), hazel (*Corylus avellana*), hornbeam (*Carpinus betulus*), maple (*Acer*), oak (*Quercus*), ornamental cherry (*Prunus*), poplar (*Populus*), sallow (*Salix*) and *Sorbus*. The larvae often occur on such plants in gardens, parks and nurseries from April to June but cause only slight damage. In their later developmental stages they will often devour the larvae of other moths. Individuals (up to 30mm long) are bright green with three white lines along the back, a yellowish line along each side and yellowish intersegmental bands; the pinacula are small and black, each at least partly edged with white (Fig. **841**). Adults (25–32mm wingspan) are mainly whitish-grey to yellowish-grey, with a pinkish tinge and a more-or-

841 Larva of dun-bar moth, *Cosmia trapezina*.

less darkened central band on each fore wing (Fig **842**). They occur in July and August, eggs eventually being deposited in batches on the bark of host trees.

842 Dun-bar moth, *Cosmia trapezina*.

Hydraecia micacea (Esper)
Rosy rustic moth

A minor pest of robust herbaceous plants, including ornamentals such as *Chrysanthemum, Dahlia, Iris*, snapdragon (*Antirrhinum*) and sunflower (*Helianthus*). Holarctic. Widely distributed in Europe but especially common in coastal areas.

DESCRIPTION **Adult:** 40–45 mm wingspan; fore wings reddish-brown, somewhat darker centrally; hind wings pale with a dark crossline. **Larva:** 40–45 mm long; slender-bodied, dull pinkish and translucent, with a slightly darker dorsal line and several brownish-black pinacula, each bearing a pinkish-brown hair; head and prothoracic plate yellowish-brown (Fig. **843**).

LIFE HISTORY Adults are most numerous in the autumn, depositing eggs close to the ground on the stems of grasses and certain other plants. The eggs hatch in the following spring, usually in late April or May. The young larvae then bore into the stems of suitable host plants, tunnelling downwards into the crowns, roots or rhizomes. Larval development is completed in July or August, fully fed individuals pupating in the soil a few centimetres below the surface, and adults emerging a few weeks later.

843 Larva of rosy rustic moth, *Hydraecia micacea*.

DAMAGE Attacked plants are weakened and may wilt and die. Infestations are usually most severe on or near field headlands and on weedy sites.

CONTROL Infested plants or stems should be removed and the larvae destroyed; the chances of attack can be reduced by keeping sites weed free.

Gortyna flavago (Denis & Schiffermüller)
Frosted orange moth

An uncommon pest of herbaceous ornamentals, including *Chrysanthemum, Dahlia*, foxglove (*Digitalis purpurea*), hollyhock (*Althaea rosea*), lupin (*Lupinus*), pot marigold (*Calendula*) and mullein (*Verbascum*); attacks also occur on woody hosts such as elder (*Sambucus*), lilac (*Syringa vulgaris*) and willow (*Salix*). Holarctic. Widely distributed in Europe.

844 Frosted orange moth, *Gortyna flavago*.

DESCRIPTION **Adult:** 32–42 mm wingspan; fore wings yellow, marked with brown, purplish-brown and orange-yellow; hind wings pale ochreous (Fig. **844**). **Larva:** 30–35 mm long; pale yellow with very large, black pinacula and black spiracles; head yellowish-brown; prothoracic and anal plates blackish-brown. **Pupa:** 17–20 mm long; elongate, yellowish-brown; cremaster with a short pair of divergent spines.

LIFE HISTORY Adults occur from late August to October, eggs then being deposited in groups at the base of suitable host plants. Larvae feed from late March or April to July or August, each tunnelling within a stem and, if necessary, moving to an adjacent plant to complete its development. Fully fed larvae pupate within the feeding gallery, just above ground level. There is one generation each year.

DAMAGE Shoots of infested plants wilt and may die.

CONTROL Infested stems containing larvae or pupae should be destroyed.

Spodoptera littoralis (Boisduval)
larva = Mediterranean climbing cutworm
A mainly tropical and subtropical species but recently established in southern Europe; sometimes introduced into northern Europe on imported *Chrysanthemum* cuttings and on other ornamentals such as *Hibiscus* and *Kalanchoe*; a potentially serious pest of various glasshouse-grown crops, especially chrysanthemums.

DESCRIPTION **Adult:** 40 mm wingspan; fore wings blackish-brown with lighter markings, and often with a purplish sheen; hind wings mainly white (Fig. **845**). **Larva:** 40–45 mm long; pale brown to blackish-brown, finely speckled with white; dorsal and subdorsal lines orange-brown, the latter interrupted on the first and eighth abdominal segments by black patches and on the other segments by yellow spots, each bordered above with a black patch; yellow spots and black patches also present on the second and eighth thoracic segments; subspiracular line broad and pale reddish-brown; dark forms are mainly blackish-brown (Fig. **846**). **Pupa:** 16–20 mm long; reddish-brown.

LIFE HISTORY Eggs are deposited in large groups on the underside of leaves of host plants or on nearby surfaces and then coated with scales from the female's abdomen. They hatch within a few days at normal glasshouse temperatures. The larvae

845 Adult of Mediterranean climbing cutworm, *Spodoptera littoralis.*

846 Mediterranean climbing cutworm, *Spodoptera littoralis.*

then attack the plants, feeding voraciously and producing noticeably wet faecal pellets. Development is completed in 2–4 weeks; the larvae then pupate in flimsy, silken cocoons, adults emerging about two weeks later. Breeding is continuous under suitably warm conditions. In northern Europe, the pest is unlikely to survive the winter out-of-doors or in unheated structures.

DAMAGE Young larvae 'window' the leaves; older individuals bite right through the leaf tissue, causing considerable defoliation; they also attack the flowers and stems.

CONTROL Suspected infestations should be reported to Plant Health authorities.

Spodoptera exigua (Hübner)
Small mottled willow moth

This polyphagous, mainly tropical and subtropical species is an introduced pest in the Netherlands, attacking various glasshouse-grown ornamentals. It also occurs elsewhere in northern Europe (including the British Isles) as a rare immigrant or as an accidentally introduced species on plants such as *Chrysanthemum*. Breeding is continuous under suitable conditions but the pre-adult stages are susceptible to cold and damp, and this reduces the likelihood of their survival out-of-doors in northern Europe. The moths (28–32 mm wingspan) are yellowish-grey, mottled with yellowish-brown, with the stigmata characteristically ochreous or pinkish, and the hind wings pearly white with a darker border. The larvae (30–35 mm long when fully grown) vary from green or olive-green to purplish-black, with a paler or darker dorsal region and the spiracles set in pinkish, yellowish or orange patches. In some parts of the world this insect is an often destructive and notorious pest; the larva is commonly known in America as the 'beet army-worm'.

Earias clorana (Linnaeus)
Cream-bordered green pea moth

A widely distributed but local, wetland species breeding mainly in willow beds, especially on osier (*Salix viminalis*), but also attacking various amenity and ornamental willows. The larvae attack the terminal shoots, feeding during July and August within closely webbed clusters of leaves. Larvae (*c.* 17 mm long when fully fed) are greenish-white to greyish-white, marked with reddish-brown and orange; there are also large, dark-brown pinacula on the second and eighth abdominal segments (Fig. **847**). Pupation occurs in boat-shaped, parchment-like cocoons spun on the shoots (Fig. **848**). The tortrix-like adults (20–24 mm wingspan), which appear in the following May and June, have pea-green fore wings and white hind wings (cf. *Tortrix viridana*, p. 249).

Nycteola revayana (Scopoli)
syn. *undulana* (Hübner)
Oak nycteoline moth

This species is associated mainly with oak (*Quercus*) but, at least in mainland Europe, will also attack certain kinds of poplar (*Populus*) and willow (*Salix*), including amenity trees. Moths occur from late summer onwards, hibernating during the winter months and finally depositing eggs in the spring. The larvae (up to 20 mm long) are mainly green, marked intersegmentally with yellow and finely

847 Larva of cream-bordered green pea moth, *Earias clorana*.

848 Pupal cocoon of cream-bordered green pea moth, *Earias clorana*.

clothed in long, whitish hairs; the spiracles are yellowish with black rims. Larval development continues from May to July, fully fed individuals then pupating in pale green, boat-shaped cocoons formed on the twigs or on the underside of the leaves. Adults, which appear a few days later, are extremely variable in appearance, ranging from greyish-white, through brownish to blackish (Fig. **849**); they are relatively small (13–15 mm wingspan) and superficially similar in appearance to certain members of the family Tortricidae (q.v.).

Polychrysia moneta (Fabricius)
Delphinium moth

A minor garden pest of *Delphinium*, globe flower (*Trollius europaeus*) and monkshood (*Aconitum*).

Widely distributed in Europe but most numerous in southerly areas.

DESCRIPTION **Adult:** 38–45 mm wingspan; fore wings pale gold, with light and dark brown markings and a silver reniform stigma; hind wings brownish-black (Fig. **850**). **Larva:** 35–40 mm long; dark green with a darker dorsal line, a white line along each side and white pinacula; abdominal segments humped; three pairs of prolegs. **Pupa:** plump; dark reddish-brown with the head and wing pads green.

LIFE HISTORY Adults occur from mid-June to early July, depositing eggs on open flowers or amongst the buds. Larvae feed from July onwards, attacking the buds, flowers and young leaves, and tying the shoots together with silk; older larvae feed in exposed situations and may also attack the developing seeds. Some larvae complete their development during the summer, descending the foodplant to pupate in yellow cocoons spun amongst the lower leaves. These individuals produce a partial second generation of moths in the autumn. Other larvae, along with those resulting from eggs laid by second-generation adults, overwinter in hollow stems, either above or below ground level, and complete their development in the following spring, pupating in May or June.

DAMAGE Defoliation, bud damage and loss of flowers may be important locally and can affect the marketability of commercially grown plants.

CONTROL Infested buds and shoot tips should be picked off and burnt. Application of insecticides throughout the egg-laying period is also recommended.

849 Oak nycteoline moth, *Nycteola revayana*.

850 Female delphinium moth, *Polychrysia moneta*.

Autographa gamma (Linnaeus)
Silver y moth
An important horticultural pest, attacking various glasshouse-grown and outdoor plants, including azalea (*Rhododendron*), carnation (*Dianthus caryophyllus*), *Chrysanthemum*, Himalayan poppy (*Meconopsis*), *Pelargonium* and snapdragon (*Antirrhinum*). Holarctic. A major migratory species, in Europe spreading annually from its permanent breeding grounds around the Mediterranean to many western and northern areas.

DESCRIPTION **Adult:** 35–40 mm wingspan; fore wings greyish-brown to velvet-black, suffused with whitish-grey, often tinged with purplish and each bearing a silver, γ-shaped mark; hind wings light brown, with a broad brownish-black or blackish border (Fig. **851**). **Larva:** 35–45 mm long; green to

851 Silver y moth, *Autographa gamma*.

blackish-green, with pale irregular lines along the back and a whitish or yellowish spiracular line; three pairs of abdominal prolegs (Fig. **852**). **Pupa:** 12–19 mm long; black or blackish-brown, and unpunctured; cremaster bulbous and bearing curved, hook-tipped spines.

LIFE HISTORY This insect is unable to survive the winter in north-western Europe, except in heated glasshouses. Nevertheless, annual infestations are commonplace, the first immigrant moths usually arriving in May or June. Eggs are laid singly or in small groups on the leaves of host plants, hatching in 1–2 weeks according to temperature. The larvae feed mainly at night, attacking the leaves, buds and flowers. Individuals are fully grown in about a month. They then spin flimsy silken cocoons on the foodplant and pupate, adults emerging a week or two later. The moths sometimes fly during the daytime but are usually most active at dusk, feeding avidly on the nectar secreted by flowers such as buddleia and honeysuckle. Breeding is continuous under favourable conditions; however, most adults reared in the autumn do not breed locally but migrate south to more favourable areas.

DAMAGE Leaf damage is often extensive but tends to be less important than direct damage to buds and flowers, that can result in blind shoots or the production of malformed blooms.

CONTROL Glasshouse plants should be inspected regularly and any larvae found destroyed. If required, apply a contact insecticide.

85

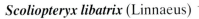

852 Larva of silver y moth, *Autographa gamma*.

85

853 Herald moth, *Scoliopteryx libatrix*.

Scoliopteryx libatrix (Linnaeus)
Herald moth

A generally common pest of poplar (*Populus*), sallow and willow (*Salix*). Widely distributed in Europe.

DESCRIPTION **Adult:** 50–60 mm wingspan; fore wings purplish-brown to greyish-brown, more-or-less tinged with reddish-orange; hind wings light brown (Fig. **853**). **Larva:** 50–60 mm long; slender and smooth-bodied; yellowish-green to dark green, with three dark dorsal lines; head green.

LIFE HISTORY Adults hibernate in barns, hollow trees, and various other situations, reappearing in the following spring. The larvae feed on poplars and willows from May or June onwards, attacking the young leaves and shoots which they often spin together with silk. They complete their development in the summer, young adults emerging in August. In favourable parts of mainland Europe there are two generations annually.

DAMAGE Defoliation on young trees and nursery stock can be extensive but attacks on older plants are usually of little or no significance.

CONTROL Damage is often not seen until larvae are relatively large; they can then be picked off by hand and destroyed. Application of an insecticide is rarely justified.

Order **TRICHOPTERA** (caddis flies)

1. Family **LIMNEPHILIDAE**

A large family of caddis flies, associated mainly with slow-moving water. Antennae of adults are about as long as the fore wings and each has a bulbous basal segment. Larval cases often very large.

Halesus radiatus (Curtis)

A generally common species, sometimes a pest of aquatic ornamentals such as white water-lily (*Nymphaea alba*) and yellow water-lily (*Nuphar lutea*). Widely distributed in Europe.

DESCRIPTION **Adult:** 42 mm wingspan; fore wings broad with a rounded apex, pale yellow striated with dark grey. **Larva:** 20–22 mm long; head brown; thorax yellowish-brown to dark brown, spotted with black; abdomen whitish to dark brown, with gill filaments on the first to seventh segments. **Case:** 30 × 6 mm; formed from small pieces of plant tissue, with up to three thin sticks cemented along the sides.

LIFE HISTORY Eggs are laid in the autumn on plants or stones above the water of suitable aquatic habitats, the egg mass being protected by a greenish gelatinous coating. After egg hatch, the larvae enter the water to begin feeding, each forming an elongated case within which to shelter. Further material is added to the case as the larva grows, the insect remaining within the habitation throughout its development, protruding the anterior end of the body when feeding or walking about but retracting rapidly back inside if disturbed. The larvae feed throughout the winter, completing their development in the following summer. Pupation occurs within the sealed-off case, usually after its attachment to a submerged plant or stone, the pupa breaking free of the case and floating to the surface immediately before the appearance of the adult.

DAMAGE Buds, leaves, stalks and roots of aquatic plants are all attacked, tissue often becoming extensively tattered and torn.

CONTROL Ornamental ponds, pools or water tanks should be cleared of debris and weeds, ideally in the spring immediately before replanting water-lilies and other aquatic ornamentals. Goldfish and trout readily devour caddis fly larvae, and their presence in ornamental ponds will often be sufficient to prevent plant damage. The use of insecticides is not recommended.

Limnephilus marmoratus Curtis

A widely distributed, often abundant species, breeding in most kinds of water and sometimes damaging to ornamental aquatic plants, including water-lilies (Nymphaeaceae). The larvae (up to 28 mm long) inhabit cases measuring *c.* 27 × 8 mm, each shelter composed of short lengths of plant stalks, arranged either transversely or obliquely. Adults (30–35 mm wingspan) have relatively narrow, mainly yellowish, brownish-marked, parchment-like fore wings; they occur in the autumn.

Order **HYMENOPTERA** (ants, bees, sawflies and wasps)

1. Family **PAMPHILIIDAE**

A small group of primitive, fast-flying, flattened, broad-bodied sawflies; antennae long and thin, with 18–24 segments. Larvae lack abdominal prolegs, and are often gregarious and web-forming.

854 Larva of social pear sawfly, *Neurotoma saltuum*.

Neurotoma saltuum (Linnaeus)
syn. *flaviventris* (Retzius in Degeer)
Social pear sawfly

Although mainly a pest of pear (*Pyrus communis*), infestations of this widespread but local, southerly-distributed pest occur on various other rosaceous plants, including *Cotoneaster*, hawthorn (*Crataegus*), medlar (*Mespilus germanica*) and ornamental cherry (*Prunus*). The unmistakeable larvae feed in communal webs in June and July, rapidly stripping the branches of foliage; fully fed individuals are 20–25 mm long, yellowish-orange with a shiny black head and a pair of prominent anal cerci (Fig. **854**). They overwinter in the soil and pupate in the spring, adults appearing in May and June.

CONTROL Occupied larval tents should be cut out and burnt.

855 Larva of *Pamphilius varius*.

Pamphilius varius (Lepeletier)

A minor pest of birch (*Betula*), the solitary, green-bodied larvae (*c.* 20mm long) (Fig. **855**), which have very small thoracic legs and no prolegs, inhabiting a rolled leaf. They are sometimes present on young trees during the summer, causing minor damage to the foliage. Fully grown individuals overwinter in the soil, pupating in the following spring. The stout-bodied adults occur in May and June. Species of *Pamphilius* are also associated with other hosts, including aspen (*Populus tremula*), oak (*Quercus*), rose (*Rosa*) and sallow (*Salix*).

2. Family ARGIDAE

Slow-moving, heavily built, moderate-sized sawflies with a short ovipositor; antennae divided into a scape, pedicel and fused flagellar segment, the latter sometimes distinctly bifid like a tuning fork.

856 Larva of large rose sawfly, *Arge ochropus*.

Arge ochropus (Gmelin in Linnaeus)
Large rose sawfly

A local and often destructive pest of rose (*Rosa*), the larvae occurring on both wild and cultivated bushes. Eurasiatic. Widely distributed in mainland Europe; in Britain generally uncommon (cf. *Arge pagana*, p. 334).

DESCRIPTION **Adult:** 7–10 mm long; head and thorax black but with the pronotum and tegulae yellow; abdomen yellow; legs yellow with the apices of the tibiae and tarsi black; wings yellowish. **Larva:** 22–28 mm long; head black or orange; body bluish-green marked with yellow along the back, and covered with numerous black verrucae, on most segments the latter usually forming one relatively indistinct and two distinct transverse rows (cf. *Arge pagana*, p. 334); the last two segments bear only small verrucae; anal plate black (Fig. **856**); five pairs of abdominal prolegs (cf. *Arge nigripes*, p. 335).

LIFE HISTORY Adults emerge in late May or June and may then be found on or near rose bushes. Eggs are deposited in double rows along the young vegetative shoots or flower stalks, the female penetrating the tissue with her short, stubby ovipositor to form distinctive oviposition scars (Fig. **857**); up to 20 eggs are inserted into each affected shoot. The eggs hatch a few weeks later, the larvae attacking the expanded foliage. At first they graze on the lower epidermis but, later, they bite completely through the leaf tissue, either in the middle or at the edge. Fully fed larvae drop to the ground and eventually pupate in brown, double-walled cocoons formed in the soil. There is often just one generation annually but, if conditions are favourable, a second brood of larvae may occur in the autumn.

857 Large rose sawfly oviposition scar in shoot of *Rosa*.

DAMAGE Shoots containing oviposition scars become blackened and distorted, heavy attacks affecting growth and flowering. Larvae cause extensive defoliation, commonly skeletonizing the leaves.

CONTROL On a small scale, egg-laying females or young larvae can be killed by hand. If necessary, apply a contact insecticide.

Arge pagana (Panzer)
syn. *stephensii* (Leach)
Variable rose sawfly

A locally common and important pest of rose (*Rosa*). Eurasiatic. Present throughout much of mainland Europe; in Britain most abundant in southern, south-eastern and south-western England.

DESCRIPTION **Adult:** 7–9 mm long; head, thorax and legs mainly black; abdomen mainly yellow; wings blackish (Fig. **858**). **Larva:** 22–25 mm long; head black or orange; body bluish-green, suffused above with yellowish-green; body bearing numerous, usually very prominent, shiny, black verrucae; the verrucae form eight rows along the body, arranged into three distinct transverse rows per segment but with just one row on the last (cf. *Arge ochropus*, p. 333); anal plate black (Fig. **859**); five pairs of abdominal prolegs (cf. *Arge nigripes*, p. 335).

LIFE HISTORY Adults occur from May to October, depositing eggs in the young shoots as described under *Arge ochropus* (p. 333). The larvae feed voraciously from June onwards, there being two main broods annually. When fully grown, the larvae enter the soil to pupate in brown, double-walled cocoons.

DAMAGE Leaf skeletonization is often extensive, seriously affecting shoot growth and the appearance of bushes.

CONTROL As for *Arge ochropus* (p. 333).

858 Variable rose sawfly, *Arge pagana.*

859 Larva of variable rose sawfly, *Arge pagana.*

Arge berberidis Schrank

Absent from the British Isles but widely distributed in central and southern Europe; a well-known pest of barberry (*Berberis*), the larvae sometimes defoliating ornamental bushes. There are two generations annually, adults flying in May and in August, and larvae occurring during the summer and again in the autumn. Pupation occurs in dark brown cocoons spun on or in the ground beneath infested bushes. Adult females are 7–10 mm long and dark metallic blue with smoky wings. The larvae are up to 18 mm long, relatively plump and whitish, clearly marked with yellow patches and numerous black verrucae; the head and anal plate are black (Fig. **860**).

860 Larva of *Arge berberidis.*

Arge nigripes (Retzius in Degeer)

Widely distributed on rose (_Rosa_) and sometimes a pest of cultivated bushes. Oviposition occurs in the leaf margins, close to the tips of the main serrations, the position of each egg being indicated by a distinct swelling (Fig. 861). The larvae feed gregariously from late May onwards, often causing noticeable defoliation. Individuals complete their development in about a month, each then pupating on the ground in a double-walled cocoon; there are either one or two generations annually. Larvae (up to 30 mm long) are greenish and translucent, the body bearing numerous small, mainly black, verrucae (Fig. 862); unlike _Arge ochropus_ (p. 333) and _Arge pagana_ (p. 334), there are seven pairs of abdominal prolegs. Adults (9–11 mm long) are mainly black; they occur in April and May, with members of a second generation appearing in July or early August.

861 _Arge nigripes_ egg pouches in leaf of _Rosa_.

862 Larva of _Arge nigripes_.

Arge ustulata (Linnaeus)

A widely distributed and often common species on birch (_Betula_) and willow (_Salix_), including ornamentals. Larvae (up to 22 mm long) are dark green and shiny, with a pair of white lines down the back; there are numerous pairs of black verrucae on the body, those on the thorax being largest and most conspicuous (Fig. 863). They occur from July to September or October, feeding along the edge of the leaves and causing noticeable, but usually only minor, defoliation. Adults (10–11 mm long) are bronzy-black with yellowish wings.

863 Larva of _Arge ustulata_.

3. Family CIMBICIDAE

Medium-sized to large, stout-bodied, fast-flying sawflies, with distinctly clubbed antennae.

864 Female large birch sawfly, *Cimbex femoratus*.

Cimbex femoratus (Linnaeus)
syn. *sylvarum* (Fabricius)
Large birch sawfly
A common but minor pest of birch (*Betula*) including, occasionally, nursery and garden trees. Present throughout Europe.

DESCRIPTION **Adult:** 20–22 mm long; head, thorax and abdomen black but sometimes partly, if not entirely, reddish-brown or yellow; antennae and tarsi yellowish-orange; wings mainly clear, with brownish margins (Fig. **864**). **Larva:** 50 mm long; head pale yellow with distinct black eyes; body green, marked with yellow and with a distinct black, blue-centred dorsal line; body also distinctly wrinkled and covered with white verrucae; spiracles black (Fig. **865**).

LIFE HISTORY Adults occur from May to July, eggs then being laid in the petioles of birch leaves. The larvae feed from July to September, remaining curled up on the underside of an expanded leaf and, in spite of their size, often escaping detection. When fully fed, they pupate in large (*c.* 10 × 20 mm), barrel-shaped, reddish-brown cocoons spun on the host plant, adults emerging in the following spring.

DAMAGE Larvae cause slight defoliation but damage is of little or no significance.

865 Larva of large birch sawfly, *Cimbex femoratus*.

4. Family DIPRIONIDAE

Slow-moving, stout-bodied sawflies, the males with strongly bipectinated antennae; antennae of females distinctly toothed. Larvae are associated with coniferous plants.

Neodiprion sertifer (Geoffroy in Fourcroy)
syn. *rufus* (Latreille)
Fox-coloured sawfly
A locally common and destructive pest of pine (*Pinus*), often attacking young ornamental and amenity trees. Eurasiatic. Widespread in Europe; also present in North America.

DESCRIPTION **Adult female:** 8–10 mm long; mainly brownish to reddish-yellow, partly marked with black. **Adult male:** 7–9 mm long; mainly black, the underside partly brownish-red; legs brownish-yellow; antennae strongly bipectinated. **Larva:** 23–25 mm long; head shiny black; body dirty greyish-green with a diffuse, black, lateral stripe and small, black verrucae (Fig. **866**).

LIFE HISTORY This species overwinters in the egg stage. The larvae commence feeding in May, forming small groups at the tips of the needles immediately below the opening buds. If disturbed, the larvae jerk the anterior end of the body over their backs, members of the group typically acting in unison.

The larvae feed gregariously on the older needles throughout their development, often causing extensive damage. When fully grown, in late June or July, they enter the soil to pupate, each in a tough, oval cocoon. Adults appear in September and October, eggs then being laid in rows in the edges of the needles.

DAMAGE Needles produced in the previous year are chewed down to the basal stalks, leaving unsightly lengths of bare wood.

CONTROL If practical, apply an insecticide to control the young larvae before damage becomes extensive. On a small scale, the larvae can be picked off by hand and destroyed.

Diprion pini (Linnaeus)
Pine sawfly

A locally common pest of pine (*Pinus*), sometimes associated with ornamentals. Widely distributed in central and northern Europe.

DESCRIPTION **Adult female:** 10 mm long; head and thorax mainly dark brown; abdomen pale yellowish-brown marked with black; antennae serrated (Fig. **867**). **Adult male:** 8–9 mm long; blackish-brown with partly pale brown legs; antennae strongly bipectinated. **Larva:** 25 mm long; head brown; body pale yellowish to yellowish-green, but darker dorsally, with a row of black spots along each side (Fig. **868**).

LIFE HISTORY Adults are active in May or June, females eventually depositing rows of eggs in slits made in the previous year's needles. Larvae emerge 2–3 weeks later and then feed communally on the needles. If disturbed, members of the colony adopt a threatening posture, grasping the leaf edge with the thoracic legs and extending the abdomen upwards in the shape of an 'S'. Individuals are fully fed in July. They then spin cocoons on the host plant or on the ground, a second generation of adults appearing in late July or August. Larvae of the second generation feed during the late summer; in September or early October, when fully grown, they enter the soil or leaf litter to spin cocoons. Pupation occurs in the following spring.

DAMAGE Larvae consume complete needles and may also graze on the bark of the young shoots. Attacked shoots are often completely stripped of needles; small trees can be completely defoliated.

CONTROL Apply an insecticide to kill the young larvae.

866 Larvae of fox-coloured sawfly, *Neodiprion sertifer*.

867 Female pine sawfly, *Diprion pini*.

868 Larva of pine sawfly, *Diprion pini*.

5. Family **TENTHREDINIDAE**

The main family of sawflies with, in females, a characteristic, saw-like ovipositor; antennae with 7–15 (but usually 9) segments. Larvae are mainly free-living, often feeding gregariously on the leaves of trees, shrubs and herbaceous plants they possess 6–8 pairs of abdominal prolegs which, unlike those of lepidopterous larvae, lack crochets.

Heterarthrus vagans (Fallén)

An often abundant pest of alder (*Alnus*), including ornamentals and trees planted as windbreaks. Widespread in central and northern Europe.

DESCRIPTION **Adult:** 3–5 mm long; mainly blackish, the abdomen orange-yellow below; legs yellowish; wings smoky. **Larva:** 6–9 mm long; head small, brown, and pointed anteriorly; body flattened and tapering back from a relatively broad thoracic region; whitish and translucent, with the green gut contents clearly visible; distinctive black markings present between the thoracic legs (except in final instar); legs very small, the anal pair fused (Fig. **869**).

LIFE HISTORY Adults occur in the spring, depositing eggs in the leaves of alder. The larvae then form distinctive brownish blotch mines in the leaves. There are commonly two or three mines on each infested leaf (Fig. **870**); these commence their development separately close to the leaf edge (cf. *Fenusa dohrnii*, p. 352), but may eventually coalesce to form one large chamber. When feeding, the larvae remove all the tissue between the upper and lower epidermis, and eject frass through a small slit made in the wall of the gallery; occupied mines occur in June and July, and the larvae are clearly visible if a mined leaf is held up to the light. Pupation occurs within the mine in a brown, flat, disc-like cocoon. A second generation of adults appears in the summer, with a second and usually larger brood of larvae feeding in the autumn.

DAMAGE Infested leaves are extensively disfigured, spoiling the appearance of specimen trees. Heavy attacks in the autumn lead to premature leaf-fall.

CONTROL Some control can be achieved by applying an insecticide recommended for use against leaf-mining insects; on a small scale, infested leaves can be picked off by hand and burnt.

869 Larva of *Heterarthrus vagans*.

870 Mines of *Heterarthrus vagans* in leaf of *Alnus*.

Heterarthrus aceris (Kaltenbach)

Locally common on field maple (*Acer campestre*) and sycamore (*Acer pseudoplatanus*). Adults (3.5–4.5 mm long) are mainly black with white legs; they occur in May and June, eggs being laid singly in the tips of the leaf lobes. The whitish larvae mine the leaves during the summer, each forming a large, brown blotch (Fig. **871**). When fully fed, each larva cuts out a circular section from the upper cuticle of its feeding gallery, to which it attaches a flat, disc-like cocoon. These habitations drop to the ground before leaf-fall, the occupants overwintering and eventually pupating shortly before the emergence of the adults. By flexing their bodies, individuals are capable of propelling their cocoons: hence the colloquial name 'jerking disc sawfly'. Although usually of little significance as a pest severe outbreaks of this mainly parthenogenetic sawfly have occurred, with extensive damage reported on amenity trees.

Heterarthrus microcephalus (Klug)

A generally common species, associated with various kinds of sallow and willow (*Salix*) including, occasionally, nursery and ornamental trees. The whitish larvae (Fig. **872**) feed mainly from July or August to September or October, forming brown blotch mines on the leaves (Fig. **873**). Adults (3–5 mm long) are black and shiny, with pale yellow legs; they occur from May to July.

Protemphytus carpini (Hartig)
syn. *glottianus* Cameron
Geranium sawfly

This widely distributed species is associated with *Geranium*, including cultivated forms, the larvae often causing noticeable defoliation. Larvae (up to 12 mm long) are mainly olive-green to greyish-black but paler below. There are two generations annually. Adults (6–8 mm long) are mainly black and shiny; they occur from May to June and from July to August.

Protemphytus pallipes (Spinola)
syn *grossulariae* Klug
Viola sawfly

An extremely widespread pest of *Viola*; often especially numerous on cultivated plants. Adults (6–8 mm long) are mainly black; the larvae (up to 10 mm long) are greenish-grey, but paler below, with a brown head. Adults occur from April to September and larvae from May onwards, there being three or more generations annually.

871

871 Mines of *Heterarthrus aceris* in leaf of *Acer.*

872 Larva of *Heterarthrus microcephala.*

873

873 Mine of *Heterarthrus microcephala* in leaf of *Salix.*

Allantus cinctus (Linnaeus)
Banded rose sawfly

An often common pest of rose (*Rosa*) and often present on cultivated bushes. Widely distributed in Europe; also present in North America.

DESCRIPTION **Adult:** 7–10 mm long; mainly shiny black, female with a distinct white or creamish band on the fifth abdominal segment; body elongate; wings pale yellow with brown veins. **Larva:** 12–15 mm long; head pale yellowish-brown; body greyish-green above, with numerous white verrucae on the back, whitish-green below (Fig. **874**).

LIFE HISTORY Adults occur in May or early June, flying strongly in sunny weather. Eggs are laid in slits made in the mid-rib on the underside of leaves of host plants, usually one or two per leaf. The eggs swell considerably after laying, causing conspicuous bulges on the upper surface of the leaves. They hatch about two weeks later. The larvae feed from July onwards, curling into a ball on the underside of a leaf when at rest and dropping to the ground if disturbed. At first they graze on the underside of the leaves, the upper surface remaining intact. Later, however, they bite right through the lamina, usually feeding along the leaf edge. The larvae are fully fed in about three weeks; individuals then burrow into decaying wood or the pith of pruned shoots to pupate, each in a flimsy, semitransparent, greenish cocoon. Adults emerge a few weeks later. Larvae of a second generation feed from August to September or October; they then spin cocoons but do not pupate until the following spring.

DAMAGE Although loss of leaf tissue can disfigure host plants and cause concern, the extent of damage is usually insufficient to affect plant growth.

CONTROL Apply a contact insecticide, ideally against young larvae of the first generation.

Allantus togatus (Panzer)

Widely but locally distributed on oak (*Quercus*) and, less frequently, birch (*Betula*) and sallow (*Salix*). The larvae feed on the leaves in the summer

874 Larva of banded rose sawfly, *Allantus cinctus*.

875 Larva of *Allantus togatus*.

and early autumn; they are sometimes found on young trees but do not cause significant damage. Individuals (*c.* 15 mm long) are dirty greenish to greyish-white, with several small, white verrucae on the back and an orange-yellow head; when at rest they curl into a ball on the underside of the leaves (Fig. **875**), dropping to the ground if disturbed. Fully fed individuals overwinter as prepupae in rotting wood. Adults (8–9 mm long) are mainly black, partly marked with yellowish-white and with partly cloudy wings; they occur from June to August.

Allantus viennensis (Schrank)

This species occurs in mainland Europe but not in the British Isles, the larvae infesting wild and cultivated varieties of rose (*Rosa*). Eggs are laid in the young shoots, usually some distance from the tip. The young larvae feed gregariously on the youngest leaves, forming distinctive circular holes in the laminae. In their final feeding stage they move downwards, often invading other branches and tending to occur singly; they will then devour tissue around the leaf margins. When at rest, larvae curl into a ball. Larvae are fully fed in 2–3 weeks. They then pupate in snags or broken branches on the host plant. Individuals of the last generation overwinter, most adults appearing in the spring but the emergence of some being delayed until the summer. The larvae are relatively large (up to 20 mm long) with an orange-yellow head and a green to bluish-green body, the first, second and fourth fold of each segment marked with prominent white verrucae (Fig. **876**). Larvae may be found from June onwards with, in favourable situations, up to three generations annually.

876 Larva of *Allantus viennensis*.

Apethymus filiformis (Klug)

syn. *abdominalis* (Lepeletier); *autumnalis* Forsius
A minor, locally common pest of oak (*Quercus*). Widely distributed in central and northern Europe.

DESCRIPTION **Adult:** 9–11 mm long; shiny black with a mainly yellow abdomen; antennae black; legs yellow. **Larva:** 15–19 mm long; head dark grey above but yellow below, covered in white, waxy powder; body light green, also mealy coated (Fig. **877**). **Prepupa:** yellowish-green and very shiny, with a mainly yellow head.

877 Larva of *Apethymus filiformis*.

LIFE HISTORY Unlike all other known European species of the family Tenthredinidae, adults of the genus *Apethymus* occur only in the late summer or autumn, individuals of *Apethymus filiformis* (cf. *Apethymus serotinus*, p. 342) occurring from September to November. Eggs laid in the autumn on host plants hatch in the following May or early June. Larvae are then common on young oak leaves in June, often curling into a ball when at rest. They usually complete their development in July, fully grown individuals moulting to an active, non-feeding prepupal stage and entering the soil to pupate but without forming a cocoon. Adults emerge a few weeks later.

DAMAGE Larvae graze the foliage but damage is not significant.

341

Apethymus serotinus (Müller)

syn. *braccatus* (Gmelin in Linnaeus); *tibialis* (Panzer)

This widely distributed but local species is also associated with oak (*Quercus*). The larvae feed on the foliage from May to July, commonly resting fully exposed on the upper surface of expanded leaves. Fully grown individuals are 15–18 mm long, with a shiny black head and the much wrinkled body dark grey above and pale grey to yellowish-grey at the sides and below (Fig. **878**). Adults (8–10 mm long) are mainly black with the hind tibiae white basally and the antennae with the sixth to eighth segments usually white. They occur from August to October.

878 Larva of *Apethymus serotinus*.

Endelomyia aethiops (Fabricius)

syn. *nigricolle* (Cameron); *rosae* (Harris)
Rose slug sawfly

An often common pest of rose (*Rosa*). Widespread throughout Europe; also present in North America.

DESCRIPTION **Adult:** 4–5 mm long; black with smoky wings; legs mainly black, with the tibiae and knees of the fore and middle legs yellowish-white. **Larva:** 15 mm long; head yellowish-brown; body translucent-yellow, the gut contents imparting a greenish tinge (Fig. **879**).

LIFE HISTORY This species is mainly partheno-genetic. Adult females occur from May to June, depositing eggs on the underside of rose leaves in slits cut close to the edges. The eggs hatch in 1–2 weeks and the larvae then feed on the upper surface, grazing the tissue but leaving the lower epidermis more-or-less intact. Larvae are fully fed in 3–4 weeks; they then enter the soil, usually in late June or early July, to construct silken cocoons in which to overwinter. Individuals pupate in the following spring, shortly before the emergence of adults, but development is sometimes delayed for a further year.

DAMAGE Infested leaves become extensively blanched (Fig. **880**), the damaged areas eventually turning brown and shrivelling up. Severely infested bushes appear scorched and their growth is checked.

879 Larva of rose slug sawfly, *Endelomyia aethiops*.

880 Rose slug sawfly damage to leaves of *Rosa*.

CONTROL Infested leaves can be removed by hand and the larvae destroyed. If necessary, apply a contact insecticide.

881 Female oak slug sawfly, *Caliroa annulipes*.

Caliroa annulipes (Klug)
Oak slug sawfly
larva = oak slugworm
A generally common pest of oak (*Quercus*); infestations also occur on lime (*Tilia*) and, less frequently, beech (*Fagus sylvatica*), birch (*Betula*) and sallow (*Salix*). Widespread in central and northern Europe.

DESCRIPTION **Adult:** 7–8 mm long; body black; wings blackish with an iridescent sheen and distinctly subhyaline apically; female with base of the hind tibiae and basitarsi white (Fig. **881**). **Larva:** 10–12 mm long; head blackish-brown, with the anterior third yellowish-brown; body pear-shaped, pale yellowish, covered in yellowish slime, shiny and translucent, with the gut contents clearly visible (Fig. **882**); young larva translucent-whitish with a black head (Fig. **883**).

LIFE HISTORY Adults appear in the spring, the females depositing eggs in the leaves of lime, oak and other host plants. The larvae feed in groups on the underside of the leaves during May and June, eventually entering the soil and pupating in tough, blackish cocoons. Larvae of a second generation feed in late July and August. In favourable situations there may be a further generation in the autumn.

DAMAGE Foliage is grazed from below, the upper epidermis remaining intact but turning brownish or whitish. Leaves are not distorted and attacks have little or no effect on plant growth, although the presence of damaged leaves on nursery or ornamental trees may be undesirable.

CONTROL If necessary on small trees, infested leaves can be removed by hand and burnt.

882 Oak slugworm, *Caliroa annulipes*.

883 Young oak slugworm, *Caliroa annulipes*.

Caliroa cerasi (Linnaeus)

syn. *limacina* (Retzius in Degeer)

Pear slug sawfly

larva = pear and cherry slugworm

A locally common pest of rosaceous trees and shrubs, including almond (*Prunus dulcis*), hawthorn (*Crataegus*), japonica (*Chaenomeles speciosa*), ornamental cherry (*Prunus*), ornamental pear (*Pyrus*) and rowan (*Sorbus aucuparia*). Eurasiatic. Present throughout Europe; accidentally introduced into many other parts of the world, including Africa, Australasia, North and South America.

DESCRIPTION Adult: 4–6mm long; black and shiny. **Larva:** 8–10mm long; greenish-yellow to orange-yellow but covered in olive-black, shiny slime; body pear-shaped, tapering towards the hind end; head and legs inconspicuous (Fig. **884**).

LIFE HISTORY Adults appear in late May and June. Eggs are then laid in small slits cut into the underside of leaves, often several on the same leaf. The larvae rest and feed whilst fully exposed on the upper surface of expanded leaves, grazing away the epidermis but without biting through the lamina. Individual are fully grown in July. They then pupate in small, black cocoons formed in the soil about 10cm below the surface, adults emerging a week or two later. There are usually two, and sometimes three, generations each year, the larvae occurring throughout the summer and early autumn. Reproduction is parthenogenetic, males being very rare.

DAMAGE Leaf grazing is often extensive and very disfiguring to ornamentals (Fig. **885**). Severe infestations cause premature leaf-fall and affect the growth of plants in the following season.

CONTROL Spray in June with a contact insecticide.

Caliroa cinxia (Klug)

This species occurs on oak (*Quercus*) in various parts of mainland Europe, and in central and southern England. The larvae graze on the underside of the leaves, damage (Fig. **886**) being identical to that caused by *Caliroa annulipes* (p. 343). *Caliroa cinxia* is single brooded, infestations being most

884 Pear and cherry slugworm, *Caliroa cerasi*.

885 Pear and cherry slugworm damage to leaves of *Sorbus aucuparia*.

886 Young larvae of *Caliroa cinxia*.

evident in the autumn from September to October. Adults are mainly black, and distinguished from those of *Caliroa annulipes* by their less cloudy wings and (in females) by the less extensive area of white on the hind legs; the larvae of both species are also similar but those of *Caliroa cinxia* have a uniformly reddish-brown head.

887 Larva of Solomon's seal sawfly, *Phymatocera aterrima*.

Phymatocera aterrima (Klug)
syn. *robinsoni* (Curtis)
Solomon's seal sawfly

A destructive pest of Solomon's seals, especially *Polygonatum multiflorum*. Widely distributed in central and southern Europe. In Britain first reported in London in 1846; now common throughout much of England and Wales.

DESCRIPTION **Adult:** 8–9 mm long; stout-bodied, black and shiny, with moderately long antennae and smoky, iridescent wings. **Larva:** 18–20 mm long; head black but inconspicuous; body greyish-white and wrinkled, bearing numerous small, black, spinose verrucae; thoracic legs black (Fig. **887**); young larva greyish-yellow with a prominent black head (Fig. **888**).

LIFE HISTORY Adults occur in May and June. Eggs are deposited in the petioles of Solomon's seals. They hatch in June, the larvae then feeding in company on the underside of the expanded leaves, forming elongate holes between the major veins or devouring the leaf edges (Fig. **889**). Fully grown larvae enter the soil to overwinter in silken cocoons, pupation taking place in the following spring.

DAMAGE The larvae cause considerable defoliation, the foliage of heavily infested plants being totally destroyed.

CONTROL Spray with a contact insecticide to kill the young larvae.

888 Young larvae of Solomon's seal sawfly, *Phymatocera aterrima*.

889 Solomon's seal sawfly damage to foliage of *Polygonatum multiflorum*.

Rhadinoceraea micans (Klug)

Iris sawfly

A locally distributed pest of mainly waterside *Iris*, occurring on wild yellow flag (*Iris pseudacorus*) and also cultivated species such as *Iris laevigata* and the butterfly iris (*Iris spuria*). Most common in central Europe; in Britain found mainly in the southern half of England.

DESCRIPTION **Adult:** 7–8 mm long; body black and stout; wings slightly smoky. **Larva:** 20 mm long; head black; body bluish-grey to dirty greenish-yellow, with pale verrucae on the back and sides (Fig. **890**).

LIFE HISTORY Adults fly in May and June, depositing eggs on the leaves of irises. The larvae feed from late May onwards, most completing their development by the end of July. They then enter the soil and spin silken cocoons, adult sawflies emerging in the following year.

DAMAGE Young larvae bite out V-shaped wedges along the leaf edges; older larvae graze away longitudinal sections of the leaf margins, feeding from the tips downwards and eventually reducing the foliage to ragged stumps. Larvae will also feed on the flower buds.

CONTROL Spray with a contact insecticide as soon as larvae are seen but only if there is no likelihood of contaminating ponds or other aquatic habitats.

89

890 Larva of iris sawfly, *Rhadinoceraea micans*.

89

891 Adult of *Periclista lineolata*.

Periclista lineolata (Klug)

A locally common but minor pest of oak, especially English oak (*Quercus robur*). Widely distributed in central and northern Europe.

DESCRIPTION **Adult:** 5–6 mm long; mainly shiny black with a short greyish or whitish pubescence; tegulae, edge of pronotum, knees and tibiae whitish; wings hyaline and iridescent, with black veins and each fore wing with a black stigma (Fig. **891**). **Larva:** 17 mm long; head black; body green with numerous black, mainly Y-shaped spines along the back and sides, and a series of smaller spines (some Y-shaped and some simple) above the legs (Fig. **892**).

89

892 Larva of *Periclista lineolata*.

LIFE HISTORY Adults occur in April and May, eggs being laid close to the veins of unfurling leaves (Fig. **893**). The larvae feed on the underside of oak leaves from mid-May to mid-June. When fully grown they moult to a mobile prepupal stage, which is devoid of spines, and enter the soil to pupate in a silken cocoon. There is just one generation annually.

DAMAGE Larvae bite out large holes in the leaves. Infestations are most common on young trees but damage caused is of little or no significance.

893

893 *Periclista lineolata* egg pouches in young leaves of *Quercus.*

894

894 Adult of *Periclista albida*

Periclista albida (Klug)
syn. *melanocephala* (Fabricius)
A widely distributed species, also associated with oak (*Quercus*). The adult females (5–7 mm long) have a mainly reddish-yellow thorax, a yellowish, brown-spotted abdomen and a black head and black antennae; although the wing veins are black, the stigma on each fore wing is yellow (Fig. **894**). The larvae are green with a black head, the body bearing numerous, mainly black, Y-shaped, thick-stemmed spines (Fig. **895**). Adults occur in April and May, the larvae feeding on the underside of the expanded leaves of oak from May to June. Pupation takes place in the ground in silken cocoons, individuals overwintering as prepupae and pupating in the spring.

895

895 Larva of *Periclista albida.*

Ardis brunniventris (Hartig)

syn. *bipunctata* (Cameron); *sulcata* (Cameron)
A locally common and often destructive pest of rose (*Rosa*). Widely distributed in Europe; also present in North America.

DESCRIPTION **Adult:** 5.5–6.5 mm long; black with pronotum, tegulae and legs partly yellowish-white. **Larva:** 12 mm long; brownish-white with a pale brown head; legs poorly developed (Fig. **896**).

LIFE HISTORY Adults occur from May onwards, the period of emergence being very protracted (often giving the impression of two generations) and also varying considerably from year to year. Eggs are laid in leaf tissue on the young terminal buds, the eggs hatching a few days later. Each young larva feeds briefly on the leaf tissue but soon bores into the shoot tip, typically one per infested shoot. Feeding continues within the shoot for about three weeks, the larva burrowing downwards for a few centimetres and expelling considerable quantities of wet, black frass through the original entry hole. When fully grown the larva vacates the shoot and drops to the ground, leaving behind a characteristic exit hole at the base of the feeding gallery (Fig. **897**). Larvae overwinter in earthen cells and eventually pupate in the following spring.

DAMAGE Infested shoot tips wilt and die, affected tissue turning black. Growth of young bushes is greatly affected, loss of terminal shoots resulting in the development of lateral shoots. Damage is especially severe on nursery rootstock.

CONTROL If practical, infested shoots should be cut off and burnt. Spraying with an insecticide is of limited value unless several applications are made throughout the egg-laying period.

Blennocampa pusilla (Klug)

Leaf-rolling rose sawfly
Generally common on wild and cultivated rose (*Rosa*); often an important pest in nurseries and gardens. Present throughout Europe.

DESCRIPTION **Adult:** 3.0–4.5 mm long; mainly black with whitish knees, tibiae and tarsi (Fig. **898**).

896 Larva of *Ardis brunniventris*.

897 *Ardis brunniventris* larval exit hole in shoot of *Rosa*.

898 Leaf-rolling rose sawfly, *Blennocampa pusilla*.

Larva: 8–10mm long; head brown; body pale green with short, spiny hairs on the back (Fig. **899**).

LIFE HISTORY Adults are active during May and June, females depositing eggs in the underside of rose leaves, close to the leaf margin. Females also use their ovipositor to probe the tissue on either side of the mid-rib; this causes the lamina to roll tightly inwards to form a protective tube in which the larvae will eventually feed (Fig. **900**); in some instances leaves in which eggs have not been laid also become rolled. The larvae feed from late May or June to July or August. They then enter the soil to overwinter in cocoons, pupating in the following spring. There is just one generation annually.

DAMAGE The presence of rolled, drooping leaves (Fig. **901**) spoils the appearance of bushes and the quality of nursery stock; severe attacks also reduce plant vigour. Certain cultivars, including various climbing roses, are especially liable to be attacked but damage is uncommon on standard-grown bushes.

CONTROL On a small scale, infested foliage can be picked off and burnt. Insecticides are of limited value against this pest because of the difficulty of reaching the larvae inside their protective leaf rolls.

899

899 Larva of leaf-rolling rose sawfly, *Blennocampa pusilla.*

900

900 Larval habitations of leaf-rolling rose sawfly on *Rosa.*

901

901 Leaf-rolling rose sawfly damage to foliage of *Rosa.*

Monophadnoides geniculatus (Hartig)
Geum sawfly

A generally common pest of *Filipendula* and *Geum*; also attacks *Rubus*. Eurasiatic. Widely distributed in Europe.

DESCRIPTION **Adult:** 5–6 mm long; stout-bodied and mainly black. **Larva:** 14 mm long; head greenish-yellow; body green to dark green, with numerous branched spines.

LIFE HISTORY Adults occur in May and early June, females depositing eggs on the underside of leaves of geum and other hosts. The larvae feed from late May to early July. They then enter the soil and spin cocoons, individuals eventually pupating and adults emerging in the following spring. There is just one generation annually.

DAMAGE Larvae bite out large, irregular holes in the foliage; such damage is often extensive, and heavily infested plants are seriously disfigured and weakened.

CONTROL Apply a contact insecticide as soon as larvae are seen.

Claremontia waldheimii (Gimmerthal)
syn. *subcana* (Zaddach); *subserrata* (Thomson)

Larvae of this widely distributed species are very similar in appearance to those of the previous species but slightly larger (up to 15 mm long). They also infest cultivated *Geum*, often causing extensive defoliation. Adults (5.5–6.5 mm long) are shiny black with mainly white tibiae. They occur in May and June, larvae feeding during the summer.

Metallus gei (Brischke)
Geum leaf-mining sawfly

Generally common on wild and cultivated *Geum*, and frequently a pest in gardens and nurseries. Widely distributed in Europe.

DESCRIPTION **Adult female:** 3.5–4.5 mm long; body mainly black but the abdomen partly brown; hind legs mainly white. **Larva:** pale greenish-white with a dark head, prothoracic plate and anal plate.

LIFE HISTORY Adult females of this partheno-genetic species occur in May or early June, with a second generation in July and August. The larvae are leaf miners, each forming an expansive, pale brown blotch on the upper surface of a fully expanded leaf of geum. The larvae feed from June to July and from August to September or October, fully grown individuals eventually pupating in cocoons in the soil. In contrast with many other leaf miners, the larval frass does not accumulate in the mine but is ejected through a small slit in the wall of the gallery.

DAMAGE Infested plants are disfigured by the conspicuous mines. There are often several mines in each infested leaf and, if attacks are severe (with the mines occupying the greater part of the upper surface of leaves), plants are seriously weakened.

CONTROL Application of an insecticide active against leaf-mining insects may be worth while; on a small scale, infested leaves can be picked off by hand and burnt.

Scolioneura betuleti (Klug)
syn. *betulae* (Zaddach)

A generally common pest of birch (*Betula*) and often injurious to young amenity trees and nursery stock. Widely distributed in central and northern Europe.

DESCRIPTION **Adult:** 4–5 mm long; shiny black and slender-bodied, with mainly reddish-yellow legs. **Larva:** 10 mm long; white with a brown head, a large prothoracic plate and a large plate ventrally on the first thoracic segment; distinct black markings usually present ventrally on the remaining two thoracic and first (sometimes also second) abdominal segments, plus a pair of marks ventrally on the penultimate abdominal segment; black markings also present along the sides; legs banded with black (Fig. **902**).

LIFE HISTORY Adults occur in May and June, and from July to September. Eggs are laid at the edge of the leaves of birch, the larvae subsequently mining and forming brownish, semitransparent blotches. There may be one or several mines per leaf, the galleries often uniting and the larvae then feeding gregariously. Fully grown larvae eventually vacate the leaf mine to pupate in the soil. Occupied mines occur in June and early July, and again in the autumn.

DAMAGE Attacked leaves may contain one or several mines which, if numerous, can destroy much of the leaf tissue; heavily infested trees appear brown.

CONTROL Some control may be achieved by applying an insecticide recommended for use against leaf-mining insects; on a small scale, infested leaves can be picked off by hand and burnt.

902 Larva of *Scolioneura betuleti.*

903 Larva of *Messa hortulana.*

Messa hortulana (Klug)

A locally distributed, leaf-mining and mainly parthenogenetic sawfly associated with black poplar (*Populus nigra*), infestations occurring occasionally on ornamental trees. The larvae (up to 10 mm long) are whitish and shiny, with a brownish-black head, a blackish prothoracic plate and black plates between the thoracic legs (Fig. **903**). They occur from May or June to July, each feeding within a brown, blister-like, frass-filled blotch which develops from the leaf margin, and causes slight distortion of the lamina (Fig. **904**). If attacks are heavy, several eggs may be deposited in the same leaf, mines then eventually merging to occupy most if not all of the lamina. Larvae are fully fed in about four weeks. They then vacate their mines to pupate in cocoons formed in the ground. The adult females are 4.0–4.5 mm long, stout-bodied and mainly black, with white tegulae and yellowish-white knees, tibiae and tarsi. They usually emerge in May, there being just one generation annually.

904 Mine of *Messa hortulana* in leaf of *Populus.*

Messa nana (Klug)
syn. *quercus* (Cameron)

Infestations of this widely distributed, mainly par-thenogenetic species occur on birch, especially downy birch (*Betula pubescens*). The larvae (up to 8mm long) are whitish to yellowish-white, with a brown head and prothoracic plate, and black plates between the thoracic legs. They form brown, frass-filled blotches on the leaves (Fig. **905**), feeding from July or August to September. Adults are similar to those of the previous species; they appear in May and June.

905 Mine of *Messa nana* in leaf of *Betula*.

Profenusa pygmaea (Klug)
Oak leaf-mining sawfly

This widely distributed, locally common leaf miner is associated with English oak (*Quercus robur*) and sessile oak (*Quercus petraea*), including young trees. The larvae mine the upper side of the leaves in the summer and early autumn, forming prominent, brown blotches (Fig. **906**). These mines are dis-figuring but they do not affect plant growth. When fully fed the larvae vacate the mines to overwinter in the soil, adults emerging in the following year. Adult females (3–4mm long) are black with white tegulae, knees, tibiae and tarsi. They occur from May to July, but males are unknown, reproduction being entirely parthenogenetic. The larvae (up to 8 mm long) are white with a light brown head, a black prothoracic plate and black plates between the thoracic legs; abdominal prolegs are lacking.

906 Mine of oak leaf-mining sawfly on *Quercus*.

Fenusa dohrnii (Tischbein)
syn. *melanopoda* Cameron; *westwoodi* Cameron

A generally common pest of alder (*Alnus*). Holarc-tic. Widespread in central and northern Europe.

DESCRIPTION **Adult:** 3–4mm long; body black and shiny, the abdomen stumpy; wings blackish. **Larva:** 8–10mm long; whitish with a light brown head and prothoracic plate (Fig. **907**); a large black plate present between the first pair of legs and a small black spot between the second and third pairs, these markings absent after the last moult.

LIFE HISTORY The larvae occur from July to October. They mine within the leaves of alder, the galleries commencing within the lamina, often close

907 Larva of *Fenusa dohrnii*.

to the mid-rib (Fig. **908**) (cf. *Heterarthrus vagans*, p. 338); there are often several mines in the same leaf. There are usually two generations annually, adults appearing in May or June and in August; in favourable conditions there may be a third generation. Reproduction is parthenogenetic.

DAMAGE The feeding galleries disfigure the foliage and, if numerous, can cause significant defoliation.

CONTROL Some control can be achieved by applying an insecticide recommended for use against leaf-mining insects; on a small scale, infested leaves can be picked off by hand and burnt.

908 Mines of *Fenusa dohrnii* in leaf of *Alnus cordata*.

Fenusa pusilla (Lepeletier)
syn. *pumila* (Klug)

A locally common and sometimes important pest of birch, especially downy birch (*Betula pubescens*). Holarctic, but probably introduced into North America. Present throughout Europe.

DESCRIPTION **Adult:** 2.5–3.5 mm long; mainly shiny black, the legs with brownish tibiae and tarsi; antennae short. **Larva:** 6–8 mm long; flattened and tapered from front to rear; head brownish; body white and translucent, with a greenish tinge, finally becoming yellowish-white; before final moult with a black dumbbell-shaped ventral mark on the first thoracic segment and blackish or brownish dots on the following two, three or four segments; prolegs are lacking on the anal segment (Fig. **909**).

909 Larva of *Fenusa pusilla*.

LIFE HISTORY Adults occur mainly in May and June, and in July and August. Eggs are laid in leaves of birch, usually towards the mid-rib in the tissue between two major lateral veins, the females tending to select the young leaves on the new shoots. The larvae feed in kidney-shaped blotches, infested leaves commonly containing several mines which either remain separated throughout their development or eventually unite into a common gallery which may occupy the whole leaf (Fig. **910**). Fully grown larvae pupate in the soil without forming cocoons. Occupied mines occur from June to October, there being two or more generations per year.

DAMAGE Infested leaves are disfigured by the larval mines, heavily infested foliage turning completely brown and dropping prematurely to reduce the vigour of affected shoots.

CONTROL As for *Fenusa dohrnii*.

910 Composite mine of *Fenusa pusilla* in leaf of *Betula pendula* 'Dalecarlica'.

Fenusa ulmi Sundewall
Elm leaf-mining sawfly

A generally common pest of elm (*Ulmus*). Holarctic but probably introduced into North America. Widely distributed in Europe.

DESCRIPTION **Adult:** 3–4 mm long; black and shiny, with short antennae and slightly smoky wings. **Larva:** 10 mm long; white to yellowish-white, with a pale brown head (Fig. **911**); legs banded with brown; a ventral plate on the first and a black dot on several of the following segments.

LIFE HISTORY This insect is mainly partheno-genetic, adult females occurring from late April to June and in August. The larvae feed in blister mines formed in the leaves of elm from June to July, and in September. The mines are brown and tend to extend between the major veins, without crossing the mid-rib; there are usually several in each infested leaf.

DAMAGE The mines cause slight distortion of the leaf lamina but, even when numerous, have little or no effect on growth.

CONTROL If necessary on young trees, infested leaves can be picked off by hand and burnt.

911 Larva of elm leaf-mining sawfly, *Fenusa ulmi.*

Macrophya punctumalbum (Linnaeus)

In continental Europe, an important pest of ash (*Fraxinus excelsior*) growing as shade trees; infestations also occur on lilac (*Syringa*) and privet (*Ligustrum vulgare*) and are sometimes of significance in nurseries. Present throughout Europe; locally common in Britain, especially in southern England. An introduced pest in North America.

DESCRIPTION **Adult female:** 7–8 mm long; mainly black with two white marks on the prothorax, a white scutellum, and white marks on the sides of the abdomen; legs mainly black with partly white tibiae and bright red hind femora (Fig. **912**). **Larva:** 16 mm long; dull green with a yellowish-brown head (Fig. **913**); young larva yellowish-green to whitish green, with a green head.

912 Adult female of *Macrophya punctumalbum.*

913 Larva of *Macrophya punctumalbum.*

914 *Macrophya punctumalbum* damage to leaves of *Syringa emodi*.

915 Female lesser antler sawfly, *Cladius difformis*.

LIFE HISTORY Adults occur in May and June, but males are extremely rare and reproduction is normally parthenogenetic. Eggs are deposited in the leaves of host plants from June onwards, larvae feeding throughout the summer and becoming fully grown by the autumn. Individuals then enter the soil and eventually pupate in dark brown cocoons, adults appearing in the following spring.

DAMAGE Larvae cause considerable defoliation, attacked leaves becoming extensively holed and often totally destroyed (Fig. **914**). Adults are also responsible for damage, mascerating leaves with their jaws.

CONTROL On nursery stock, apply a contact insecticide as soon as damage is seen.

Cladius difformis (Panzer)
Lesser antler sawfly

Widespread and common on wild and cultivated rose (*Rosa*), the larvae also feeding on meadowsweet (*Filipendula ulmaria*), marsh cinquefoil (*Potentilla palustris*) and strawberry (*Fragaria*). Present throughout Europe; also occurs in North America, but probably introduced.

DESCRIPTION **Adult:** 5–7 mm long; body black; legs yellowish-white; antennae of male with characteristic, long projections on the two basal segments and shorter projections on the third and fourth segments; female with slight projections on the first and second antennal segments (Fig. **915**). **Larva:** 10–12 mm long; somewhat flattened and distinctly hairy; head yellowish-brown; body translucent, yellowish to pale green, with darker subdorsal stripes (Fig. **916**); young larva paler with a blackish head (Fig. **917**).

916 Larva of lesser antler sawfly, *Cladius difformis*.

917 Young larva of lesser antler sawfly, *Cladius difformis*.

LIFE HISTORY Adults occur in May, depositing eggs singly in the leaf petioles. After egg hatch the larvae browse on the underside of leaves, most often attacking fully expanded leaflets. They are fully grown in four or five weeks and then pupate in thin, double-walled, brownish cocoons formed on the leaves or amongst debris on the ground. Adults emerge two or three weeks later, usually in late July or August. Larvae of a second generation feed during August and September, eventually over-wintering in cocoons and pupating in the following spring.

DAMAGE At first, larvae 'window' the leaves, the upper epidermis remaining intact; older larvae make holes right through the lamina and also browse on the leaf edges. Such damage is disfiguring but of significance only if extensive.

CONTROL Apply a contact insecticide as soon as larvae are seen.

Cladius pectinicornis (Geoffroy in Fourcroy)
Antler sawfly
This abundant sawfly occurs widely in Europe and parts of Asia, the larvae often attacking cultivated rose (*Rosa*). Adults of *Cladius pectinicornis* are distinguished from those of *Cladius difformis* (with difficulty in females) by the longer and increased number of antennal projections. Both species have a similar life-cycle and cause similar damage.

Priophorus morio (Lepeletier)
syn. *brullei* Dahlbom; *tener* Zaddach
Small raspberry sawfly
A widespread and locally common species, some-times infesting rowan (*Sorbus aucuparia*), including ornamental trees in gardens, but associated mainly with cane fruits, especially raspberry (*Rubus idaeus*). The larvae feed on the expanded leaves from late May or early June onwards but do not cause significant damage. Fully grown individuals (*c.* 12mm long) are translucent-whitish but mainly black dorsally, with numerous whitish hairs arising from pale verrucae and the head black and shiny (Fig. **918**). Adults are about 5–7mm long and mainly black with pale legs. There are two or more generations annually.

918 Larva of small raspberry sawfly, *Priophorus morio*.

919 Poplar sawfly, *Trichiocampus viminalis*.

Trichiocampus viminalis (Fallén)
syn. *luteicornis* (Stephens)
Poplar sawfly
An often common pest of poplar (*Populus*); some-times also attacking willow (*Salix*). Present through-out the whole of Europe; also occurs in North America.

DESCRIPTION **Adult:** 7–9mm long; body mainly yellow with the head and dorsal part of thorax black; antennae relatively long, black above and yellowish below; wings subhyaline, the fore wings each with a brown stigma (Fig. **919**). **Larva:** 20mm long; head black and shiny; body varying from glassy-green to greenish-yellow or orange, each side bearing a row of prominent black patches; body also somewhat flattened and hairy, the hairs arising from pale verrucae (Fig. **920**).

LIFE HISTORY Adults occur from May to August, eggs being deposited in rows within the petioles of poplar leaves. After egg hatch, the oviposition sites appear as characteristic depressions running along the length of the petiole and these often betray the presence nearby of the young larvae. The larvae occur in groups, sheltering beneath the leaves during the daytime; they also feed whilst lying alongside one another (Fig. **921**). The larvae continue to feed in groups until the final stages of their development. Individuals then wander over the trunks of host trees in search of suitable pupation sites. They eventually spin double-walled cocoons in bark crevices or amongst debris on the ground, adults emerging shortly afterwards. Larvae occur throughout the summer and early autumn, there being two or more generations annually.

DAMAGE Young larvae graze on the lower epidermis, the upper surface of the leaf turning brown, but older individuals bite through the entire lamina, sometimes causing significant defoliation.

CONTROL On a small scale, infested leaves can be picked off by hand and burnt. Alternatively, apply a contact insecticide as soon as larvae are seen.

920 Larva of poplar sawfly, *Trichiocampus viminalis.*

921 Young larvae of poplar sawfly, *Trichiocampus viminalis.*

Hemichroa crocea (Geoffroy in Fourcroy)

A sporadically abundant pest of alder (*Alnus*) and birch (*Betula*). Widely distributed in Europe; also occurs in North America.

DESCRIPTION **Adult female:** 5–8mm long; reddish-yellow with antennae, part of thorax and legs black. **Larva:** 20mm long; head pale brown to black; body mainly greyish-green, the first thoracic segment yellowish-orange, with a black longitudinal line above the spiracles and two lines of black spots above the legs.

LIFE HISTORY Adults of this mainly parthenogenetic species appear from May onwards. Eggs are laid in alder or birch leaves, along either side of a petiole. After egg hatch, the larvae move onto the lamina to feed. Older larvae feed in company along the leaf edges but they become less gregarious as they mature. When fully grown, the larvae enter the soil to pupate, each in a dark brown, single-walled cocoon. This species is usually double-brooded but, in favourable conditions, there may be a third generation.

DAMAGE Young larvae form distinctive sigmoid-shaped holes in the leaf laminae. Older larvae devour most of the leaf tissue, apart from the major veins, and are capable of causing considerable defoliation; greatest damage is caused to alder, especially red alder (*Alnus rubra*).

CONTROL On a small scale, infested leaves can be picked off by hand and burnt. Alternatively, apply a contact insecticide as soon as damage is seen.

Platycampus luridiventris (Fallén)

The unusual, extremely flattened, woodlouse-like larvae of this locally common species are occasionally noted on alder (*Alnus*) but are not important pests; the larvae also occur on birch (*Betula*) and hazel (*Corylus avellana*). They feed on the underside of the leaves from July to October, biting irregular holes through the lamina between the major veins. When at rest, they lie flat against the lower surface of a leaf, blending well with their surroundings. Fully grown larvae are 10–12 mm long and yellowish-green, with pairs of small black spots towards the sides of most abdominal segments; the lateral borders of each segment are rounded and fringed with white hairs; the head is yellowish-green with a pair of yellowish-brown patches, rising sharply upwards from behind and sloping gently downwards towards the mouthparts (Fig. **922**). The mainly black-bodied, orange-legged adults (5–6 mm long) occur in May and June.

922 Larva of *Platycampus luridiventris*.

Dineura stilata (Klug)

A generally common but minor pest of hawthorn (*Crataegus*) and sometimes present on nursery stock and garden hedges. Widely distributed throughout central and northern Europe.

DESCRIPTION **Adult:** 5–6 mm long; head black; thorax black, with reddish marks; abdomen black and yellow, and relatively narrow; legs yellowish. **Larva:** 12–14 mm long; head yellowish-green; body green, semitranslucent, slightly flattened and distinctly tapered from front to rear; head and body bear short, whitish hairs (Fig. **923**).

923 Larva of *Dineura stilata*.

LIFE HISTORY The larvae feed on hawthorn from August to October, lying pressed flat against the leaf and grazing away the upper epidermis but leaving the lower surface intact. They often occur in small groups, sheltering beneath the leaves during the daytime. When fully fed, larvae enter the soil to spin single-walled cocoons; individuals pupate in the following spring, the adults emerging in late May or June.

DAMAGE Infested leaves become noticeably discoloured where the upper epidermis is removed but damage has no effect on plant growth.

924 Recently moulted larva of *Dineura testaceipes*.

Dineura testaceipes (Klug)

A widespread pest of rowan (*Sorbus aucuparia*), the pale green larvae (Fig. **924**) feeding on the leaves during the summer months. Minor infestations sometimes occur on ornamental trees, the leaves being disfigured by pale patches, but the insect is more numerous on wild hosts.

925

925 Larva of *Dineura viridorsata*.

Dineura viridorsata (Retzius in Degeer)

Although most abundant in birch woods, this widely distributed sawfly is also a common pest of ornamental and amenity birches (*Betula*). The larvae (*c.* 16 mm long when fully grown) are greenish-white, tapered and flattened, with a brownish head (Fig. **925**). They rest during the daytime on the underside of the leaves, where the cast skins of earlier instars are often present. Feeding takes place at night, the larvae grazing away the upper epidermis of the leaves to produce distinct whitish blotches; such damage is often extensive but is of little or no significance. When fully grown, individuals enter the soil to overwinter in black, thin-walled cocoons, pupating in the following spring. The adults, which occur in late May or June, are 5–8 mm long and mainly black with a yellow abdomen.

926

926 Columbine sawfly, *Pristiphora alnivora*.

Pristiphora alnivora (Hartig)
Columbine sawfly

A common and often important pest of columbine (*Aquilegia vulgaris*). Widely distributed in central Europe; now well established in southern England, where it was first reported in 1946.

DESCRIPTION **Adult:** 4.5–5.5 mm long; mainly black with pale brown legs (Fig. **926**). **Larva:** 10 mm long; head greenish-yellow to blackish; body green and shiny.

LIFE HISTORY Adults occur from April or May onwards, there being three or more generations annually. The larvae feed on the leaves of host plants, biting out large holes in the edges of leaves whilst lying with the abdomen curled beneath the leaf lamina. They are fully fed within a few weeks, individuals then dropping from the host plant to pupate in brownish-orange cocoons. During the summer months, pupal development is rapid and adults emerge within about two weeks.

DAMAGE Defoliation is often very extensive, affecting both the appearance and vigour of infested plants.

CONTROL Apply a contact insecticide as soon as damage is seen.

Pristiphora testacea (Jurine)

syn. *betulae* (Retzius in Degeer)

A locally common pest of birch (*Betula*). Widely distributed in central and northern Europe but absent from more northerly areas.

DESCRIPTION **Adult:** 5–7 mm long; mainly black with orange-yellow tegulae and an orange-yellow abdomen; wings hyaline, the fore wings each with a black stigma (Fig. **927**). **Larva:** 16 mm long; head black; body yellowish-green and shiny, with large orange-yellow lateral patches on the second and third thoracic segments; a small, orange-yellow mark on the sides of the first and large lateral patches on the second to eighth abdominal segments: legs pale green with black claws (Fig. **928**).

LIFE HISTORY Adults occur from late April or May to June, and in July and August. Eggs are deposited in the lamina of birch leaves, each close to a major vein. The larvae feed gregariously on the leaf edges during the summer months, a second generation appearing in the autumn. Pupation occurs in the soil in brownish cocoons.

DAMAGE Larvae bite out large pieces from the leaves but attacks are rarely of significance, except on small trees.

CONTROL If necessary on small trees, apply a contact insecticide; alternatively, infested leaves can be picked off by hand and burnt.

927 Adult female of *Pristiphora testacea*.

928 Larva of *Pristiphora testacea*.

Pristiphora abietina (Christ)

syn. *pini* (Retzius in Degeer)

Gregarious spruce sawfly

Widely distributed and locally common on fir (*Abies*) and spruce (*Picea*). Although regarded mainly as a forest pest, damage can also occur on Norway spruces (*Picea abies*) intended for the Christmas tree market, especially when these are being raised near established spruce plantations. The larvae (up to 15 mm long) are green with a yellowish or brownish head. They occur in May and June, feeding gregariously on the needles of the new shoots. Infestations occur mainly at the tops of the plants, sometimes causing distortion and shoot death, and thereby reducing the quality of the trees.

CONTROL If attacks occur on young trees, apply a contact insecticide as soon as larvae are seen.

Pristiphora geniculata (Hartig)

Associated with rowan (*Sorbus aucuparia*) and locally important as a pest of forest and ornamental trees; widespread in mainland Europe, from Italy northwards, but uncommon in the British Isles. Also present in North America. The larvae feed gregariously from June onwards, sometimes causing noticeable defoliation. Fully grown larvae (15–18 mm long) are mainly yellow, marked with black. There are up to two generations annually. Adults (6.5–7.5 mm long) are mainly black; they occur in May and June and in July and August.

CONTROL If attacks occur on young trees, apply a contact insecticide as soon as larvae are seen.

Pristiphora punctifrons (Thomson)
syn. *viridana* Konow

Minor infestations of this widespread species occur occasionally on cultivated rose (*Rosa*), causing slight damage to the foliage. The mainly black to dull greyish-yellow adults (4–5 mm long) (Fig. **929**) occur from April to May or June. The larvae (up to 11 mm long) are green and shiny, with a few small blackish plates dorsally on the thoracic segments, pale yellowish thoracic legs and a dirty-yellowish, blackish-marked head; they feed on the edges of the leaves in May and June, typically resting with the tip of the abdomen curled downwards (Fig. **930**). Fully fed individuals enter the soil to pupate, adults appearing in the following spring.

929

929 Adult female of *Pristiphora punctifrons*.

930

930 Larva of *Pristiphora punctifrons*.

Pristiphora wesmaeli (Tischbein)
Larch sawfly

A widely distributed pest of larch (*Larix*), especially young trees; mainly of importance in forestry plantations but sometimes also present on ornamentals. The mainly yellow, black-marked adults (5.0–6.5 mm long) occur from May to June or early July, eggs being deposited in young needles at the tips of new shoots. The larvae (12–18 mm long) are green with a yellowish-green head (Fig. **931**). They feed during the summer, often stripping the needles from the terminal shoots.

931

931 Larva of larch sawfly, *Pristiphora wesmaeli*.

Amauronematus leucolaenus (Zaddach)
syn. *saarineni* Lindqvist
Locally common on various species of sallow and willow, including common sallow (*Salix atrocinerea*), creeping willow (*Salix repens*), eared sallow (*Salix aurita*) and grey sallow (*Salix cinerea*). Eurasiatic. Widespread in central and northern Europe.

DESCRIPTION **Adult:** 5.5–6.5 mm long; mainly black with pale legs (Fig. **932**). **Larva:** 18 mm long; elongate, green and shiny, with small black verrucae; the tracheae conspicuous (Fig. **933**). **Prepupa:** green to yellowish-green, with small, black verrucae.

LIFE HISTORY Adults of the genus *Amauronematus* are active early in the spring, *Amauronematus leucolaenus* occurring from March to April or early May. Eggs are deposited in the leaves of willows, the larvae feeding mainly in April and early May. They lie stretched out on the leaves, blending well with their surroundings, often several on each infested shoot. When fully grown the larvae moult to a non-feeding prepupal stage and enter the soil. Here they overwinter, each in a dark brown cocoon, pupation occurring shortly before the adults emerge.

DAMAGE Larvae cause some defoliation but, since this occurs relatively early in the season, plants are well able to compensate and growth is not affected.

Phyllocolpa leucapsis (Tischbein)
A generally common but minor pest of common sallow (*Salix atrocinerea*), eared sallow (*Salix aurita*) and pussy willow (*Salix caprea*). Present throughout central and northern Europe.

DESCRIPTION **Adult:** 3.5–5.0 mm long; mainly black, with a short, broad abdomen, the legs brownish marked with white or yellow. **Larva:** 10 mm long; head reddish-brown; body bluish-green to greyish, with a pair of small, black cerci; young larva mainly yellowish (Fig. **934**).

LIFE HISTORY Adults occur from May onwards, depositing eggs in the leaves of certain willows. The larvae develop during June and July, each within a folded leaf-edge which forms a flattened pouch on the lower surface (Fig. **935**); a second brood appears later in the season. Fully grown individuals construct brown cocoons in the soil and then pupate, those of the second generation over-

932 Adult of *Amauronematus leucolaenus*.

933 Larva of *Amauronematus leucolaenus*.

934 Larva of *Phyllocolpa leucapsis*.

wintering within their cocoons and producing adults in the spring.

DAMAGE Heavily infested bushes are disfigured but growth is not impaired.

Phyllocolpa leucosticta (Hartig)
syn. *sharpi* (Cameron)
Infestations of this species also occur widely on broad-leaved willows (*Salix*), the larvae feeding within elongate folds along the leaf edges (Fig. **936**). The young larva is whitish and later becomes greenish-white with a blackish to brownish head. Larvae are slightly larger than those of the previous species; adults (4.5–5.5 mm long) are distinguished by their entirely reddish-yellow hind femora.

935 Larval habitation of *Phyllocolpa leucapsis* on leaf of *Salix*.

Pontania proxima (Lepeletier)
syn. *flavipes* (Cameron); *gallicola* (Cameron)
Willow bean-gall sawfly
An often abundant pest of crack willow (*Salix fragilis*) and white willow (*Salix alba*). Widespread throughout central and northern Europe.

DESCRIPTION **Adult:** 3.0–4.5 mm long; black and shiny, with whitish-yellow legs. **Larva:** 7–10 mm long; head brownish-black and shiny; body yellowish-green, the legs whitish with brown claws.

LIFE HISTORY Adults first appear in May, eggs being deposited in the leaf buds. As attacked leaves unfurl, galls develop on the laminae, each housing a single larva. The mature galls, which project equally from both sides of the leaf, are red above and yellowish-green below, with the surface roughened by numerous small protuberances and ridges (Fig. **937**). During its development, the larva expels frass through a small hole formed in the lower wall of the gall. Fully fed larvae pupate in cocoons spun in the ground or in bark crevices, a second generation of adults emerging in July. These deposit eggs in the surface of expanded leaves, to produce larvae which usually complete their development in October and then overwinter.

DAMAGE Galls spoil the appearance of infested plants but, although often numerous, have little or no effect on tree growth.

CONTROL Chemical treatment is not worth while but, if necessary on young plants, galled leaves can be picked off by hand and burnt.

936 Larval habitation of *Phyllocolpa leucosticta* on leaf of *Salix*.

937 Galls of willow bean-gall sawfly, *Pontania proxima*.

Pontania bridgmanii (Cameron)
Sallow bean-gall sawfly

A widely distributed and generally common sawfly, the larvae developing in relatively flat, dark green, bean-shaped galls on the leaves of common sallow (*Salix atrocinerea*), grey sallow (*Salix cinerea*) and pussy willow (*Salix caprea*). The galls are similar in appearance to those of *Pontania proxima* but are usually expanded more above the leaf than below; also, their surface is smooth and somewhat pubescent, especially below; galls on *Salix caprea* are noticeably larger and less hairy than those on other hosts. The life-cycle is similar to that of *Pontania proxima* (p. 363).

938 Gall of sallow pea-gall sawfly, *Pontania pedunculi.*

Pontania pedunculi (Hartig)
syn. *baccarum* (Cameron); *bellus* (Zaddach)
Sallow pea-gall sawfly

This common and widely distributed species forms pale greenish, densely hairy, pea-shaped galls, attached to the mid-rib on the underside of leaves of broad-leaved willows, including common sallow (*Salix atrocinerea*), grey sallow (*Salix cinerea*) and pussy willow (*Salix caprea*) (Fig. **938**) (cf. *Pontania viminalis*, p. 365). Each gall measures about 7 mm across and encloses a single larva (Fig. **939**), occupied galls occurring in two main generations from late May or early June onwards. Fully grown larvae (10–12 mm long) are whitish with a brown head, pupation occurring in cocoons spun amongst debris on the ground. Adults are 3.5–4.5 mm long, and mainly black with white legs; they occur from April to June and from July to August.

939 Larva of sallow pea-gall sawfly, *Pontania pedunculi.*

Pontania triandrae Benson

This species is essentially similar to *Pontania proxima* (p. 363) but infests the leaves of almond willow (*Salix triandra*). The characteristically smooth, glabrous galls are at first pale green but soon become red above (Fig. **940**) and pale yellowish-

940 Galls of *Pontania triandrae*, viewed from above.

green below (Fig. **941**); in common with those of *Pontania proxima* they are developed equally on both sides of the lamina and there are two generations annually. The galls are often associated with plants cultivated in willow beds but they do not normally cause damage. However, if present in large numbers on the top-most growth of tall canes, the extra weight of the galls can result in shoots becoming bent.

941 Galls of *Pontania triandrae*, viewed from below.

Pontania viminalis (Linnaeus)
syn. *saliciscinereae* (Retzius); *vollenhoveni* (Cameron)
Willow pea-gall sawfly
A generally common, double-brooded species, forming pea-shaped galls on the leaves of purple willow (*Salix purpurea*) and, less commonly, crack willow (*Salix fragilis*) and osier (*Salix viminalis*); the galls are pinkish to orange-yellow, with a glabrous, somewhat warty surface (Fig. **942**) (cf. *Pontania pedunculi*, p. 364). The adults (4–5mm long) occur in May and June, and in July and August, the galls appearing on the underside of the leaves from late May onwards. The larvae (12–13 mm long) are whitish-green, each with a pale greenish-brown head (Fig. **943**).

942 Galls of willow pea-gall sawfly, *Pontania viminalis*.

Croesus septentrionalis (Linnaeus)
Hazel sawfly
A common and often serious pest of ornamental trees and shrubs, especially alder (*Alnus*), birch (*Betula*), hazel (*Corylus avellana*), poplar (*Populus*), rowan (*Sorbus aucuparia*) and willow (*Salix*). Eurasiatic. Widespread in Europe; also present in parts of Asia.

DESCRIPTION **Adult:** 8–10mm long; head and thorax black; abdomen with the basal two and the apical two or three segments black, the rest reddish-

943 Larva of willow pea-gall sawfly, *Pontania viminalis*.

brown; hind basitarsus and tip of hind tibia greatly expanded; wings mainly clear but apex of each fore wing cloudy (Fig. **944**). **Larva:** 22 mm long; head shiny black; body yellowish to bluish-green, variably marked with orange-yellow and with prominent black patches along the sides (Fig. **945**).

LIFE HISTORY Adults appear in May and June, depositing eggs in the leaf veins of host plants. Eggs hatch about two weeks later. The larvae then feed gregariously along the leaf edge, grasping onto the leaf with their thoracic legs and arching the body over the head. If disturbed, they thrash their bodies violently in the air. Fully fed larvae pupate in the soil, each spinning an elongate, brown cocoon a short distance below the surface. Larvae occur from mid-June onwards, those of the second generation, in late summer and autumn, being especially numerous.

DAMAGE The larvae are voracious feeders and cause considerable defoliation, often completely stripping the foliage from small trees.

CONTROL Apply a contact insecticide as soon as larvae are seen; on a small scale, infested leaves can be picked off by hand and burnt.

Croesus latipes (Villaret)

A widespread but usually uncommon species, the larvae attacking birch (*Betula*). Infestations occur from May or June onwards, the larvae feeding gregariously and causing noticeable defoliation; damage sometimes occurs on nursery trees. Fully grown individuals are 15–20 mm long and shiny black to shiny brownish-black, with the legs and underside of the last few abdominal segments pale yellowish (Fig. **946**). Adults are similar to those of *Croesus septentrionalis* but slightly smaller (7.5–10 mm long), with the cloudy apex of the fore wings less extensive and the hind femora of females reddish below. They occur from May to June and, in favourable districts where there are two generations, from July to August.

944 Hazel sawfly, *Croesus septentrionalis*.

945 Larva of hazel sawfly, *Croesus septentrionalis*.

946 Larva of *Croesus latipes*.

Nematus melanaspis Hartig
syn. *maculiger* Cameron
Gregarious poplar sawfly

A generally common pest of poplar (*Populus*); infestations also occur on birch (*Betula*) and willow (*Salix*). Widely distributed throughout central and northern Europe.

DESCRIPTION **Adult:** 6–8 mm long; pale yellowish-white, the thorax and upper side of abdomen mainly black; head brownish, with a dark facial mark; antennae short; legs mainly pale, the hind legs partly black. **Larva:** 18–21 mm long; head shiny black; body mainly pale green to dull green, with the first thoracic segment orange; three black stripes along the back, a double series of shiny, black verrucae along the sides and another series of verrucae above the legs; anal plate black (Fig. **947**).

LIFE HISTORY Adults occur from May to June and from July to August, eggs being deposited along the major veins on the underside of leaves of poplar. The larvae feed gregariously from June onwards, members of the second generation completing their development in the autumn. Fully grown larvae pupate in the soil in tough, brown cocoons.

DAMAGE Infested leaves are reduced to a network of veins, heavy infestations affecting the appearance and vigour of host plants.

CONTROL Apply a contact insecticide as soon as larvae are seen; on a small scale, infested leaves can be picked off by hand and burnt.

947 Larva of gregarious poplar sawfly, *Nematus melanaspis*.

948 Larva of *Nematus nigricornis*.

Nematus nigricornis Lepeletier

A minor pest of poplar, especially black poplar (*Populus nigra*), and willow (*Salix*). Widely distributed in central and northern Europe.

DESCRIPTION **Adult:** 6–8 mm long; head black and hairy; thorax black, marked with red; abdomen mainly yellow, with a black back; legs reddish. **Larva:** 15–18 mm long; head black; body relatively plump and mainly whitish-green to pale yellowish-green, with several shiny black verrucae along the back and sides; thoracic legs mainly black (Fig. **948**).

LIFE HISTORY Adults occur in two main generations, from May to June and from July to August. Eggs are deposited in the petioles of leaves of poplar or willow. The larvae feed gregariously along the edges of leaves, typically lying with the tip of the abdomen curled downwards on one side of the lamina. Larvae occur from June to September, fully grown individuals eventually pupating in cocoons formed in the soil.

DAMAGE The larvae devour all but the major leaf veins, infested branches being significantly defoliated.

CONTROL As for *Nematus melanaspis*.

Nematus pavidus Lepeletier

Lesser willow sawfly

An abundant and often destructive pest of willow (*Salix*); infestations also occur on alder (*Alnus*) and poplar (*Populus*). Widely distributed in Europe.

DESCRIPTION **Adult:** 6–7 mm long; head and thorax black; abdomen mainly yellow (male usually with black crossbands); antennae black; legs mainly yellowish; wings hyaline with blackish veins, the stigma on each fore wing brownish-black. **Larva:** 18–20 mm long; head shiny black; body green with the first thoracic and last three abdominal segments mainly orange; three black stripes along the back and a single series of shiny black verrucae along the sides and another series of black verrucae above the legs (Fig. **949**).

LIFE HISTORY Adult sawflies occur in two main generations, from April to June and from August to September. Eggs are deposited in dense groups on the underside of willow leaves. Each female lays either fertilized or unfertilized eggs, the former eventually giving rise to females and the latter developing parthenogenetically to produce males. After eggs have hatched, the larvae feed along the edge of the leaves, with the hind part of the body arched upwards. They may be found at any time from May to October and are voracious feeders, large numbers typically occurring together on the same leaf or branch. Fully grown caterpillars eventually pupate in the soil in tough, brownish cocoons.

DAMAGE Larvae rapidly skeletonize the leaves to a mere network of veins, heavy infestations affecting the appearance and vigour of plants. Attacks are often very damaging in osier beds.

CONTROL As for *Nematus melanaspis* (p. 367).

94

949 Larva of lesser willow sawfly, *Nematus pavidus*.

95

950 Larva of willow sawfly, *Nematus salicis*.

Nematus salicis (Linnaeus)

Willow sawfly

A local pest of willow, especially crack willow (*Salix fragilis*) and white willow (*Salix alba*). Widely distributed in central and northern Europe.

DESCRIPTION **Adult female:** 8–10 mm long; head and thorax black; abdomen and legs mainly yellow; wings hyaline with black veins. **Larva:** 25–30 mm long; head shiny black; body bluish-green with the thoracic and last three abdominal segments brownish-orange; numerous black verrucae arranged in rows along the body, and (predominantly on the green segments) five rows of black marks along the back (Fig. **950**).

LIFE HISTORY Adults occur most commonly from May to June and from August to September, depositing eggs on the underside of willow leaves. Young larvae feed gregariously but as development progresses they eventually disperse, older individuals usually occurring singly or in pairs. When fully grown, the caterpillars enter the soil to pupate in brownish-black, double-walled cocoons.

DAMAGE Larvae devour the tissue of fully expanded leaves but cause far less damage than those of *Nematus pavidus*.

CONTROL As for *Nematus melanaspis* (p. 367).

Nematus spiraeae Zaddach

Aruncus sawfly

A common and destructive pest of goat's beard (*Aruncus dioicus*). Widely distributed in central and northern Europe; in Britain, first recorded in 1924 and now often an abundant garden pest.

DESCRIPTION **Adult:** 5–6mm long; yellowish-brown, darker above and with a brownish-black head, thorax and antennae; tegulae pale; wings clear with brown veins (Fig. **951**). **Egg:** 1mm long; whitish and capsule-shaped. **Larva:** 15–20mm long; head brownish-green to brown; body green and translucent, with short, pale hairs arising from pale, inconspicuous verrucae (Fig. **952**).

LIFE HISTORY Adults of this parthenogenetic sawfly emerge in late April or May, eventually depositing eggs on the underside of leaves of host plants. Eggs hatch in about a week. The larvae feed in groups on the leaves during May and June, rapidly devouring the tissue between the major veins. Individuals are fully grown in 4–5 weeks; they then enter the soil to pupate in silken cocoons. A second generation of adults appears in late July or August, larvae then feeding in August and September; in some seasons there may be a partial third generation. Larvae of the final brood over-winter in their cocoons, pupating early in the following spring.

DAMAGE Defoliation is often extensive, the leaves of plants becoming skeletonized with only the major veins remaining.

CONTROL Apply a contact insecticide as soon as larvae are seen.

951 Aruncus sawfly, *Nematus spiraeae*.

952 Larva of aruncus sawfly, *Nematus spiraeae*.

Nematus tibialis Newman

False acacia sawfly

A minor pest of false acacia (*Robinia pseudoacacia*). Widely distributed in mainland Europe and in England, having been introduced from North America along with its foodplant in the early nineteenth century.

DESCRIPTION **Adult:** 6–7mm long; yellow, marked above with black; antennae black; legs mainly yellow, the hind tibiae and tarsi black (Fig. **953**).

953 False acacia sawfly, *Nematus tibialis*.

Larva: 12 mm long; head brownish-green marked with black; body green and shiny (Fig. **954**).

LIFE HISTORY Adults appear in May and June, depositing eggs in association with the young growth of false acacia. The young larvae feed on expanded leaves, each forming a small hole through the lamina and resting along the cut edge (Fig. **955**); later, the larvae devour more of the lamina, becoming fully fed after two or three weeks. They then enter the soil and pupate in tough, dark brown cocoons, adults emerging shortly afterwards. A second generation occurs in the late summer with, in favourable seasons, a partial third developing in the autumn.

DAMAGE Holed or partially devoured leaves attract attention on small ornamental trees and nursery stock but damage is rarely of any significance.

CONTROL On nursery stock and small trees, infested leaves can be picked off by hand and burnt. Such treatment on older trees is not worth while.

Nematus bergmanni Dahlbom
syn. *curtispina* (Thomson)
A generally common species, the larvae feeding on the leaves of willow, especially crack willow (*Salix fragilis*) and white willow (*Salix alba*). Infestations occur from May to October, but the larvae are rarely numerous and do not cause extensive damage. Individuals (up to 18 mm long) are green with a broad, translucent-white or pinkish-white dorsal stripe and a pair of darker subdorsal stripes; the head is a pale yellowish-brown, marked with brownish-black, and the tapered body terminates in a pair of reddish cerci (Fig. **956**). Adults (6–8 mm long) are mainly black with the paler parts of the body green; they occur from April onwards.

954 Larva of false acacia sawfly, *Nematus tibialis*.

955 Young larva of false acacia sawfly on leaf of *Robinia*.

956 Larva of *Nematus bergmanni*.

957 Larva of *Nematus umbratus*.

958 Adult of *Nematus umbratus*.

Nematus umbratus Thomson
syn. *collinus* Çameron; *similis* (Forsius)
A gregarious species, associated with birch (*Betula*); sometimes damaging to young nursery or amenity trees but usually uncommon. The larvae (up to 25 mm long) are blackish-green to whitish-green, shiny and translucent (Fig. **957**). They feed gregariously from May onwards, causing considerable damage to the foliage. There are two generations annually, the mainly yellowish-orange adults (5.5–7.5 mm long) (Fig. **958**) occurring from May to June and from late July to early September.

959 Larva of *Nematus viridescens*.

Nematus viridescens Cameron
A generally common species, associated with birch (*Betula*); sometimes present on nursery trees and young ornamentals but not an important pest. Adults occur from April onwards, the mainly green larvae (Fig. **959**) feeding on the leaves during the summer and autumn. There are two or more generations annually. Adults (6–8 mm long) are mainly green (Fig. **960**).

960 Adult of *Nematus viridescens*.

96

6. Family **CYNIPIDAE** (gall wasps)

Small, gall-forming insects with the gaster (part of abdomen behind the 'waist') flattened from side to side.

961 Bedeguar gall wasp, *Diplolepis rosae.*

Diplolepis rosae (Linnaeus)
Bedeguar gall wasp

Generally common on wild rose (*Rosa*) and sometimes present on cultivated bushes. Present throughout Europe; also occurs in North America.

DESCRIPTION **Adult:** 3.5–4.5 mm long; black with a mainly red gaster; tip of body with a distinct spine; legs partly reddish; fore wings mainly dark, each with a darker subapical patch (Fig. **961**). **Larva:** 5–6 mm long; whitish with a pale yellowish-brown head (Fig. **962**). **Pupa:** 5 mm long; white with purplish-white eyes (Fig. **963**).

LIFE HISTORY Adult wasps emerge in the spring from May onwards, depositing eggs in the unopened buds. Affected tissue begins to swell into a large, compact gall containing up to 60 hard, wooden, cherry-stone-like cells, each containing a single larva. The galls are surrounded by a dense, sticky mass of branched, moss-like filaments and measure up to 10 cm across (Fig. **964**); they change in colour from green, through pink and bright red to reddish-brown, darkening further during the winter months. Larvae develop throughout the summer, eventually pupating in their cells in the following spring shortly before the appearance of the adult wasps. Males are unknown, reproduction being entirely partheno-genetic. Old weather-worn galls (Fig. **965**) remain attached to host plants long after the emergence of the adults, and are especially noticeable during the winter and early spring.

DAMAGE The galls have little effect on bush growth but, if present in large numbers, they can be disfiguring.

CONTROL If necessary, galls should be cut out and burnt.

962 Larva of bedeguar gall wasp, *Diplolepis rosae.*

963 Pupa of bedeguar gall wasp, *Diplolepis rosae.*

964 Young gall of bedeguar gall wasp on *Rosa*.

965 Old gall of bedeguar gall wasp on *Rosa*.

Diplolepis eglanteriae (Hartig)
Rose smooth pea-gall cynipid

A local but often overlooked species, forming smooth, pea-like galls (4–5mm in diameter) on the underside of the leaves of rose (*Rosa*) (Fig. **966**). The galls, which sometimes bear slight surface depressions or small tubercules, also occur occasionally on the upper surface. They are at first pale green, appearing in July and maturing in September or October, by which time they have turned rose-red; each gall, which encloses a single larva, then drops to the ground. The occupant pupates within the larval chamber, the adult emerging in the following May or June. Reproduction is mainly parthenogenetic, males being very rare. Infestations are associated mainly with wild, rather than cultivated, bushes and are unimportant.

966 Galls of rose smooth pea-gall cynipid, *Diplolepis eglanteriae*.

Diplolepis nervosa (Curtis)
Rose spiked pea-gall cynipid

This generally common species is essentially similar to *Diplolepis eglanteriae* but usually produces pea-like leaf galls characterized by the presence of one or more long, thorn-like spines (Fig. **967**). Young galls are yellowish-green, becoming flushed with pink and, finally, brown.

967 Gall of rose spiked pea-gall cynipid, *Diplolepis nervosa*.

Neuroterus quercusbaccarum (Linnaeus)
syn. *lenticularis* (Olivier)
Oak leaf spangle-gall cynipid

A generally abundant gall wasp, associated with deciduous oak (*Quercus*) and often present on young trees. Widely distributed in Europe.

DESCRIPTION **Adult (asexual) female:** 2–3 mm long; black and shiny, with brownish-yellow legs; wings mainly hyaline; ovipositor long and curved. **Adult (sexual) female:** 1.5–2.0 mm long; black with yellowish legs; wings hyaline to subhyaline; ovipositor short. **Adult male:** similar to sexual female but with a long petiole and no ovipositor.

LIFE HISTORY In summer, mated female wasps deposit large numbers of eggs in the tissue on the underside of expanded oak leaves, inducing the formation of characteristic spangle galls. The disc-like galls are slightly hairy (the hairs stellate) and yellowish to yellowish-white in colour (Fig. **968**). Each gall contains a central chamber within which a single larva develops. The galls, often 100 or more on a leaf, occur from July onwards, maturing in October when about 6 mm in diameter. They then fall to the ground and swell as they take up moisture. The larvae, which represent the unisexual generation of the species, overwinter within the galls and pupate in the following spring. Asexual female wasps then emerge, depositing unfertilized eggs in male catkins. Characteristic currant-like galls develop in strings on the catkins (Fig. **969**), each changing from green through pink to red; they measure about 4 mm in diameter when mature. Larvae destined to become either males or females occur singly within these galls, completing their development in June. Adult wasps eventually emerge and, after mating, the females initiate the familiar generation of spangle galls.

DAMAGE Although spangle galls may occur in vast numbers and can cause spotting of the foliage, visible from above (Fig. **970**), infested trees are seldom, if ever, harmed. The currant galls are also unimportant.

968 Spangle galls of oak leaf spangle-gall cynipid, *Neuroterus quercusbaccarum*.

969 Currant galls of oak leaf spangle-gall cynipid, *Neuroterus quercusbaccarum*.

970 Oak leaf spangle-gall cynipid damage to leaf of *Quercus*.

Neuroterus albipes (Schenck)

syn. *laeviusculus* Schenck
Oak leaf smooth-gall cynipid

Spangle galls formed on the leaves of oak (*Quercus*) by this cynipid wasp are smooth, irregularly saucer-shaped with a slight central knob. They vary from creamish-white to reddish (Fig. **971**), developing from July to October and finally dropping to the ground when about 4 mm in diameter. The galls are usually present in relatively small numbers, sometimes occurring in company with the more abundant species *Neuroterus quercusbaccarum* (p. 374). Larvae of the bisexual generation develop within so-called Schenck's galls, which occur on the margins of leaves in May and June. These galls are oval (*c.* 2 × 1 mm), green and smooth, although slightly hairy when young.

971 Spangle galls of oak leaf smooth-gall cynipid, *Neuroterus albipes*.

Neuroterus numismalis (Geoffroy in Fourcroy)

syn. *politus* Hartig; *vesicator* (Schlechtendal)
Oak leaf blister-gall cynipid

The very characteristic golden-brown spangle galls formed by this species, develop on the underside of the leaves of oak (*Quercus*) from August onwards. They measure up to 3 mm across, have a distinct central pit and a dense silky covering of short hairs (Fig. **972**). The galls often occur in very large numbers, sometimes more than 1,000 appearing on a single leaf. The galls also cause a distinct discoloration of the upper surface of infested leaves. Larvae of the bisexual generation occur in May and June, within greenish or greyish, hemispherical, partly ribbed blister galls (*c.* 3 mm across); these are formed on either surface of young leaves.

972 Galls of oak leaf blister-gall cynipid, *Neuroterus numismalis*.

Neuroterus tricolor (Hartig)

syn. *fumipennis* Hartig
Oak leaf cupped-gall cynipid

This wasp forms spangle galls on the underside of the leaves of oak (*Quercus*) from July onwards. The galls are yellowish-green, 3 mm in diameter and button-like, each with a slightly raised rim bearing reddish hairs, and a slight but noticeable central elevation. The bisexual generation develops in shiny, whitish, yellowish or greenish, pea-like galls, each with a temporary coating of reddish hairs which drop off when about 5 mm long. These galls, each up to 6 mm in diameter, develop on the underside of leaves during May and June, arising from the mid-rib or major veins; they often occur singly but may coalesce and cause noticeable distortion of infested leaves.

Andricus kollari (Hartig)
Marble gall wasp

A common and often abundant pest of oak, especially English oak (*Quercus robur*) and sessile oak (*Quercus petraea*), and often of importance on young trees. Widespread in Europe.

DESCRIPTION **Adult (asexual) female:** 4–5 mm long; reddish-yellow with gaster partly black; legs pale; wings hyaline but tinged with red. **Larva:** plump, whitish.

LIFE HISTORY This cynipid forms large (up to 28 mm in diameter), smooth, green to reddish, marble-like galls on young oak trees (Fig. **973**). Each gall, which develops from the base of a bud from late spring onwards, surrounds a small central cavity and contains a single larva. The galls mature in August or September, adult female wasps emerging during the autumn or early in the following spring. These wasps deposit eggs in the axillary buds of Turkish oak (*Quercus cerris*), to produce an often overlooked bisexual spring generation of larvae. (In areas where Turkish oak does not exist, the insect is represented only by the asexual generation.) Mated adult females from this early-spring generation then initiate the familiar unisexual, summer generation of marble galls. Deserted marble galls, which soon become woody (Fig. **974**), may persist for several years and are often a common sight on scrub-oaks. The galls are commonly invaded by parasites and inquilines, these generally emerging in the spring through a series of small exit holes; parasitized galls may also remain small and become prematurely woody.

DAMAGE Infestations cause considerable distortion on young trees and can reduce the marketability of nursery stock. The galls are also disfiguring, especially when numerous.

CONTROL Rarely necessary but marble galls on nursery trees can be picked of and burnt.

Andricus curvator Hartig
Oak bud collared-gall cynipid

Infestations of this species occur abundantly on oak (*Quercus*). Female wasps of the asexual generation occur in February or March, initiating galls which eventually develop on the major veins or petioles of the young leaves. The galls cause considerable distortion of the foliage (Fig. **975**) and often attract

973 Galls of marble gall wasp, *Andricus kollari*.

974 Old galls of marble gall wasp, *Andricus kollari*.

975 Galls of oak bud collared-gall cynipid, *Andricus curvator*.

attention when present on small trees. Each gall, which contains a single larva in a large chamber, is pale green to light brown and measures 5–8mm across. The larvae feed during April and May, developing into male or female wasps which emerge in June. After mating, females of the bisexual generation deposit eggs between the scales of the leaf buds, where small, inconspicuous 'collared galls' are eventually formed. Such galls are commonly initiated in buds already hosting the 'larch-cone gall' generation of *Andricus fecundator* (see below). Development within the galls commences in the summer and extends throughout the following year, the complete life-cycle thus occupying two years. Unlike leaf galls of the asexual generation, the galls produced by larvae of the bisexual generation usually pass unnoticed.

976 Gall of larch-cone gall cynipid, *Andricus fecundator*.

977 Old galls of larch-cone gall cynipid, *Andricus fecundator*.

Andricus fecundator (Hartig)
Larch-cone gall cynipid

This species is responsible for transforming the buds of oak (*Quercus*) into artichoke-like larch-cone galls (Fig. **976**). Each gall, which begins its development in June, is up to 20mm long at maturity, and contains a single larva. Adult female wasps emerge from the galls in the following spring but their emergence may be delayed for up to three years. These wasps lay eggs in male flower buds on oak, a bisexual generation then developing in May to June inside hairy, catkin galls. The resulting adults mate, females then depositing fertilized eggs in leaf buds and thereby initiating the next round of larch-cone galls. Attacks are most common on scrub-oaks but also occur on nursery trees. The dead remains of old larch-cone galls (Fig. **977**) often persist on host plants for several years.

CONTROL Rarely necessary but larch-cone galls on nursery trees can be picked off and burnt.

Andricus quercuscalicis (Burgsdorf)
Acorn cup gall cynipid

A widespread and common pest in continental Europe and, having appeared in southern England in the 1960s, now also well established in England and Wales. Adult females (Fig. **978**) appear in the early spring, depositing eggs on the flower initials of Turkish oak (*Quercus cerris*), a bisexual genera-

978 Asexual female of acorn cup gall cynipid, *Andricus quercuscalicis*.

tion then developing in galls on the catkins. Later, after the production of males and females, eggs are deposited in the acorn primordia of English oak (*Quercus robur*). Characteristic 'knopper' galls (usually one per infested acorn) then develop from the acorn-cup tissue, each containing a single, whitish larva. During the summer the galled tissue expands into irregular, green, sticky outgrowths (Fig. **979**) which become brown and woody by the autumn (Fig. **980**); any acorns surviving within these infested cups are also malformed. Mature galls fall from the tree in October, still associated with the peduncle, the occupants pupating before the onset of winter, each in a small pupal cell within the shelter of the gall. Infestations of this wasp are sometimes heavy and can totally prevent acorn production on some trees; however, there appears to be no real effect on tree growth.

CONTROL Affected acorn cups on young trees can be removed and burnt but control measures on established trees are not worth while.

979 Young galls of acorn cup gall cynipid, *Andricus quercuscalicis*.

980 Mature galls of acorn cup gall cynipid, *Andricus quercuscalicis*.

Cynips divisa Hartig
syn. *verrucosa* (Schlechtendal)
Oak bud red-gall cynipid

A locally distributed cynipid, forming smooth, whitish-yellow to bright red, woody, thick-walled galls on the underside of the leaves of oak (*Quercus*). Each gall arises from a major vein (Fig. **981**) and encloses a small cavity within which a small, whitish larva develops. The galls represent the asexual generation and occur from July onwards; adults appear in October. Although sometimes numerous on young trees, the galls do not distort the foliage and are often overlooked. A sexual generation occurs in the spring, larvae developing in small insignificant galls associated with the young leaves or terminal buds.

981 Galls of oak bud red-gall cynipid, *Cynips divisa*.

Cynips longiventris Hartig
syn. *substituta* Kinsey
Oak leaf striped-gall cynipid

Asexual galls formed by this locally common species also occur on the underside of expanded leaves of oak (*Quercus*). They are irregular in shape, 7–8 mm across and slightly flattened, with a hard, roughened wall; each encloses a small larval cavity. The galls are whitish-yellow, more-or-less marked with red (Fig. **982**), and often appear striped. They develop during the summer months, reaching maturity in October. The sexual generation develops in the spring in adventitious buds on old oak trees.

982 Gall of oak leaf striped-gall cynipid, *Cynips longiventris*.

Cynips quercusfolii Linnaeus
syn. *taschenbergi* (Schlechtendal)
Oak leaf cherry-gall cynipid

A common and widely distributed cynipid, inducing the formation of cherry-like galls on the underside of the leaves of oak, mainly English oak (*Quercus robur*) and sessile oak (*Quercus petraea*). The smooth, rounded galls, each 15–20mm in diameter, are at first green or yellowish-green but eventually become yellowish-brown, flushed with red. The galls arise from the major veins but, in spite of their size, they do not cause distortion, even when several occur on the same leaf. On young trees, the galls are sometimes present in considerable numbers. Each gall contains a single larva, which develops within a small central cavity surrounded by spongy tissue (Fig. **983**). The galls reach maturity by October but remain attached to the leaves after leaf-fall. The adult wasps, although fully formed, delay their escape until mid-winter. These wasps (Fig. **984**) initiate a bisexual generation in dormant buds, adult males and females eventually appearing in June. Fertilized eggs are then deposited in leaf veins to give rise to the next unisexual cherry-gall generation.

983 Galls of oak-leaf cherry-gall cynipid, *Cynips quercusfolii*.

984 Asexual female of oak-leaf cherry-gall cynipid, *Cynips quercusfolii*.

Biorhiza pallida (Olivier)
syn. *aptera* (Fabricius)
Oak-apple gall wasp
Oak-apple galls, caused by this widespread cynipid, are often common on scrub-oaks (*Quercus*). The galls develop in spring, following egg laying in the base of axillary and terminal buds by newly emerged, wingless females. Affected buds swell rapidly into smooth, slightly irregular, whitish to brownish-yellow, spongy galls about 25–40mm in diameter (Fig. **985**). These galls, which contain several larvae, each within its own internal chamber, become extensively tinged with pinkish-red as maturity is reached, fully-winged adults of both sexes appearing in June or July. Vacated galls eventually darken and their remains often persist on trees long after the emergence of the original occupants. After mating, females of the summer generation enter the soil and give rise to a unisexual brood which develops inside root galls. Larvae of this generation complete their development in about 16 months. The wingless females then appear and ascend the trunks to initiate the next crop of oak-apple galls.

985 Young gall of oak-apple gall wasp, *Biorhiza pallida*.

7. Family **FORMICIDAE** (ants)

Ants, such as the generally common and well-known garden species *Lasius niger* (Linnaeus), are sometimes harmful to ornamental plants but are of only minor importance. They often ascend plants to collect nectar from flowers, occasionally also damaging young buds of trees and shrubs in their quest for moisture. More frequently, the presence of ants on plants is an indication that the leaves, shoots or branches are infested by honeydew-excreting pests such as aphids and scale insects; ants sometimes construct earthen shelters on plants to protect colonies of aphids, notably certain species of *Cinara* on conifers (p. 44 *et seq.*) and the rose root aphid, *Maculolachnus submacula* (p. 48). Subterranean activities by ants can be harmful, damage most often being restricted to the accidental disturbance of seedlings and established plants as soil around the roots becomes loosened; seriously affected plants wilt and die, damage being especially severe in hot, dry conditions. Ants are sometimes troublesome on lawns, especially on those freshly constructed in light, sandy soils inhabited by the yellow meadow ant (*Lasius flavus* (Fabricius)), this usually being the damaging species. Ants will also remove seeds from pots, seedboxes and seedtrays but losses are rarely significant.

CONTROL Eradication of ants' nests can be achieved by breaking them open and drenching them with a dilute insecticide solution. Alternatively, especially when nests are inaccessible, insecticide-treated baits may be deposited in areas where worker ants are known to forage.

8. Family VESPIDAE (social wasps)

Vespula spp.

Social wasps (Fig. **986**) are unimportant pests of ornamental plants but they can be a nuisance in gardens and nurseries, commonly establishing their colonies in banks, buildings, hollow trees and walls. Wasps frequently visit fences, sheds, shrubs and trees to scrape off pieces of dead wood which are formed into a papery material used in the construction of their nests; they also gnaw tissue from the woody stems of ornamentals such as *Dahlia*, injured plants then collapsing. Although sometimes injurious, wasps are also beneficial, especially during the spring and early summer when they collect large numbers of harmful caterpillars and other insect pests which they then feed to their developing brood.

986

986 Worker social wasp, *Vespula* sp.

9. Family ANDRENIDAE

A relatively large group of often very hairy, solitary, burrowing bees; the tongue short, ovate and pointed; the abdomen somewhat flattened dorsoventrally.

Andrena fulva (Müller in Allioni)
syn. *armata* (Gmelin in Linnaeus)
Tawny burrowing bee

An often abundant species, nesting in the soil and sometimes causing concern when its burrows are formed in lawns. Widely distributed in Europe.

DESCRIPTION **Adult female:** 12–14 mm long; thorax and abdomen clothed in bright brown hairs, those on the abdomen distinctly reddish; hairs on the face, legs, at tip of abdomen and on the underside of the body black. **Adult male:** 12–14 mm long; body clothed in mainly brown hairs.

LIFE HISTORY Adults appear in March and April, when they are active in warm, sunny weather and forage on various spring flowers. Mated females excavate deep (15–30 cm long) burrows in the soil, the excavated soil forming volcano-like mounds on the surface. Each burrow consists of a main tunnel, with a cell at the bottom, and a series of short, lateral branches, each also ending in a single cell. Into each cell the female places a quantity of nectar and pollen, upon which she then lays an egg; when all the cells have been provisioned, she seals the burrow with soil and flies away. After the eggs have hatched, the entombed larvae feed on their supplies of food; fully fed larvae eventually pupate, adult bees emerging in the following spring.

DAMAGE Heaps of excavated soil can be a nuisance, especially when fine lawns are invaded by large numbers of the bees attracted by particularly favourable nesting conditions.

CONTROL On fine lawns, excavated mounds of soil should be dispersed with a broom before mowing. Soil insecticides can provide protection on regularly infested sites (if applied during the period of nest building – late March to May) but their use is rarely justified, the nuisance value of the bees being outweighed by their value as pollinators.

10. Family **HALICTIDAE** (solitary bees)

Megachile centuncularis (Linnaeus)
Common leaf-cutter bee
Generally common, the adult females sometimes damaging the leaves of rose (*Rosa*) and other ornamental plants, including *Laburnum*, lilac (*Syringa vulgaris*), privet (*Ligustrum vulgare*), *Rhododendron* and snowy mespilus (*Amelanchier laevis*); petals of *Geranium* may also be damaged. Widely distributed in Europe.

DESCRIPTION **Adult female:** 10–12 mm long; black-bodied; head and thorax clothed in golden-brown hairs; abdomen clothed above with black hairs but banded with pale hairs, especially basally; pollen-collecting hairs on underside of abdomen orange-red and projecting beyond the sides to form an apparent fringe; legs pale-haired; wings smoky.

LIFE HISTORY Adults are active in June and July. They occur commonly in gardens, foraging during the daytime on various flowers from which they collect both nectar and pollen. When ready to breed, the females form elongate burrows in decaying wood, soft brickwork or light soil; each then collects several fresh leaf fragments from nearby rose bushes or other suitable plants, and carries these into her burrow to form a series of thimble-like brood cells. When completed, each cell is provisioned with a mixture of nectar and pollen, and an egg deposited on the surface; the cell is then capped with a leaf fragment and another cell constructed above it. A fully furnished nest usually contains a series of about six cells placed end to end; the burrow is then sealed with wood-pulp or

987 Common leaf-cutter bee damage to leaves of *Rosa*.

soil and abandoned. Larvae feed on their store of food from late summer onwards. They complete their development in the following spring and then pupate, adult bees emerging in June.

DAMAGE Attacked leaves have large, regular, oblong or semicircular pieces removed from the lamina (Fig. **987**), such damage often causing concern. Although plants are disfigured, growth is not affected.

CONTROL Application of a pesticide to discourage attacks is sometimes advocated but, because of the value of bees as pollinators, this is undesirable.

MITES

1. Family **PHYTOPTIDAE** (gall mites)

Phytoptus avellanae Nalepa
Filbert bud mite

Associated with hazel (*Corylus avellana*), inducing the formation of greatly enlarged buds ('big buds') (Fig. **988**). The mites breed within the shelter of the galled buds throughout the autumn and winter. In the early spring, female mites migrate from the galls to invade the leaves, where further eggs are laid. The eggs produce active protonymphs that feed and eventually moult into more-or-less sedentary deutonymphs; these, unlike adults and protonymphs, are distinctly flattened and have few abdominal tergites. In late June or July the summer deutonymphs moult into 'normal' adults. These invade new terminal buds which then swell and become very noticeable from September onwards. Such galls are often abundant in hedgerows but are also often numerous on ornamentals and nursery stock, affecting both the appearance and potential structure of young plants. The causal mites are whitish and elongate (*c.* 0.3 mm long), with numerous abdominal tergites and sternites, and a pair of short, anterior (frontal) setae and a pair of longer posterior (dorsal) setae on the prodorsal shield.

CONTROL Galled buds should be picked off and burnt during the autumn and winter, before female mites emerge in the spring.

988 Gall of filbert bud mite, *Phytoptus avellanae*.

989 Galls of *Phytoptus tetratrichus* on leaf of *Tilia*.

Phytoptus tetratrichus Nalepa
This species forms tight upward leaf-roll galls along the edge of the leaves of various kinds of lime (*Tilia*) (Fig. **989**). The galls cause slight distortion but are less noticeable than those formed on lime by the midge *Dasineura tiliamvolvens* (p. 159).

383

2. Family **ERIOPHYIDAE** (gall mites, rust mites)

Many different species of eriophyid mite are associated with ornamental plants, some forming galls or bronzing the leaves but others existing merely as inquilines or as leaf vagrants. Most species mentioned in the following account are indigenous to northern Europe but some are relatively new arrivals, having been introduced along with their host plants from other parts of the world. The nomenclature for eriophyid mites is very confusing, with many species referred to in the literature under two or more specific if not subspecific names. Confusion at generic level is also widespread (see footnote on p. 385). As a result it is often difficult, if not impossible, to assign correctly an observed plant symptom to any particular species or subspecies of mite. These problems are exacerbated by the occurrence in eriophyid galls of inquilines, which are not themselves the gall-forming species, and by the presence of several races or subspecies which may produce different symptoms on the same host and, sometimes, similar galls on different hosts; further difficulties arise in those eriophyids exhibiting deuterogeny (two or more structural female forms — a summer form or 'deutogyne' and a winter form or 'protogyne'). In this book, to reduce confusion and to aid correct cross-reference to names used in other publications, eriophyid mites are catalogued under their main host plants rather than in any taxonomic order. Also, under the host-plant entries, the various mites are treated alphabetically, without regard to their systematic hierarchy. Apart from occasional reference to colour, size and form, taxonomic descriptions of the mites have been excluded. N.B. Although most, if not all, of the more important pest species are included, the following account is far from exhaustive, there being many other eriophyids on ornamental plants, especially shrubs and trees. Also, various eriophyid galls are known that cannot be attributed with certainty, to any particular mite species.

Acer

Aculops acericola (Nalepa)
Sycamore gall mite
Although sometimes considered a gall-forming species, this mite is merely an inquiline in erinea produced on sycamore (*Acer pseudoplatanus*) by *Eriophyes psilomerus*, p. 386 (q.v.). The mites are *c.* 0.12 mm long with about 23 abdominal tergites; microscopically, therefore, the two species are readily distinguishable.

990 Young galls of maple bead-gall mite on leaves of *Acer campestre*.

991 Mature galls of maple bead-gall mite on leaf of *Acer campestre*.

Artacris cephaloneus (Nalepa)*
Maple bead-gall mite
A generally abundant species, responsible for the development of pimple-like bead galls on the leaves of maple (*Acer*) and sycamore. Female mites, which overwinter in sheltered situations on the shoots, become active in the early spring. They then invade the expanding foliage, each initiating the development of large numbers of galls. These galls occur from April onwards, changing from green to red as they mature (Figs. **990**, **991**). The

* See footnote on p. 385.

galls are often present in very large numbers, especially on young trees, but are harmless. Breeding continues in the galls throughout the summer, although many galls will be found on examination to be empty. The causal mites are *c.* 0.16 mm long and slender-bodied, with about 65 abdominal tergites and sternites and a pair of backwardly directed setae arising from tubercules on the hind margin of the prodorsal shield.

992 Galls of sycamore leaf gall mite on *Acer pseudoplatanus*.

Artacris macrorhynchus (Nalepa)*
Sycamore leaf gall mite

Widespread and generally common, inducing the development of elongate (2–4 mm long), dark red galls on the upper surface of the leaves of sycamore (Fig. **992**). The galls, which occur throughout the spring and summer, are distinctly longer than those of the previous species.

Eriophyes eriobius Nalepa

Infestations of this species induce the development of erinea on the underside of the leaves of maple. The galls commence as whitish patches (Fig. **993**) which later become flushed with purple. The upper surface of affected leaves is slightly discoloured but damage caused is unimportant. The causal mites (*c.* 0.2 mm long) are pale yellowish with about 68 abdominal tergites and sternites.

993 Galls of *Eriophyes eriobius* on underside of leaf of *Acer campestre*.

Eriophyes eriobius pseudoplatani Nalepa

This subspecies is associated with sycamore, forming yellowish to brownish erinea on the underside of the leaves. The galls occur from May onwards, and tend to be concentrated alongside and at the junctions of the major veins (Fig. **994**); cf. galls formed by *Eriophyes psilomerus* (p. 386). Although infested leaves are often extensively galled, the foliage is not distorted.

994 Galls of *Eriophyes eriobius pseudoplatani* on underside of leaf of *Acer pseudoplatanus*.

* The specific name *'cephaloneus'* (sometimes cited as *'macrorhynchus cephaloneus'*) is often restricted to the mites producing bead galls on maple; the specific name *'macrorhynchus'* (sometimes cited as *'macrorhynchus aceribus'*) is then applied to the mites on sycamore, without recognizing differences between the two types of gall found on the latter host.

Eriophyes macrochelus (Nalepa)

Maple leaf gall mite

An often abundant mite on ornamental maples, producing distinctive pouch galls on the leaves. The galls appear from May onwards, commencing as greenish warts but later becoming more-or-less red (Fig. **995**); they frequently arise at the junction of the major veins and are often very brightly coloured but eventually turn black. On heavily infested leaves, the galls coalesce and cause considerable distortion (Fig. **996**). Mites inhabiting the galls are *c.* 0.16 mm long and cylindrical, with about 60 abdominal tergites and sternites, and a relatively small pair of backwardly directed setae arising from tubercules on the hind margin of the prodorsal shield.

CONTROL Heavily infested leaves on young ornamentals should be removed and destroyed.

995 Young galls of maple leaf gall mite, *Eriophyes macrochelus.*

996 Maple leaf gall mite damage to foliage of *Acer.*

Eriophyes psilomerus Nalepa

The leaves of sycamore are often disfigured by large, irregular erinea induced by this generally common gall mite. Each gall appears as a pale green to brownish blister on the upper surface of the leaf, with the underside densely clothed in whitish to purplish hairs (Fig. **997**). Such galls occur from May onwards, gradually darkening as they mature. Although infested foliage looks unsightly, tree growth is not affected. The causal mites are *c.* 0.18 mm long with about 65 abdominal tergites and sternites (cf. *Aculops acericola*, p. 384).

997 Galls of *Eriophyes psilomerus* on underside of leaf of *Acer pseudoplatanus.*

Aesculus

Tegonotus carinatus Nalepa
Horse chestnut rust mite
This widespread and generally common species is free-living on the underside of the leaves of horse chestnut (*Aesculus hippocastanum*). When present in large numbers, the mites cause extensive bronzing and premature leaf-fall, infestations sometimes being of importance on nursery trees. The mites are *c.* 0.19 mm long and rather flattened, with few abdominal tergites; there are distinct overwintering and summer forms: protogynes and deuterogynes, respectively. The latter are thought to aestivate within bark crevices on the previous year's shoots when leaves harden during the early summer.

998 Galls of alder erineum mite, *Acalitus brevitarsus*, viewed from above.

Alnus

Acalitus brevitarsus (Fockeu)
Alder erineum mite
A generally common pest of various kinds of alder (*Alnus*), infested leaves becoming distorted by large, irregular, blister-like erinea. The upper surface of each gall is pale, somewhat warty and shiny (Fig. **998**); below, the galled tissue is densely coated with whitish to pale brown, multi-headed hairs (Fig. **999**). The galls develop from June to October, changing in colour from pale green, through pale yellow, to reddish-brown. The inhabitants are *c.* 0.16mm long, with about 60 abdominal tergites and sternites, and a pair of backwardly directed setae arising from tubercules close to the hind margin of the prodorsal shield. This species is deuterogenous.

999 Galls of alder erineum mite, *Acalitus brevitarsus*, viewed from below.

Acaricalus paralobus Keifer
Alder leaf rust mite
One of several free-living eriophyid mites responsible for bronzing of the foliage of alder. Affected foliage becomes dull and noticeably discoloured, heavy infestations reducing the vigour of young trees and nursery stock.

Eriophyes inangulis (Nalepa)

This mite induces the development of prominent swellings in the angles between the mid-rib and the major veins of leaves of alder, the position of each gall demarcated above by a discoloured, shiny swelling and below by a small patch of whitish to reddish-brown hairs (Fig. **1000**). The galls develop from May onwards, changing from green, through yellow and red, to brown.

1000 Galls of *Eriophyes inangulis* on leaf of *Alnus*.

Eriophyes laevis (Nalepa)
Alder bead-gall mite

A deuterogenous species, forming small, compact pimple-like galls on the upper surface of the leaves of alder. The galls often occur in vast numbers, sometimes several hundred on a leaf, and may cause significant distortion; attacks on established trees are of little importance but damage to young nursery stock may affect plant vigour. The galls develop from June to October, and vary in colour from green, through yellow, to purplish-brown (Fig. **1001**). The causal mites are relatively large (*c.* 0.28 mm long) with about 65 abdominal tergites and sternites, and a pair of short, backwardly directed setae arising from tubercules in front of the hind margin of the prodorsal shield.

1001 Galls of alder bead-gall mite, *Eriophyes laevis*.

Betula

Acalitus rudis (Canestrini)
syn. *rudis calycophthirus* (Nalepa) [in part]
Birch witches' broom mite

Abundant on birch (*Betula*) in association with various symptoms. In some cases, buds are invaded; these then fail to open and, instead, become greatly enlarged and cone-like (Fig. **1002**). Shoot growth on affected branches is disrupted but, because infestations are most frequently established on mature trees, damage caused is insignificant. The mites also inhabit witches' broom galls (Fig. **1003**) and were once considered to be their initiators; however, they are now generally accepted as merely inquilines, the growths being induced by fungi (*Taphrina* spp.).* The mites are *c.* 0.19 mm long, whitish and slender-bodied, with about 65 abdominal tergites and sternites, and a pair of backwardly directed setae arising from tubercules close to the hind margin of the prodorsal shield.

* Species of *Taphrina* are also responsible for witches' brooms on certain other trees and for various leaf-curl galls, including the well-known peach leaf curl (caused by *Taphrina deformans*); these galls are sometimes mistaken for pest-induced deformities, especially if invertebrates (such as mites or small insects) are sheltering within them.

1002 Cone gall of birch witches' broom mite, *Acalitus rudis.*

Aceria leionotus (Nalepa)
syn. *laevis lionotus* (Nalepa)
Birch bead-gall mite
An often common species, inhabiting small (*c.* 1mm diameter), red, pimple-like galls on the upper side of the leaves. Although sometimes numerous, the galls cause little or no distortion.

1003 Witches' broom galls on *Betula.*

Aceria longisetosus (Nalepa)
syn. *rudis longisetosus* (Nalepa)
This mite inhabits whitish, often reddish-tinged erinea on the underside of the leaves (Fig. **1004**). The galls may also develop on the upper surface of the foliage and are capable of causing noticeable distortion.

1004 Galls of *Aceria longisetosus* on underside of leaf of *Betula.*

Carpinus

Eriophyes macrotrichus (Nalepa)
Hornbeam leaf gall mite
A southerly-distributed species which induces an interveinal furrowing on the leaves of hornbeam (*Carpinus betulus*), the affected foliage becoming distinctly crinkled and discoloured (Fig. **1005**). Although damage is disfiguring and sometimes appears of some significance, plant growth is not affected. The causal mites are *c.* 0.16mm long and relatively plump.

1005 Hornbeam leaf gall mite damage to leaf of *Carpinus*.

Chrysanthemum

Epitrimerus alinae Liro
Chrysanthemum leaf rust mite
Free-living on glasshouse-grown *Chrysanthemum*, the stems of which become russeted, especially around the uppermost leaf petioles; such damage causes the leaves to wilt and may result in premature leaf-fall. The adult mites are 0.16mm long with about 45 abdominal tergites and a greater number of sternites.

Crataegus

Epitrimerus piri (Nalepa)
syn. *pyri* (Nalepa)
Pear rust mite
The foliage of hawthorn (*Crataegus*) is sometimes bronzed following the development of infestations of this generally common mite. The pest sometimes occurs on nursery stock but is more important in pear orchards, russeting both the foliage and the fruit.

Eriophyes pyri crataegi (Nalepa)
See under *Sorbus*, p. 400.

Phyllocoptes goniothorax (Nalepa)
Hawthorn leaf gall mite
This mite causes a tight downward leaf-edge rolling on hawthorn (_Crataegus_) (Fig. **1006**) and is often present on cultivated plants. The rolled edges vary in colour from pale green to yellowish-green, often with a reddish tinge, and sometimes result in noticeable distortion of the leaf lamina. Galling is apparent from spring to autumn and is often extensive, but not of significance, on unclipped hedges. The causal mites are whitish, _c._ 0.18 mm long, with about 54 abdominal tergites and sternites and a pair of small forwardly directed setae arising from tubercules in front of the hind margin of the prodorsal shield. Forms of this mite also occur in whitish erinea on the underside of the leaves. Erineum-producing forms of _Phyllocoptes goniothorax_ also occur on crab apple (_Malus_) and on rowan (_Sorbus aucuparia_). These are often regarded as subspecies: _Phyllocoptes goniothorax malinus_ (Nalepa) and _P. goniothorax sorbeus_ (Nalepa) respectively.

1006 Galls of hawthorn leaf gall mite, _Phyllocoptes goniothorax_.

Cytisus

Eriophyes genistae (Nalepa)
Broom gall mite
A locally common pest in parks and gardens, breeding throughout the spring and summer in galled buds of broom (_Cytisus_) and greenweed (_Genista_). Young buds are invaded in the spring, each then developing into a tight cluster of fleshy, pale green lobes covered with a downy coating of whitish hairs (Fig. **1007**). The galls measure about 20–30 mm across at maturity and often persist on infested plants for several years. Infested shoots are stunted and, if attacks persist, bushes can be severely disfigured. Mites inhabiting the galls are brownish, _c._ 0.13 mm long, with about 70 abdominal tergites and sternites, and a pair of backwardly directed setae arising from tubercules on the hind margin of the prodorsal shield.

CONTROL Galled buds should be cut out and burnt as soon as they appear; severely infested bushes should be grubbed and destroyed.

1007 Gall of broom gall mite, _Eriophyes genistae_.

Euonymus

Eriophyes convolvens (Nalepa)
Spindle leaf-roll gall mite

This mite causes an upward leaf-edge rolling on spindle (*Euonymus*) (Fig. **1008**), attacks often being very extensive. Galling is sometimes noted on cultivated bushes but, although unsightly, does not affect plant growth. The mites are whitish and relatively small (*c.* 0.11 mm long), with about 60 abdominal tergites and sternites, and a pair of small, forwardly directed, convergent, setae arising from tubercules in front of the hind margin of the prodorsal shield.

1008 Galls of spindle leaf-roll gall mite, *Eriophyes convolvens.*

Fagus

Aceria stenaspis (Nalepa)
syn. *stenaspis plicans* (Nalepa)

A locally common pest of beech (*Fagus sylvatica*), the mites causing considerable malformation and stunting of young leaves, and also death of the opening buds (Fig. **1009**). The mites, which breed within the shelter of the galled tissue throughout the spring and summer, are whitish and *c.* 0.14 mm long, with about 75 abdominal tergites and sternites. Damage frequently occurs on garden hedges but is often overlooked. A species of *Acaricalus* is also associated with leaf deformation on beech but this may be merely an inquiline in buds damaged by *Aceria stenaspis*.

CONTROL Affected shoots should be cut out and burnt as soon as leaf distortion is seen.

1009 *Aceria stenaspis* damage to young shoots of *Fagus.*

Aceria stenaspis stenaspis (Nalepa)

This subspecies breeds in marginal leaf-roll galls formed on the extreme edge of infested leaves, the tissue rolling over the upper surface to form a tight, hair-lined tube 1–2 mm in diameter (Fig. **1010**); distortion and discoloration may also spread onto the lamina of affected leaves. The galls occur from late April or May onwards, changing from green to brown as they mature.

1010 Galls of *Aceria stenaspis stenaspis* on leaf of *Fagus*.

Eriophyes nervisequus (Canestrini)
Beech leaf-vein gall mite

A common species, forming hairy, white to brownish ridges along the major lateral veins on the upper surface of leaves of beech (Fig. **1011**). The galls are most often noticed in June, during the early stages of development before the hairs darken. The mites inhabiting these galls are *c.* 0.13 mm long with about 60 abdominal tergites and sternites, and a pair of backwardly directed setae arising from tubercules on the hind margin of the prodorsal shield.

1011 Galls of beech leaf-vein gall mite, *Eriophyes nervisequus*.

Eriophyes nervisequus fagineus Nalepa
Beech erineum gall mite

This locally common subspecies is responsible for the formation of conspicuous erinea between the major veins on the underside of the leaves. The galls develop throughout the summer months, commencing as pale patches of enlarged, club-shaped hairs, amongst which various stages of the mite may be found; the erinea soon turn red (Fig. **1012**) and finally brown. Heavy infestations on young leaves may cause the laminae to curl at the edges.

1012 Galls of beech erineum gall mite, *Eriophyes nervisequus fagineus*.

Fraxinus

Aculus epiphyllus (Nalepa)
Ash rust mite

This generally common mite causes extensive bronzing of the foliage of ash (*Fraxinus excelsior*) and is often present on nursery trees. In severe cases, the foliage becomes brittle and distorted (Fig. **1013**), and the tips of new shoots turn black and die (Fig. **1014**) (cf. damage caused by the ash bud moth, *Prays fraxinella*, p.000). Loss of terminal growth often results in the forking of young trees and this can be a serious problem in nurseries. The pale, yellowish, pear-shaped mites are *c.* 0.15 mm long with about 30 abdominal tergites, and a pair of very short, backwardly directed setae arising from widely spaced tubercules on the hind margin of the prodorsal shield (cf. *Tegonotus collaris*, p.000). When present in large numbers, the mites are clearly visible against the darker background of infested leaves or bud scales.

CONTROL Heavily affected nursery stock should be grubbed and burnt. Application of an acaricide in early spring can reduce the extent of damage but may not be entirely effective.

Eriophyes fraxinivorus Nalepa
Ash inflorescence gall mite

A widespread species, causing a distinctive galling of the inflorescences of ash. The mites overwinter in bark crevices. In the spring, they invade the emerging inflorescences, depositing eggs and producing a succession of overlapping generations throughout the summer. The mites cause considerable distortion, the flower clusters swelling into a series

1013 Ash rust mite damage to leaves of *Fraxinus*.

1014 Ash rust mite damage to young shoots of *Fraxinus*.

of brownish lumps up to 20mm across (Fig. **1015**); galled pedicels may also coalesce. The galled inflorescences remain attached to trees throughout the year and are especially obvious after leaf-fall. Infestations are often heavy on mature trees but any effect on vegetative growth appears to be slight. Adult mites are *c*. 0.18mm long with about 65 abdominal tergites and sternites, and a pair of backwardly directed setae arising from tubercules on the hind margin of the prodorsal shield.

Tegonotus collaris Nalepa

Often reported on bronzed foliage of ash, usually in company with *Aculus epiphyllus* (p. 394). The mites are *c*. 0.16mm long and rather stumpy, with about 13 broad, roof-like abdominal tergites and a pair of posteriorly directed setae arising from tubercules on the hind margin of the prodorsal shield.

1015

1015 Galls of ash inflorescence gall mite, *Eriophyes fraxinivorus*.

Juglans

Eriophyes erineus (Nalepa)
Walnut leaf gall mite

A generally abundant mite, responsible for the formation of erinea on the leaves of walnut (*Juglans*). The galls appear as large, reddish-tinged blisters on the upper surface of the expanded leaves (Fig. **1016**); the lower surface of the galls is coated with a felt-like mass of whitish hairs, within which the mites breed. Galling is often extensive on both nursery stock and mature trees but is of little or no importance. The causal mites are *c*. 0.22 mm long and whitish, with a pair of moderately long, posteriorly directed setae arising from tubercules on the hind margin of the prodorsal shield.

1016

1016 Galls of walnut leaf gall mite, *Eriophyes erineus*.

Malus

Aculus schlechtendali (Nalepa)
Apple rust mite

Although more frequently reported as a pest in apple orchards, infestations of this generally abundant species also occur on crab apple (*Malus*) and are occasionally of significance on young trees. The mites are *c*. 0.17mm long and yellowish-brown; they are deuterogenous, breeding throughout the spring and summer on the underside of leaves, the protogynes sheltering during the winter in bark crevices and beneath bud scales. Heavy infestations lead to bronzing and shrivelling of the foliage, and may also affect the growth of new shoots.

Eriophyes pyri (Pagenstecher)
Pear leaf blister mite
See under *Sorbus*, p. 400.

Phyllocoptes goniothorax malinus (Nalepa)
Apple leaf erineum mite
See under *Crataegus*, p. 391.

Populus

Aceria dispar (Nalepa)

Associated with aspen (*Populus tremula*), the mites infesting the young, lateral shoots; affected leaves are rolled and crinkled, the internodes of heavily infested shoots becoming noticeably shortened to form brush-like clumps of small, deformed leaves.

Phyllocoptes populi (Nalepa)
Poplar erineum mite

Associated with aspen and black poplar (*Populus nigra*), forming erinea on the underside of the leaves; the upper surface of affected leaves is disfigured by the development of pallid, blister-like areas (Fig. **1017**). These galls should not be confused with similar distortions caused on poplar by the generally common fungal disease *Taphrina populina*, which produces bright yellow blisters on the underside of the leaves; in the case of the disease the upper surface of an infected leaf, although noticeably distorted, remains green.

1017 Galls of poplar erineum mite, *Phyllocoptes populi*.

1018 Plum rust mite damage to leaf of *Prunus*.

Prunus

Aculus fockeui (Nalepa & Trouessart)
Plum rust mite

Although mainly a pest in damson and plum orchards, infestations of this generally common species may also occur on various other kinds of *Prunus*, including ornamentals. Affected foliage becomes distinctly bronzed and may develop a characteristic yellowish flecking (Fig. **1018**); small blotches sometimes also appear on the new wood of the young shoots. Damage of greatest significance is caused in the early spring by the overwintered mites feeding on the newly developing tissue. This species is deuterogenous, with distinct over-wintering and summer forms. Protogynes are *c.* 0.17mm long with about 30 abdominal tergites and 50 sternites; deutogynes are *c.* 0.16mm long with about 32 abdominal tergites and sternites.

1019 Galls of plum leaf gall mite, *Eriophyes padi*.

Eriophyes padi (Nalepa)
Plum leaf gall mite

The erect, dark red leaf galls produced by this locally common species occur on blackthorn (*Prunus spinosa*) and certain other species of *Prunus*, and are sometimes of significance on nursery stock. The galls, which are very noticeable (Fig. **1019**), tend to occur in tight clusters towards the mid-section of the base of the lamina but cause only slight distortion; they do not affect tree growth. Occupants of the galls are *c.* 0.22 mm long with about 55 abdominal tergites and sternites.

1020 Galls of plum pouch-gall mite, *Eriophyes similis*.

Eriophyes similis (Nalepa)
Plum pouch-gall mite

A widespread and locally common pest, often established on blackthorn and sometimes causing considerable distortion of the leaves. Infestations may also appear on other kinds of *Prunus*, including nursery trees, the pale greenish to yellowish or reddish, pouch-like galls (unlike those of *Eriophyes padi*, q.v.) tending to occur around the periphery of the leaves (Fig. **1020**). The causal mites are *c.* 0.2 mm long with about 50 abdominal tergites and sternites.

CONTROL If necessary, infested leaves on young plants should be cut off and burnt.

1021 Azalea bud and rust mite damage to young shoot of *Rhododendron* 'Mollis'.

Rhododendron

Aculus atlantazaleae (Keifer)
Azalea bud and rust mite

Although formerly restricted to North America, this pest is now established in various parts of Europe, including southern England and the Netherlands. Infestations occur on certain kinds of azalea (*Rhododendron*), especially 'Mollis' varieties. The pale yellowish-brown mites (*c.* 0.17 mm long) feed on the leaf bases and buds, causing considerable distortion (Fig. **1021**). They also occur on the expanding or expanded leaves, infestations resulting in noticeable bronzing of the foliage (Fig. **1022**).

CONTROL Apply an acaricide at the first signs of damage, selecting a product recommended for use against eriophyid mites.

1022 Azalea bud and rust mite damage to leaves of *Rhododendron* 'Mollis'.

Robinia

Vasates allotrichus (Nalepa)
False acacia rust mite

Infestations of this mite occur on false acacia (*Robinia pseudoacacia*), puckering and rolling the leaves; heavily infested ·foliage also becomes blackened and brittle (Fig. **1023**). The mite is often responsible for extensive damage to nursery and specimen trees, affected shoots or branches developing a dull, sickly appearance. The pale yellowish adults are *c.* 0.15 mm long and stumpy, with about 45 abdominal tergites and a slender pair of backwardly directed setae arising from tubercules on the hind margin of the prodorsal shield (cf. *Vasates robiniae*).

CONTROL On a small scale, heavily affected foliage may be cut out and burnt; on young trees, application of a spring acaricide may reduce the extent of damage.

Vasates robiniae (Nalepa)

This mite also infests false acacia, causing discoloration and a marginal leaf rolling (Fig. **1024**). Adults are similar to those of *Vasates allotrichus* (q.v.) but have fewer (*c.* 25) abdominal tergites and stronger prodorsal-shield setae.

1023 False acacia rust mite damage to leaf of *Robinia*.

1024 *Vasates robiniae* damage to leaves of *Robinia*.

Salix

Aculops tetanothrix (Nalepa)
Willow leaf gall mite

Galls formed by this locally common mite occur on willow, especially crack willow (*Salix fragilis*). They occur on the leaves from May onwards, and are often present in considerable numbers as irregular, hairy, yellowish-green to red swellings, each 2–4 mm in diameter (Fig. **1025**). Occupants of these galls are *c.* 0.19 mm long and spindle-shaped, with about 60 abdominal tergites and a larger number of sternites.

1025 Galls of willow leaf gall mite, *Aculops tetanothrix*.

Aculus truncatus (Nalepa)

This mite inhabits distinctive leaf-edge galls formed on the upper surface of the leaves of purple willow (*Salix purpurea*). On suitable hosts, galling is often extensive and may affect a considerable proportion of the leaves. Each gall becomes distinctively reddened throughout its length (Fig. **1026**) and contains numerous whitish mites (cf. galls formed on *Salix viminalis* by the midge *Rhabdophaga marginemtorquens*, p. 164).

1026 Galls of *Aculus truncatus* on leaves of *Salix purpurea*.

Eriophyes triradiatus (Nalepa)
Willow witches' broom gall mite

This widely distributed and generally common species occurs in witches' broom galls (Fig. **1027**) on various kinds of willow, including ornamentals such as weeping willow (*Salix vitellina* var. *pendula*). Although usually abundant in the galls, breeding within them throughout the summer, the mites are not considered to be involved in their initiation or development. Adult females are *c.* 0.17 mm long with about 80 abdominal tergites and sternites, and a pair of upwardly directed prodorsal-shield setae.

1027 Young gall of willow witches' broom gall mite, *Eriophyes triradiatus*.

Sambucus

Epitrimerus trilobus (Nalepa)
Elder leaf mite

Widely distributed in association with elder (*Sambucus*), causing considerable distortion and discoloration of affected foliage (Fig. **1028**). Infestations often occur on ornamental bushes, including nursery stock, but are most common on wild hosts.

1028 Elder leaf mite damage to leaves of *Sambucus*.

Sorbus

Aculus aucuparia Liro

A free-living species on rowan (*Sorbus aucuparia*); infestations are sometimes noted on nursery stock, the mites causing a distinct bronzing of the foliage.

Eriophyes pyri (Pagenstecher)
Pear leaf blister mite

Although most important as a pest of pear (*Pyrus*) this species or species complex is also associated with other rosaceous hosts, including various ornamentals. The elongate (*c.* 0.22 mm long), brownish mites overwinter under bud scales, becoming active in the early spring when they invade the unfurling leaves. Small, pale green to yellowish, blister-like galls are then formed on the foliage, each with a tiny aperture on its lower surface; mites breed within these chambers throughout the summer, colony development terminating in the autumn, following the production of overwintering forms. The galls, which appear as pale blisters and gradually darken as they mature, are often common on rowan (Fig. **1029**), whitebeam (*Sorbus aria*) and wild service tree (*Sorbus torminalis*); galling also occurs on crab apple (*Malus*) and on hawthorn (*Crataegus*). The pest is usually of minor importance but heavy attacks on young ornamentals can be troublesome. On certain hosts the mites are sometimes afforded subspecific status: e.g. *Eriophyes pyri sorbi* Nalepa on rowan; *Eriophyes pyri arianus* (Pagenstecher) on whitebeam; *Eriophyes pyri torminalis* Nalepa on wild service tree. The mites on hawthorn are often described as *Eriophyes pyri crataegi* (Canestrini) or treated as a distinct species: *Eriophyes crataegi* (Canestrini). Also, those on *Sorbus* are sometimes referred to as *Eriophyes sorbi* (Canestrini).

CONTROL Where practical, infested leaves should be removed and burnt. Application of an acaricide at bud-burst can reduce infestation levels but is of limited value and rarely justified.

1029 Pear leaf blister mite damage to leaves of *Sorbus aucuparia*.

Phyllocoptes goniothorax sorbeus (Nalepa)
See under *Crataegus*, p. 391.

Syringa

Eriophyes löwi Nalepa
Lilac bud mite

A widely distributed pest of lilac (*Syringa vulgaris*). The mites infest the buds, causing desiccation and a proliferation of dwarfed lateral shoots. Expanded leaves are also invaded, affected foliage becoming discoloured and the leaf edges distorted (Fig. **1030**). The mites are *c.* 0.17mm long with about 60 tergites and sternites. Infestations also occur on privet (*Ligustrum vulgare*).

1030 Lilac bud mite damage to leaves of *Syringa*.

Taxus

Cecidophyopsis psilaspis (Nalepa)
Yew gall mite

A widely distributed and generally common pest of yew (*Taxus baccata*); especially abundant on nursery stock and on regularly clipped bushes or hedges which provide an abundance of new growth. Breeding occurs throughout the year within enlarged buds ('big buds'), the mites emerging in large numbers in the spring and then invading terminal and lateral buds on the young shoots. Infested buds fail to open and, in their turn, develop into the characteristic mite-laden galls (Fig. **1031**); attacked buds and surrounding foliage also become blackened (Fig. **1032**). Attacks on young plants are especially disfiguring, shoot growth being attenuated and distorted (often reminiscent of herbicide damage); damage on mature hosts is of little or no consequence. The causal mites (*c.* 0.16mm long) are whitish with about 75 abdominal tergites and sternites; they lack setae on the prodorsal shield.

1031 Gall of yew gall mite, *Cecidophyopsis psilaspis*.

CONTROL Trimmings containing 'big buds' should not be left lying about beneath clipped bushes but should be gathered up and burnt. On a small scale, hand-picking and destruction of galls may be worth while. Chemical control measures are of limited value but an acaricide applied in the spring during the period of mite invasion will reduce mite numbers and can offer some protection to the new buds.

1032 Yew gall mite damage to young shoot of *Taxus*.

Tilia*

Aceria exilis (Nalepa)
syn. *tiliae exilis* (Nalepa)
Leaf galls formed by this mite at the junction of two major veins may be found on lime (*Tilia*) from May or June onwards. The galls develop on the upper surface of the leaf as small, hairy, greenish-yellow to brownish pimples, the position of each being marked below by pale brownish hairs; such hairs also line the gall's inner surface. The mites are similar in appearance to *Eriophyes tiliae* (see below) but have fewer (*c.* 60) abdominal tergites.

Aculus ballei (Nalepa)
An often common, free-living species on lime, sometimes causing significant leaf bronzing on young trees and nursery stock. Affected foliage appears dull and sickly (Fig. **1033**), heavy infestations affecting the vigour of host plants.

CONTROL Application of an acaricide in the spring can reduce the extent of damage but may not be entirely effective.

Eriophyes leiosoma (Nalepa)
syn. *tiliae liosoma* (Nalepa)
Lime leaf erineum mite
A generally common species inhabiting large, irregular, white, downy patches on the underside of the leaves of lime (Fig. **1034**). The galls appear from May onwards, the upper surface appearing pale green; the galls are well developed by mid-summer, affected tissue eventually turning brown. Although a considerable proportion of the leaf area may be affected, the galls are relatively shallow and usually cause little or no distortion of the lamina. However, attacks are sometimes heavy on small trees, resulting in considerable disfigurement.

CONTROL Affected leaves on small trees should be picked off and burnt but control measures on larger trees are not worth while.

Eriophyes tiliae (Nalepa)
syn. *gallarumtiliae* (Turpin); *tiliae* (Pagenstecher)
Lime nail-gall mite
An often abundant species, responsible for the elongated, tack-like galls which often occur in vast numbers on the upper surface of the leaves of large-leafed lime (*Tilia platyphyllos*) (Fig. **1035**).

* See also p. 383.

1033 *Aculus ballei* damage to leaf of *Tilia*.

1034 Galls of *Eriophyes leiosoma* on leaf of *Tilia*.

1035 Galls of lime nail gall mite on leaf of *Tilia platyphyllos*.

The galls are up to 15 mm long and vary from pale greenish to red or brown. They develop from May or June onwards but, although disfiguring the foliage of ornamental trees, appear to have little or no effect on plant growth. The causal mites are elongate with about 75 abdominal tergites and sternites.

Eriophyes tiliae lateannulatus Schulze
syn. *tiliae rudis* Nalepa

This subspecies is associated with small-leafed lime (*Tilia cordata*), producing nail galls (Fig. **1036**) that are much smaller (up to 5 mm long) than those formed on large-leafed lime by *Eriophyes tiliae*; the mites are sometimes regarded as a distinct species.

1036 Galls of *Eriophyes tiliae lateannulatus* on leaf of *Tilia cordata*.

Ulmus

Aculus ulmicola (Nalepa)
Elm bead-gall mite

An often abundant species, responsible for the development of tiny (*c.* 1 mm diameter) bead-like galls on the upper surface of the leaves of elm (*Ulmus*) (Fig. **1037**). Adult females overwinter under the bud scales, invading the underside of the unfurling leaves in the spring and eventually inducing the formation of the distinctive galls. Breeding continues within the galls throughout the late spring and early summer, at the height of their development each gall containing up to 200 or more mites. Infestations are often very extensive, vast numbers of the galls occurring on each infested leaf and the foliage of affected branches developing a distinctly roughened appearance. Heavy infestations have a detrimental effect on host plants, reducing their resistance to severe weather conditions and leaving them more susceptible to other disorders. Mites inhabiting these galls are *c.* 0.17 mm long with about 55 abdominal tergites and characteristic, two-branched feather claws.

1037 Galls of *Aculus ulmicola* on leaves of *Ulmus*.

CONTROL Heavily galled foliage on young trees should be cut off and burnt; application of an acaricide in the early spring might also be worth while but is usually impractical on established trees.

Eriophyes filiformis (Nalepa)
Elm leaf blister mite

This southerly-distributed species causes the development of pouch-like galls on the leaves of elm (Fig. **1038**), affected leaves sometimes occurring on amenity trees. The galls are much larger than those formed by the previous species but are usually less numerous; they may cause slight distortion of the foliage but shoot growth is not affected. The causal mites are *c.* 0.17 mm long, with about 90 abdominal tergites.

1038 Galls of *Eriophyes filiformis* on leaf of *Ulmus.*

Yucca

Cecidophyopsis hendersonii (Keifer)
A North American species, recently introduced into Europe on imported *Yucca* plants; in Europe, first reported on glasshouse-grown *Yucca elephantipes* in Scandinavia, the upper surface of affected fronds appearing dusted with whitish powder. The mites are relatively stumpy and lack setae on the prodorsal shield.

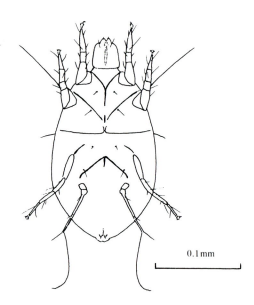

0.1 mm

1039 Female cyclamen mite, *Phytonemus pallidus*, ventral view.

3. Family **TARSONEMIDAE**

Small, elliptical, pale brown to whitish mites with a distinct head-like gnathosoma, short, needle-like chelicerae and pronounced sexual dimorphism; the hind legs of females are clawless and those of males elaborated as claspers.

Phytonemus pallidus (Banks)
Cyclamen mite

A major pest of glasshouse ornamentals, including African violet (*Saintpaulia*), *Aralia*, azalea (*Rhododendron*), *Begonia*, *Cyclamen*, *Gerbera*, *Gloxinia*, *Impatiens*, ivy (*Hedera*), *Pelargonium*, *Petunia* and *Verbena*; in favourable situations infestations also survive on outdoor plants, a distinct biological race being associated with Michaelmas daisy (*Aster*). Virtually cosmopolitan. Widely distributed in Europe.

DESCRIPTION **Adult female:** 0.25 mm long; pale brown and translucent; body oval-elongate and somewhat barrel-shaped; gnathosoma longer than broad, with the palpi directed forwards; hind legs very thin, each terminating in a long, whip-like seta (Fig. **1039**). **Adult male:** 0.2 mm long; pale brown and oval-bodied; hind legs broad, each with a very large femur bearing a rounded inner flange and terminating in a strong claw (Fig. **1040**). **Egg:** 0.125 × 0.075 mm; elliptical, semitransparent and whitish. **Larva:** whitish with hind part of body triangular; six-legged.

LIFE HISTORY In glasshouses this species is active throughout the year, breeding continuously whilst conditions remain favourable. The mites are light-shy and tend to occur on the young, succulent tissue of host plants, all stages (eggs, larvae, quiescent nymphs and adults) sheltering within leaf folds, amongst leaf hairs and between bud scales. As the tissue ages and hardens, the mites move to younger, more suitable feeding and breeding sites, commonly invading the still-furled leaves and un-opened flower buds. The mites may also spread from plant to plant, especially if leaves or shoots of adjacent hosts overlap, but they rarely if ever wander over the soil or the glasshouse staging. There are several overlapping generations annually, mites passing from egg to adult in about 2–3 weeks at temperatures of 20–25°C; the rate of development is much reduced at lower temperatures, the egg stage becoming especially protracted. Although males occur, especially during the summer months, they are usually greatly outnumbered by females and reproduction is mainly parthenogenetic. There are also several overlapping generations annually on outdoor plants, populations reaching a peak in August or September, but breeding usually ceases completely during the winter months.

DAMAGE Infested foliage becomes brittle, dis-coloured and crinkled, the margins of young leaves often rolling tightly inwards (Fig. **1041**); flower buds are also affected. Attacked plants are stunted and young growth significantly distorted; if infestations are severe, leaves, flower buds or complete plants can be killed. Mites on Michaelmas daisy (especially *Aster novi-belgii* can cause severe scarring of flower stems, affected 'flowers' being converted into rosettes of small, green leaves.

CONTROL Apply an acaricide specifically recommended for use against tarsonemid mites, and repeat as appropriate; many acaricides, although active against spider mites, are ineffective against this and related species.

1040 Hind leg of male cyclamen mite, *Phytonemus pallidus*.

1041 Cyclamen mite damage to leaf of *Aralia*.

Phytonemus confusus (Ewing)

This species occurs in England and in various parts of mainland Europe, where it was first reported in 1957, having previously been found only in North America. The mites are associated with glasshouse ornamentals, including African violet (*Saintpaulia*), azalea (*Rhododendron*), *Cissus*, *Cyclamen*, *Gloxinia*, ivy (*Hedera*) and *Pilea*. They often occur in association with *Phytonemus pallidus* (p. 404) or *Polyphagotarsonemus latus* (p. 407), but are usually present in smaller numbers; they do not cause primary damage. The mites are similar in appearance to *Phytonemus pallidus* (q.v.) but smaller; apart from size and detailed microscopical features (Fig. **1042**), they are best distinguished by the absence in males of flanges on the inner edge of the hind legs.

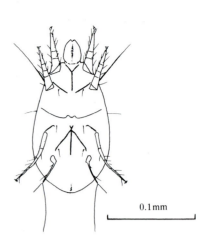

1042 Female *Phytonemus confusus*, ventral view (after Karl).

Steneotarsonemus laticeps (Halbert)

syn. *approximatus* (Banks)

Bulb scale mite

An important pest of *Amaryllis* and forced *Narcissus*; also associated with certain other members of the Amaryllidaceae, including *Eucharis*, Scarborough lily (*Vallota purpurea*) and *Sprekelia*. Present in several parts of Europe, including England, Ireland, the Netherlands and Sweden; also found in North America.

DESCRIPTION **Adult female:** 0.2 mm long; pale brown and translucent; gnathosoma broader than long; palpi directed inwards; hind legs thin, each terminating in a long seta. **Adult male:** similar to female but smaller and with the hind legs forming strong claspers. **Egg:** 0.1 mm long; oval, translucent-whitish. **Larva:** similar to adult but smaller and six-legged.

1043 Bulb scale mite damage to bulb of *Narcissus*.

LIFE HISTORY Bulbs are invaded in August and September, the mites entering the spaces between the shrinking scales to feed on the surface of the tissue, especially around the neck region. Breeding is continuous whilst conditions remain suitable, the life-cycle (which includes egg, larval and quiescent nymphal stages) being completed in about a month at 20°C but in about two weeks at bulb-forcing temperatures. Mite development is greatly protracted on infested narcissus bulbs planted out in the autumn but does increase in response to warmer conditions in the following spring and early summer; at this stage the mites occur both within the bulbs and on aerial parts of the plants but by the time of lifting in mid-summer most will have moved back into the neck region.

DAMAGE **Amaryllis**: vegetative growth from infested bulbs becomes spotted, and streaked or scarred with red; flowers are malformed and may wither. **Narcissus**: infested bulbs lack vigour, producing weakened, distorted, often sickle-shaped leaves and small, malformed flowers. Emerging foliage tends to be bright green, lacking the normal greyish bloom, and later becomes streaked with yellow and scarred, the leaf edges appearing saw-like. Heavy attacks result in a marked reduction in both crop yield and flower quality; they also cause early senescence of foliage and, sometimes, death of bulbs. Bulbs in store become very dry and, especially in the neck region, display brown streaks of dead tissue if sliced across (Fig. **1043**).

CONTROL Hot-water and fumigant treatments are effective; alternatively, infested plants can be sprayed as recommended for control of *Phytonemus pallidus* (p. 405). Forced narcissus bulbs should be drenched thoroughly with a dilute acaricide shortly after bulbs are boxed.

Polyphagotarsonemus latus (Banks)
Broad mite
A tropical and subtropical pest which, in temperate countries, infests a wide variety of glasshouse plants including ornamentals such as African violet (*Saintpaulia*), *Begonia*, *Chrysanthemum*, *Cyclamen*, *Dahlia*, *Gloxinia*, *Fuchsia*, *Gerbera*, *Hibiscus*, *Impatiens*, ivy (*Hedera*) and stock (*Mathiola*). Well established in Europe.

DESCRIPTION **Adult female:** 0.14–0.24 mm long; whitish and translucent but often greenish or yellowish; body very broad and oval (cf. *Phytonemus pallidus*, p. 404). **Adult male:** 0.11–0.17 mm long; whitish and translucent; body short and broad but tapered posteriorly; legs long, the hind pair relatively stout. **Egg:** 0.11 × 0.07 mm; flattened, smooth ventrally; several rows of large, white tubercules dorsally. **Larva:** similar to adult but smaller and six-legged.

LIFE HISTORY The mites feed mainly on the underside of leaves but will also invade unopened or unfurling buds and other plant tissue. Female mites, which greatly outnumber males, normally live for about ten days, each depositing up to 50 eggs. Breeding is rapid in warm conditions, eggs hatching in 2–3 days and larvae feeding for four days at normal glasshouse temperatures. The quiescent nymphal stage is passed entirely within the larval skin, adult males often carrying female 'nymphs' around, grasping them in their pincer-like hind legs.

DAMAGE New growth of infested plants becomes stunted and discoloured (Fig. **1044**), and often shiny, brittle and distorted (Fig. **1045**). On some hosts, such as cyclamen (Fig. **1046**), flowers are malformed and unopened buds may drop off. Heavily infested plants can be killed.

CONTROL Apply an acaricide as soon as damage is seen, and repeat at intervals of 2–3 weeks as necessary.

1044

1044 Broad mite damage to leaves of *Gloxinia*.

1045

1045 Broad mite damage to leaf of *Fatsia*.

1046

1046 Broad mite damage to flowers of *Cyclamen*.

Hemitarsonemus tepidariorum (Warburton)
Fern mite
Locally distributed in England as a pest of glass-house-grown ferns, including *Polystichum* and *Pteris*, but present mainly on *Asplenium bulbiferum*.

DESCRIPTION **Adult female:** 0.23 mm long; pale yellowish-brown; body elongate-oval, the gnathosoma longer than broad and with the palpi directed forwards. **Adult male:** 0.15–0.16 mm long; pale yellowish-brown; hind leg with a broad triangular tooth on the inner margin of the tibia and a prominent tarsal claw; tibia and tarsus both longer than femur (cf. *Phytonemus pallidus*, p. 404, and *Polyphagotarsonemus latus*, p. 407). **Egg:** 0.11–0.12 mm long; oval and whitish. **Larva:** similar to adult but smaller and six-legged.

LIFE HISTORY Infestations are encouraged by warm, dark and humid conditions, the mites breeding continuously throughout the year with a succession of overlapping generations. Eggs are laid singly or in small groups close to the tips of the fronds (within the shelter of the furled leaflets or pinnae) or at the top of the rhizome between the scales. The eggs hatch within a few days at normal glasshouse temperatures but their development is greatly protracted in cool conditions. The larvae feed for 1–2 weeks before entering the quiescent nymphal stage, adults appearing 3–4 days later. Adults and larvae feed on the youngest tissue, imbibing sap from the surface cells which then collapse. All developmental stages and both sexes occur in abundance throughout the summer but winter populations consist mainly of adult females and eggs.

DAMAGE Infested leaves are speckled with brown, attacked fronds becoming distorted and swollen. The growth of heavily infested plants is checked, the fronds turning brown and dying.

CONTROL As for *Polyphagotarsonemus latus* (p. 407).

4. Family **TETRANYCHIDAE** (spider mites)

Spider-like mites with long, needle-like chelicerae and a thumb-claw on each palp. They develop from egg to adult through larval, protonymphal and deuteronymphal stages.

Bryobia kissophila van Eyndhoven
Ivy bryobia mite
Common on wild ivy (*Hedera helix*) and often a troublesome pest on ornamental varieties. Present throughout Europe.

DESCRIPTION **Adult female:** 0.7 mm long; dark reddish-brown or red; body oval and rather flat, with spatulate dorsal setae; first pair of legs very long. **Egg:** 0.2 mm across; dark red and more-or-less spherical. **Larva:** bright reddish-orange; six-legged. **Nymph:** dark red, brown or dark green; eight-legged.

LIFE HISTORY Adult and juvenile mites are present on the upper surface of ivy foliage throughout much of the year but populations often decline during the summer, the mites then occurring on clover and usually returning to ivy in August. Eggs tend to be deposited on supporting stakes and walls rather than on host plants and, unlike *Tetranychus urticae* (p. 410), the mites do not produce webbing. Breeding is continuous throughout the year, with about 5–8 overlapping generations, the duration of each generation varying according to ambient temperatures. Although development of the mites is greatly protracted during cold winter weather, there is no diapausing stage in the life-cycle. Males are unknown and reproduction is entirely partheno-genetic.

DAMAGE Infested foliage becomes pallid and silvery, and may turn brown (Fig. **1047**).

CONTROL Apply an acaricide as soon as damage is seen, and repeat as necessary.

Bryobia cristata (Dugès)
Grass-pear bryobia mite
A polyphagous species, occurring throughout the year on various grasses and herbaceous plants. Infestations often occur on ornamentals such as *Campanula*, *Cyclamen*, *Dianthus*, gentian (*Gentiana*), *Iris*, *Polyanthus* and saxifrage (*Saxifraga*), the mites feeding mainly on the upper surface of the leaves and causing a mottling and a general silvering. In common with *Bryobia kissophila* (see above), breeding is continuous throughout the year, with many generations annually. During May the mites often disperse from their normal hosts to various trees and shrubs, including fruit trees and ornamentals such as cherry (*Prunus*), hawthorn (*Crataegus*) and rose (*Rosa*), where two summer generations occur before a return migration to herbaceous plants. The mites can be distinguished from *Bryobia kissophila* by the narrower dorsal setae and by other microscopic features.

1047 Ivy bryobia mite damage to leaves of *Hedera*.

Bryobia praetiosa Koch
Clover bryobia mite

Vast numbers of this mite occur during the spring on sunny walls, especially those of new or recently renovated buildings. Fully grown nymphs and adults of this and the previous species also invade buildings through doors and windows, to moult or to lay eggs in various cracks and crevices. Eggs deposited in the spring produce a summer generation of mites which feed on grasses and herbaceous plants (including certain ornamentals) and mature by the autumn. Eggs laid in late summer or autumn produce mites which develop throughout the winter and reach maturity in the following spring. The mites are structurally similar to *Bryobia kissophila*, p. 408.

Bryobia rubrioculus (Scheuten)
Apple and pear bryobia mite

This species, which overwinters in the egg stage and usually completes no more than three genera-tions annually, occurs mainly on fruit trees, but may also attack related ornamentals such as cherry (*Prunus*) and crab apple (*Malus*). Although the mites feed on the upper surface of leaves, they often congregate on the trunks and branches; such aggregations occur mainly in late May and June, and in August and September. The mites also cluster beneath the shoots and main branches whilst moulting from one growth stage to the next, masses of greyish-white cast skins (which are a characteristic sign of an infestation) soon accumulating on the trees. The foliage of infested trees is often mal-formed and also becomes pale, silvery and brittle; damaged leaves eventually turn brown and may drop prematurely. Oribatid mites, usually the cherry beetle mite (*Humerobates rostrolamellatus* Grandjean) (a shiny, dark red to blackish, egg-like and short-legged species, about 1 mm long) often cluster in considerable numbers on the bark of trees; they are harmless, feeding mainly on algae and lichens, and should not be mistaken for apple and pear bryobia mites.

Panonychus ulmi (Koch)
Fruit tree red spider mite

This widespread and generally common mite is an important pest of apple and various other fruit crops in many parts of the world; it also occurs on ornamental trees and shrubs such as almond (*Prunus dulcis*), cherry (*Prunus*), *Cotoneaster*, crab apple (*Malus*), flowering currant (*Ribes sanguineum*), hawthorn (*Crataegus*), Japanese quince (*Chaenomeles japonica*) and rowan (*Sorbus aucuparia*). Eggs overwinter on the spurs and smaller branches, hatching in the following spring from April to mid-June. There are then several overlapping generations throughout the summer months, mite numbers declining from September onwards as breeding ceases and winter eggs are laid. Unlike *Tetranychus urticae* (see below) the mites do not inhabit silken webs but damage caused by both species is similar, heavy infestations leading to significant leaf bronzing (Fig. **1048**) and premature leaf fall. Adult females are 0.4 mm long and dark red, with short, pale legs; they are most commonly found (along with adult males, nymphs, six-legged larvae and summer eggs) on the underside of the leaves (cf. *Bryobia* spp. pp. 408–409).

CONTROL Various acaricides are active against eggs or mobile stages; however, resistance to pesticides (especially organophosphates) is widespread. If a non-systemic material is applied, it is important to ensure good coverage of the underside of the leaves.

1048 Fruit tree red spider mite damage to leaves of *Sorbus aucuparia*.

Tetranychus urticae Koch
Two-spotted spider mite

An often abundant pest of glasshouse and outdoor plants, including ornamentals such as azalea (*Rhododendron*), buddleia (*Buddleja*), calla lily (*Richardia*), *Ceanothus*, *Chrysanthemum*, *Cytisus*, *Dahlia*, *Freesia*, *Fuchsia*, *Hydrangea*, *Impatiens*, *Ipomoea*, Mexican orange (*Choisya ternata*), mulberry (*Morus*), *Phormium tenax*, passion flower (*Passiflora*), poinsettia (*Euphorbia pulcherrima*), *Primula*, rose (*Rosa*) and *Salvia*. Cosmopolitan. Present throughout Europe.

DESCRIPTION **Adult female:** 0.5–0.6 mm long; pale yellowish or greenish, with two dark patches on the body (overwintering form orange); body oval, with moderately long dorsal setae; striae on the hysterosoma form a diamond-shaped pattern. **Adult male:** similar to female but body smaller, narrower and more pointed. **Egg:** 0.13 mm across; globular and translucent. **Larva:** pale greenish with darker markings; six-legged. **Nymph:** pale greenish with darker markings; eight-legged.

LIFE HISTORY Female mites overwinter amongst debris, in dry crevices in the soil and in other suitable shelter, often clustering in cracks in glasshouse structures, stakes and poles. They become active in March or April, invading host plants to feed and eventually deposit eggs. There are several overlapping generations of summer forms annually, mites passing through egg, larval, protonymphal and deutonymphal stages before maturing; males often omit the deutonymphal stage. Development is completed in less than two weeks at temperatures above 20°C, but is greatly protracted below 12°C, extending over almost two months at 10°C; out-of-doors, therefore, there tend to be fewer generations than in protected situations. Colonies occur mainly on the underside of the expanded leaves, the various mite stages being sheltered by fine webs of silk. When populations are large, these webs often become extensive; they may then cover complete leaves and parts of shoots and stems. During September, in response to short days (daylight of less than 14 hours), the orange winter-female forms are produced; after mating, these seek shelter for the winter. As plant vigour also declines, breeding ceases and the males and summer females all die.

DAMAGE Infested leaves are speckled with yellow and often become extensively chlorotic (Fig. **1049**), affecting both the vigour and appearance of plants; hosts may also be disfigured by webbing (Fig. **1050**). Heavy infestations, which are most likely to occur in hot, dry conditions, cause considerable stunting and can result in the eventual death of plants.

CONTROL Apply an acaricide against the egg or active stages, selecting a product that will not cause phytotoxicity and against which the pest has not become resistant; good coverage of infested plants is essential. Systemic granule treatments at potting or replanting can give long-lasting control and these are recommended for use on some hosts. Biological control is possible in glasshouses, using the predatory mite *Phytoseiulus persimilis* Athias-Henriot.

1049 Two-spotted spider mite damage to leaves of *Choisya ternata*.

1050 Two-spotted spider mite damage to leaves of *Impatiens*.

Tetranychus cinnabarinus (Boisduval)
Carmine spider mite

This subtropical species occurs throughout Europe, but, in temperate areas is confined mainly to glasshouses. Infestations develop on various herbaceous plants, including cacti, the foliage often becoming coated in masses of webbing (Fig. **1051**); affected plant tissue also becomes discoloured and infested leaves of carnation and pink (*Dianthus*) tend to curl downwards. Eggs are laid singly under leaves or on the webbing, the pattern of development following that of *Tetranychus urticae* (p. 410). There are several overlapping generations annually but, unlike the previous species, there is no winter diapause; the rate of development is also less rapid, the species being adapted to higher temperatures. The adult females are mainly red, with dark internal markings on either side of the body; the nymphal stages are green and the eggs whitish to pink (cf. those of *Tetranychus urticae*). This species crosses regularly with *Tetranychus urticae*, so, especially in glasshouses, hybrid populations can be found.

CONTROL Apply an acaricide as soon as mites are seen.

1051 Web of carmine spider mite on *Aprocactus flagelliformis*.

Eotetranychus carpini (Oudemans)

Locally common on various broad-leaved trees, including alder (*Alnus*), hazel (*Corylus avellana*), hornbeam (*Carpinus betulus*), maple (*Acer*), oak (*Quercus*) and willow (*Salix*), and occasionally noted on cultivated plants. In Europe most often reported in England, Germany and the Netherlands; also present in North and Central America.

DESCRIPTION **Adult female:** 0.4 mm long; pale green or greenish-yellow, with red eyes. **Egg:** 0.1 mm across; globular, pale green.

LIFE HISTORY Adult females hibernate in suitable shelter on host plants, including bark crevices, reappearing in the following spring. They then invade the underside of leaves to form small, compact colonies beneath, often dense, silken webs. Breeding continues from April to October, there being about six generations in a season.

DAMAGE Infested leaves become pallid, usually visible only from below although heavy infestations may cause the upper surface of leaves to become speckled with yellow (Fig. **1052**); plant growth is not noticeably affected.

1052 *Eotetranychus carpini* damage to leaf of *Carpinus*.

1053 Lime mite damage to leaves of *Tilia*.

Eotetranychus tiliarium (Hermann)
Lime mite

A sporadically important pest of lime (*Tilia*), especially on established street trees. Widely distributed in Europe; also present in the eastern United States of America.

DESCRIPTION **Adult female:** 0.4 mm long; yellowish to orange-red, with red eyes and long, narrow dorsal setae. **Egg:** 0.1 mm across; yellowish-white and globular.

LIFE HISTORY This species overwinters in the adult stage. In the following spring the mites become active, invading the newly developing leaves upon which eggs are laid. The mites then feed on the underside of the leaves, especially alongside the veins, young individuals passing through larval, protonymphal and deutonymphal stages before becoming adult. There are several overlapping generations each year, populations reaching a peak from late summer onwards; in the early autumn, female mites sometimes aggregate in vast numbers amongst webbing on the trunks and main branches.

DAMAGE The mites cause noticeable bronzing of leaves (Fig. **1053**), spoiling the appearance of specimen trees; they also cause premature defoliation, heavily infested leaves shrivelling and dying. Host trees are sometimes disfigured by glistening sheets of polythene-like webbing.

CONTROL If practical, apply an acaricide as soon as damage is seen.

Eotetranychus populi (Koch)

Associated with broad-leaved willows (*Salix*) and various species of poplar (*Populus*), including aspen (*Populus tremula*). Present as a pest in various parts of Europe, including England; also found in North America. The adults overwinter under loose bark or within bark crevices on the trunks and branches of host trees, emerging in the following spring. They then invade the underside of the young leaves and soon deposit eggs. The mites often occur on young sucker growth, individuals sheltering amongst the leaf hairs and beneath silken webs. There are several generations annually, breeding continuing so long as conditions remain favourable. Infested leaves become noticeably discoloured (Fig. **1054**), and may also tend to curl, but damage is of significance only on young trees.

1054 *Eotetranychus populi* damage to leaf of *Salix*.

Schizotetranychus schizopus (Zacher)

Infestations of this widely distributed and often common mite occur mainly on crack willow (*Salix fragilis*), white willow (*Salix alba*) and other narrow-leaved species of *Salix*. The mites feed on the underside of the leaves, especially along either side of the mid-rib, causing the foliage to become discoloured (Fig. **1055**). Damage is often severe, affecting the growth and appearance of plants. In exceptional circumstances, the trunks and branches of host plants may become coated in sheets of webbing. Eggs, which are laid along the mid-rib, are pale yellowish and somewhat flattened, with a dorsal stipe; the overwintering eggs are reddish-orange and are usually found between bud scales or in other sheltered positions on the bark. The mites are small (0.2–0.4 mm long), greenish or reddish, with blackish markings, and relatively flat-bodied.

1055 *Schizotetranychus schizopus* damage to leaf of *Salix*.

CONTROL If practical, apply an acaricide as soon as damage is seen and repeat as necessary.

Oligonychus ununguis (Jacobi)
Conifer spinning mite

A generally common and virtually world-wide pest of conifers; especially important on young spruces (*Picea*). Present throughout Europe.

DESCRIPTION **Adult:** 0.2–0.5 mm long; dark green or orange to brownish or blackish. **Egg:** greenish-brown to orange-red; spherical with a dorsal spine (stipe). **Larva:** pinkish but soon turning greenish; six-legged. **Nymph:** greenish; eight-legged.

LIFE HISTORY Eggs overwintering on the shoots hatch from April or early May onwards. The mites then feed for 2–3 weeks, passing through a larval and two nymphal stages before becoming adults. There are several, usually up to five, generations annually, the mites producing considerable quantities of webbing. Summer eggs are laid on the shoots and needles, but most winter eggs are deposited close to the base of the needles.

DAMAGE Infested needles are noticeably discoloured (Fig. **1056**), becoming mottled and yellowish and eventually turning brown. Affected needles may also drop prematurely, checking shoot growth. Heavily affected shoots become curved and they develop with shortened internodes. Spruce seedlings and transplants can be killed, especially if growing in dry soils.

CONTROL Apply an acaricide in late May, immediately after egg hatch.

1056 Conifer spinning mite damage to needles of *Picea abies.*

5. Family **TENUIPALPIDAE** (false spider mites)

Distinguished from members of the Tetranychidae (true spider mites) by the absence of a thumb-claw on the palps. Also, the mites do not produce webbing.

Brevipalpus obovatus Donnadieu

A tropical or subtropical species, accidentally introduced into Europe where it is a minor pest of glasshouse ornamentals such as *Aralia*, azalea (*Rhododendron*), *Campanula, Cissus, Gardenia* and ivy (*Hedera*). In Europe found in various countries, including Austria, England, Germany and the Netherlands.

DESCRIPTION **Adult female:** 0.25–0.30mm long; red; body flat and egg-shaped, with a reticulated pattern on the idiosoma; five pairs of short dorso-lateral setae on the hysterosoma; palps four-segmented; legs relatively short. **Egg:** 0.1×0.07 mm; bright red and elliptical.

LIFE HISTORY Mites occur on both sides of young leaves but are most numerous on the underside of older foliage where they often congregate around the margins. Eggs are deposited close to the mid-rib, often in clusters of several hundred. They hatch in 2–3 weeks, the juvenile stages feeding for 3–4 weeks before attaining maturity. Breeding continues so long as conditions remain favourable; reproduction is typically parthenogenetic.

DAMAGE The mites cause brown, necrotic areas on host plants, the discoloration commonly occurring along either side of the mid-rib or appearing as a fine rusty or bronze-like speckling over the entire leaf lamina. Heavy infestations check the growth of plants and may result in premature senescence and defoliation. On ivy, damaged leaves are often 'cupped' and reduced in size, and growth from infested buds is typically weak and pallid.

CONTROL Apply an acaricide as soon as mites are seen and repeat as necessary. Many products, although active against spider mites, are ineffective against this and related species; acaricides known to kill tarsonemid mites should be effective against false spider mites.

Brevipalpus phoenicis (Geijskes)

This mite is very similar to *Brevipalpus obovatus* and, although of most importance as a pest of citrus crops and tea plant (*Camellia sinensis*), can also damage a wide range of ornamentals; it is reported from various parts of the world, including the Netherlands.

Brevipalpus russulus (Boisduval)

A Central American species, introduced into Europe many years ago; sometimes reported as a pest of cacti (e.g. *Mammillaria*) and succulents, most often in Belgium, France, Germany and the Netherlands. Infestations check the growth of plants, often producing a general reddish-grey discoloration. The mites are similar in appearance to *Brevipalpus obovatus* but possess an additional pair of dorso-lateral setae on the hysterosoma.

Tenuipalpus pacificus Baker

This species, which is associated with orchids, is native to certain Pacific areas of the world, but has also become established elsewhere. In parts of Europe, including England, Germany and the Netherlands, it is sometimes reported as a pest of glasshouse-grown plants. The mites are readily distinguished by their three-segmented palps and the constricted, inverted-bell-shaped outline of the hysterosoma.

CONTROL As for *Brevipalpus obovatus*.

6. Family PYEMOTIDAE

Siteroptes graminum (Reuter)

syn. *cerealium* (Kirchner)

Grass and cereal mite

Reported occasionally in England and mainland Europe on carnation (*Dianthus caryophyllus*), occurring in association with the fungus *Fusarium poae* and causing a brown necrosis of petals and, sometimes, death of opening buds; this symptom is commonly known as 'bud rot' but is less frequent than a condition called 'silver top' which occurs following the same association on grasses or cereals. The mite breeds entirely parthenogenetically, passing through egg, larval and nymphal stages, all of which develop within the much swollen, sac-like body of the gravid female. Such females measure up to 2mm across and are glistening, hyaline-whitish in appearance. Infestations on horticultural plants, such as glasshouse-grown carnations, tend to occur in autumns following a spell of hot, dry weather.

7. Family ACARIDAE

Rhizoglyphus callae Oudemans

A bulb mite

An often common, usually secondary pest of bulbs, corms and tubers, including *Freesia*, *Gladiolus*, hyacinth (*Hyacinthus*), lily (*Lilium*), *Narcissus* and tulip (*Tulipa*). Present throughout Europe; also found in North America.

DESCRIPTION **Adult:** 0.7mm long; body oval, smooth, whitish and very shiny, with discrete in-

1057 Bulb mite damage to bulb of *Tulipa*.

ternal brownish patches and several long hairs projecting back beyond the tip of the hysterosoma; legs reddish-brown and stout. **Egg:** 0.2×0.1mm; hyaline to whitish, smooth and shiny. **Larva:** whitish and shiny; six-legged. **Nymph:** similar to adult but smaller. **Hypopus:** 0.3mm long; dark brown; legs short, very stout and with large claws.

LIFE HISTORY Mites occur in large numbers within diseased and otherwise unhealthy bulbs, corms or tubers. They breed continuously under suitable conditions, passing through egg, larval and two nymphal stages, completion of the life-cycle taking from one to four weeks depending upon temperature. A phoretic hypopal stage appears in some populations; these hypopi attach themselves to small insects, such as adult narcissus bulb flies (*Eumerus* spp., see p.000), and are then transported to other hosts, thereby spreading infestations.

DAMAGE **General:** although usually of secondary importance, the mites, having once gained entry, can increase the extent of damage and may lead to the complete breakdown and destruction of bulbs, corms or tubers, the internal tissue turning blackish and powdery (Fig. **1057**). **Freesia and gladiolus:** attacked roots develop dark brown streaks, and are often mined internally. Also, if healthy corms are planted into heavily infested soil, growing points and leaves are affected; these become distorted and the leaves develop ragged or saw-toothed edges.

CONTROL Soil sterilization and dip treatments will eliminate bulb mites; systemic acaricidal granules are also effective. Where appropriate, steps should be taken to solve any primary problem likely to predispose plant tissue to attack.

Rhizoglyphus robini Claparède
A bulb mite

This species also infests bulbs and tubers but appears to find diseased hosts less favourable than relatively healthy ones. The mites are very similar in appearance to *Rhizoglyphus callae*, and both species have been referred to as *Rhizoglyphus echinopus* (Fumouze & Robin); however, *Rhizoglyphus robini* is slightly the larger, and has noticeably shorter body hairs and very short dorsal idiosomal setae.

MISCELLANEOUS PESTS

WOODLICE

1. Family **ARMADILLIDIIDAE**

Woodlice with flagellar segment of the antennae divided into two sections; body strongly arched, and individuals capable of rolling into a ball.

Armadillidium nasatum Budde-Lunde
Blunt snout pillbug
Widespread and often common in heated glasshouses, occasionally causing damage to ornamental plants.

DESCRIPTION **Adult:** 21 mm long; grey with a pale median line and pale lateral patches; snout drawn into a narrow, square-sided projection (Fig. **1058**).

LIFE HISTORY Adults occur throughout the year, usually hibernating in piles of soil. Breeding occurs from April to October, there being two or more generations each year. Females deposit about 50 eggs, and these are retained in a special pouch on the underside of the thorax. The eggs hatch in about 1–2 months, young, white, 12-legged woodlice (each about 1 mm long), then escaping from the maternal pouch to begin feeding. Individuals acquire the typical adult coloration after about two weeks, and a seventh pair of legs appears after two moults. Further moults occur, even after the adult stage is attained, but individuals do not reach sexual maturity until the following year. Mortality of young woodlice is high and few individuals survive for more than a couple of months; adults, however, may survive for up to four years.

DAMAGE Woodlice produce irregular holes in leaves, especially in those close to the ground; growing points of young plants may also be destroyed. Stems of potted plants and transplants are often grazed but those of older plants are rarely attacked.

CONTROL All debris, including general rubbish, rotting wood and decaying vegetation, should be removed from within and around glasshouses. If necessary, infested sites can be sprayed with a suitable pesticide; the use of poisoned bran-based baits is also recommended.

1058

1058 Blunt snout pillbug, *Armadillidium nasatum*.

1059

1059 Common pillbug, *Armadillidium vulgare*.

Armadillidium vulgare (Latreille)
Common pillbug
This species often occurs in unheated glasshouses and garden frames, causing damage to plant roots; seedlings are frequently destroyed and the growth of older plants may also be affected if soil infestations are heavy. Adults are 18 mm long, dark slate-grey to reddish, with several pale patches along the body; the snout is slightly raised (Fig. **1059**)

2. Family ONISIDAE

Woodlice with flagellar segment of the antennae divided into three sections.

Oniscus asellus Linnaeus
Grey garden woodlouse

Widespread and common, often sheltering in considerable numbers beneath bricks, stones and pieces of rotting wood. Damage is often caused in glasshouses to the young, tender growth of ornamentals, especially carnation (*Dianthus caryophyllus*) and sweet pea (*Lathyrus odoratus*); aerial roots of orchids are also attacked. Adults are 16 mm long, shiny grey, with irregular pale patches; the body is relatively broad (Fig. **1060**).

1060 Grey garden woodlouse, *Oniscus asellus*.

3. Family PORCELLIONIDAE

Woodlice with flagellar segment of the antennae divided into two sections; body distinctly flattened.

Porcellio dilatatus Brand
This species occurs in abundance beneath seedboxes and flower pots in cold-frames and in unheated glasshouses, often attacking the roots of cultivated plants. Adults are 15 mm long, and greyish-brown with distinctive lateral stripes; the body is noticeably roughened, and the telson, which reaches just beyond the endopodites, usually has a rounded tip.

1061 Garden woodlouse, *Porcellio scaber*.

Porcellio laevis Latreille
A locally common species in manure heaps and, if introduced into glasshouses, damaging to the roots of ornamental plants, including various ferns. Adults are 18 mm long, brown, smooth and glossy; the telson has a pointed tip and the uropods are very long.

Porcellio scaber Latreille
Garden woodlouse

This widespread and common species may damage plants in cold frames and seedboxes, but is not usually a problem in glasshouses. Adults are 17 mm long, dark slate-grey, with a roughened body and a pointed telson; the basal antennal segment is usually orange (Fig. **1061**). Juveniles are often yellow or orange.

MILLEPEDES

Millepedes are vegetarian creatures that feed mainly in the soil on soft, decaying plant tissue and debris. Although possessing weak mouthparts, millepedes can cause direct damage to plant roots, seeds and germinating seedlings, attacked tissue often being invaded subsequently by rot-producing bacteria and fungi; millepedes also feed on bulbs, corms and other parts of plants, usually enlarging wounds initiated by pests such as slugs and wireworms. Millepedes are light-shy; they tend to be most numerous amongst leaf mould and vegetable rubbish, and in soil with a high organic content. Although favoured by damp conditions, millepedes often cause damage during spells of dry weather by tunnelling into plant tissue in search of moisture.

1062 Spotted millepede, *Blaniulus guttulatus*.

1. Family **BLANIULIDAE** (snake millepedes)

Blaniulus guttulatus (Bosc)
Spotted millepede
A generally abundant millepede, often causing damage to seeds, seedlings and tulip (*Tulipa*) bulbs. Attacks also occur in glasshouses, especially on carnation (*Dianthus caryophyllus*) roots, *Chrysanthemum* stools and *Cyclamen* corms. Adults are 8–18mm long, snake-like, and mainly white with a series of reddish or purplish spots along each side; the body is composed of about 50 segments, most of which bear two pairs of short legs (Fig. **1062**). Eggs are laid in groups in the soil during the spring and summer. The eggs hatch into first-stage juveniles that possess few body segments and just three pairs of legs; maturity is reached in about a year, development to adulthood occurring after several moults at which additional body segments and pairs of legs are added. This species lacks eyes but the skin is photo-sensitive, individuals soon becoming active if exposed to light.

1063 Flat millepede, *Polydesmus angustus*.

2. Family **POLYDESMIDAE** (flat millepedes)

Polydesmus angustus Latzel
Flat millepede
An often common millepede, especially amongst leaf litter; frequently associated with damage to cultivated plants in the British Isles and in various other parts of western Europe. Adults are 25mm long, flat-bodied and mainly reddish-brown, with 20 rough-textured body segments (Fig. **1063**).

3. Family **IULIDAE** (snake millepedes)

Cylindroiulus londinensis (Leach)
syn. *teutonicus* (Pocock)
A black millepede
A widely distributed and often common millepede, sometimes causing damage to cultivated plants. Adults (up to 50 mm long) are brownish-black to black, the body being of similar diameter throughout its length.

Tachypodoiulus niger (Leach)
syn. *albipes* (Koch)
A black millepede
Another common, black-bodied millepede, sometimes causing damage to nursery and garden plants. Unlike that of the previous species, the body is distinctly narrowed at the tail end (Fig. **1064**).

1064 A black millepede, *Tachypodoiulus niger*.

SYMPHYLIDS

1. Family **SCUTIGERELLIDAE**

Scutigerella immaculata (Newport)
Glasshouse symphylid
A generally common soil pest, attacking various glasshouse and outdoor plants, including seedlings and transplants. Widely distributed in Europe; also present in North America and Hawaii.

DESCRIPTION **Adult:** 5–9 mm long; white with long antennae, 12 pairs of legs and a pair of pointed anal cerci. **Egg:** white, rounded and sculptured. **Nymph:** white with from six to ten pairs of legs, additional pairs being added at each moult.

LIFE HISTORY Breeding occurs throughout the year, eggs being laid near the soil surface in batches of up to 20. Eggs hatch in 1–3 weeks, the nymphs reaching maturity several months later. Nymphs and adults feed on plant roots when the soil is warm and moist but in dry conditions they migrate to deeper levels. Individuals are capable of surviving for several years.

DAMAGE Fine roots are grazed away, weakening and sometimes leading to the death of infested seedling and young plants.

CONTROL Where damage is known to occur, an insecticide should be worked into the soil before planting.

NEMATODES

Nematodes are important pests of ornamental plants, and some species are of significance as virus vectors. Brief details of the main groups associated with ornamentals are given below.

Leaf nematodes

Leaf nematodes (*Aphelenchoides* spp.) affect a wide variety of hosts, occurring as ecto- and endo-parasites of the buds, leaves and growing points; they may also feed in the epidermal layers of green stems. The pests are often found on glasshouse plants and propagation material, where they thrive in warm, moist conditions. *Aphelenchoides ritzema-bosi* (Christie) Steiner & Buhrer is a major pest of *Chrysanthemum* and one of the most important species to attack ornamentals. Adults are about 1 mm long, and usually overwinter in the dormant buds and shoot tips of chrysanthemum stools. They are capable of surviving in dried-out leaves for several years, but survival in the soil is poor. The nematodes slither over host plants in films of water, usually entering their hosts through stomata. They reproduce sexually, development from egg to adult taking about two weeks. Affected tissue becomes yellowish and then brown or blackish, these symptoms often appearing as distinctive wedge-shaped areas between the major veins (Fig. **1065**). Seriously affected leaves eventually wither and die. In moist conditions symptoms spread rapidly up the plant, new leaves emerging from infested shoots often being distorted and thickened; heavy infestations lead to stunting of growth and the production of small, malformed flowers. Other plants attacked by this species include African violet (*Saintpaulia*), *Aster*, buddleia (*Buddleja*), *Ceratostigma*, *Cineraria*, *Crassula*, *Dahlia*, *Delphinium*, *Doronicum*, lavender (*Lavendula*), peony (*Paeonia*), *Pyrethrum*, wall-flower (*Cheiranthus cheiri*) and *Weigelia*. Various other species of leaf nematode are also pests of ornamentals. These include: *Aphelenchoides blas-tophthorus* Franklin, primarily a pest of scabious (*Scabiosa*) but also associated with plants such as *Anchusa*, *Anemone*, *Begonia*, bulbous *Iris*, *Cephalaria*, globe flower (*Trollius europaeus*), lily of the valley (*Convallaria majalis*), *Narcissus* and sweet violet (*Viola odorata*); *Aphelenchoides fragariae* (Ritzema Bos) Christie, which affects

1065 *Aphelenchoides ritzemabosi* damage to leaves of *Weigelia*.

mainly Liliaceae, Primulaceae, Ranunculaceae and many kinds of fern; *Aphelenchoides subtenuis* (Cobb) Steiner & Buhrer, which is recorded on *Phlox* and various bulbous hosts, including *Allium*, *Colchicum*, *Crocus*, *Narcissus*, squill (*Scilla*) and tulip (*Tulipa*).

CONTROL Trouble is best avoided by good hygiene, and care should be taken to avoid introducing the pests with new plants. Hot-water treatment will give effective control of leaf nematodes on chrysanthemum but is impractical on AYR crops because of the small size of the stools. Alternatively, nematicide granules can be used in chrysanthemum stool beds and as a soil treatment for certain other plants, but it is better to propagate from clean stock. Where infestations have occurred, sterilization of glasshouse soil, boxes and other equipment is essential; out-of-doors, maintenance of a weed-free fallow through the winter months will eliminate the pest.

Cyst nematodes

Cyst nematodes are associated mainly with the roots of herbaceous plants. Affected plants are weakened, and often become stunted and discoloured; they may also wilt in strong sunlight. The root system is typically attenuated, plants often developing a mass of new rootlets ('hunger roots') close to the soil surface in an attempt to compensate for the loss of normal root activity. Cyst nematodes invade the young roots of host plants as minute second-stage juveniles, the first-stage having moulted whilst still within the egg. Having gained entry to the host, each nematode settles down to feed in the centre of a young root. Feeding induces the development of giant cells which interrupt normal vascular activity within the root. Some of the nematodes develop into small worm-like males which eventually escape into the soil. Females, however, mature into whitish or yellowish, lemon-shaped (typical of the genus *Heterodera*) or rounded bodies, *c.* 0.5–1.0 mm long, which burst through the root surface but remain attached at the head end. After mating, the females die and their bodies darken into hard-walled protective cysts packed with minute, oval eggs. These cysts eventually break away from the roots and drop into the soil. Hatching of the eggs is often dependent upon the presence in the soil of chemicals exuded from the young roots of host plants, cysts commonly remaining viable in the soil for many years before eventually releasing the infective second-stage juveniles. Species most often noted on ornamental plants include: *Heterodera cacti* Filipjev & Schuuremans Stekhoven on poinsettia (*Euphorbia heterophylla*) and various cacti; *Heterodera cruciferae* Franklin on wallflower (*Cheiranthus cheiri*) and other cruciferous plants; *Heterodera fici* Kir'yanova on fig and rubber plants (*Ficus*); and *Heterodera trifolii* Goffart on carnation (*Dianthus caryophyllus*).

CONTROL Infested soil should be sterilized thoroughly. Any affected plants should be destroyed, although vegetative cuttings can be taken from them without risk of contamination. Infested cacti can also be saved, so long as all roots are removed and the bases of the aerial parts thoroughly cleaned.

Migratory nematodes

Migratory nematodes usually feed externally on the roots of plants, infestations most often occurring on rosaceous hosts growing in light, sandy soils. Attacked roots are gnarled and distorted, affected plants lacking vigour and becoming distinctly stunted. Migratory nematodes are also capable of transmitting important plant viruses. Several groups of migratory nematodes cause damage to ornamental plants. These include: dagger nematodes (*Xiphinema* spp.), often a problem on rosaceous plants, including rose (*Rosa*); needle nematodes (*Longidorus* spp.), also associated mainly with rosaceous plants; and stubby-root nematodes (*Trichodorus* spp.), which commonly affect not only woody plants but also bulbs such as *Narcissus* and tulip (*Tulipa*). Root-lesion nematodes (*Pratylenchus* spp.) are also classified as migratory species although, unlike the other groups, the adults and juveniles enter the roots of host plants to feed internally; the nematodes also breed within the host. Roots of infested plants develop short, elongate lesions which afford ideal sites for the entry of pathogenic bacteria and fungi, affected plants lacking vigour and wilting under stress. Roots often break off at the point of damage, particularly once bacterial or fungal rots have gained a hold. When conditions within the roots become unfavourable, the nematodes disperse to seek more suitable hosts elsewhere. Root-lesion nematodes affect various herbaceous ornamentals, including *Anemone*, *Begonia*, Christmas rose (*Helleborus*), *Delphinium*, hyacinth (*Hyacinthus*), lily (*Lilium*) and lily of the valley (*Convallaria majalis*). More specifically, *Pratylenchus bolivianus* Corbett infests Peruvian lily (*Alstroemeria*), notably in the Netherlands but also in England, and *Pratylenchus penetrans* (Cobb) Filipjev & Schuuremans Stekhoven is associated with the rotting of *Narcissus* bulbs, particularly in the Netherlands and in the Scilly Isles. Other species (e.g. *Pratylenchus fallax* Seinhorst and *Pratylenchus thornei* Sher & Allen) are of significance in mainland Europe on field-grown trees and shrubs, especially Rosaceae; *Pratylenchus vulnus* Allen & Jensen, although a pest of glasshouse-grown roses in northern Europe, cannot survive out-of-doors except in warmer climates.

CONTROL Soil sterilization will give some control of migratory nematodes but specialist advice should be sought, especially if viruliferous nematodes are thought to be present. Root-lesion nematodes will not survive on lifted bulbs kept dry in store, but replanting in infested land should be avoided.

Root-knot nematodes

Root-knot nematodes (*Meloidogyne* spp.) attack the roots of various trees, shrubs and herbaceous plants. Infested roots become distorted, and develop rounded or irregular galls (Fig. **1066**); the galls measure anything from 1–20 mm across and often coalesce, causing considerable distortion. *Meloidogyne hapla* Chitwood is a widely distributed, polyphagous pest in northern Europe; it attacks many different kinds of plant, including various ornamentals. Root-knot nematodes are associated mainly with light soils but most damage is caused under glass, especially in hot conditions where certain tropical and subtropical species (e.g. *Meloidogyne javanica* (Treub) Chitwood) have become established. Pot plants such as *Begonia*, *Coleus*, *Cyclamen*, *Gloxinia* and various cacti may suffer considerable damage, severely affected plants appearing discoloured, lacking vigour and wilting under stress. Root-knot nematodes may also exacerbate the deleterious effects of pathogenic bacteria or fungi. As with cyst nematodes (p. 422), host plants are invaded by second-stage juveniles; these settle down to feed in the young roots and usually reach maturity about 1–2 months later. Adult female nematodes are translucent-whitish, pear-shaped and about 0.5–1.0 mm long. They may be found within the galled tissue, often attached to a gelatinous sac containing masses of eggs. In some cases, development of the pest is parthenogenetic but in others, minute worm-like males mate with the females before eggs are laid. First-stage juveniles develop within the eggs, second-stage individuals eventually breaking free and either migrating inside the root or escaping into the soil to commence feeding elsewhere. These infective nematodes are capable of surviving in moist soil for about three months but in dry conditions they will persist for no more than a few weeks.

CONTROL If practical, infested land should be fumigated, or left fallow or put into grass for 1–2 years. In glasshouses, sterilization or drenching of infested soil with a nematicide may have some effect and help save a crop of cut flowers; however, immediately after cropping, affected plants should be destroyed. Replanting into previously infested untreated compost should be avoided. Good hygiene is essential to prevent the accidental introduction of root-knot nematodes into glasshouses on the roots of plants.

Stem and tuber nematodes

Stem nematode (*Ditylenchus dipsaci* (Kühn) Filipjev) is a major pest of herbaceous and bulbous plants, including many ornamentals, existing as several distinct races or strains; the hyacinth, narcissus, onion, phlox and tulip races are examples. Some races affect a wide variety of host plants but others are more specific, the nematodes attacking various cultivated plants and also many weeds. Plants are invaded by adults and final-stage (fourth-stage) juveniles, which move through the soil in moisture films and gain entry via wounds, lenticels or stomata in the basal parts of stems or

1066 Galls of *Meloidogyne hapla*.

leaves in contact with the ground; roots are not attacked. Adult nematodes are minute (*c*. 1.2 mm long) and thread-like. They feed and breed continuously in suitable hosts, development from egg to adult taking about three weeks at 15°C. In the absence of host plants, the nematodes are able to survive in moist soil for up to a year; however, they are far more resistant to desiccation, fourth-stage juveniles often congregating in their thousands and drying out to form yellowish, wool-like masses ('nematode wool'). These desiccated nematodes are very resistant to unfavourable conditions, remaining viable and potentially infective for several years. Damage symptoms vary from host to host but commonly include stunting, crinkling, twisting, swelling or malformation of leaves, stems, petioles and flowers; infested tissue may also split. *Aubretia*, *Campanula*, evening primrose (*Oenothera*), golden rod (*Solidago*), *Gypsophila*, *Heuchera*, sneezewort (*Helenium*), *Hydrangea*, *Phlox*, *Primula* and sweet william (*Dianthus barbatus*) are often affected. Bulbous plants such as daffodil (*Narcissus*), hyacinth (*Hyacinthus*), snowdrop (*Galanthus nivalis*) and tulip (*Tulipa*) are also important hosts. Leaves arising from infested daffodil bulbs often develop small, yellowish swellings ('spickels'); also, when infested bulbs are sliced open, brown rings of damaged tissue may be visible where the scales have been destroyed (Fig. **1067**). In tulips, flower stalks from heavily infested bulbs are bent and develop lesions which eventually split open; flowers also fail to colour properly, the petals remaining partly green. Bulbs damaged by nematodes may eventually rot; they are often invaded by secondary organisms such as small narcissus flies (*Eumerus* spp.) (pp. 170–171) and bulb mites (*Rhizoglyphus* spp.) (pp. 415–416).

The potato tuber nematode (*Ditylenchus destructor* Thorne) is similar to *Ditylenchus dipsaci* but lacks a resistant stage capable of surviving periods of desiccation. Also, unlike stem nematode, potato tuber nematode is restricted mainly to the subterranean parts of plants, producing dry, brownish or blackish lesions on bulbs, corms, roots and tubers. Various ornamentals are affected, including bulbous *Iris*, *Colchicum*, *Dahlia*, *Gladiolus* and tulip (*Tulipa*); leaves developing from damaged bulbs are weakened and often have yellow tips. Damaged tissue is frequently invaded by secondary organisms, including bacteria, fungi and mites.

CONTROL Good hygiene is essential to avoid nematode problems. Any suspect plants should be destroyed, as should plant debris which might harbour the pests. Hot-water treatment of plant material will give some control, but is unsuitable for some hosts (e.g. hydrangeas and tulip bulbs). Some control can be obtained by using systemic nematicides but soil sterilization is likely to be practical only in glasshouses or on small outdoor plots. Planting in infested soil should be avoided; suspect land should be cleared and kept free of host plants (including weeds) for several years; if necessary, expert help should be sought to overcome the problem.

1067 Stem nematode damage to bulb of *Narcissus*.

1068 Slug damage to flower of *Narcissus*.

SLUGS AND SNAILS

Slugs and snails are often important pests in gardens and nurseries, attacking seedlings and various herbaceous plants such as *Anemone*, *Campanula*, *Doronicum*, *Gladiolus*, *Hosta*, hyacinth (*Hyacinthus*), *Iris*, *Narcissus*, *Petunia*, *Primula*, *Rudbeckia*, strawflower (*Helichrysum*), sweet pea (*Lathyrus odoratus*), *Tagetes*, tulip (*Tulipa*), violet (*Viola*) and various lilies. Slugs and snails feed at night, often causing severe damage to seedlings, young shoots, foliage and flowers (Fig. **1068**); such damage is sometimes confused with that caused by caterpillars or various other pests but slime trails (if not the pests themselves) on or in the immediate vicinity of attacked plants readily betray the identity of the true culprits.

Unlike snails, which usually hibernate during the winter months, slugs breed throughout the year and will remain active in all but the coldest and

driest of conditions. Their translucent, often pearl-like eggs are deposited in groups in the soil or amongst surface vegetation. The eggs usually hatch within a few weeks but those deposited during the winter may not complete their development until the following spring. Juvenile slugs are similar in appearance to adults but smaller and usually paler. They take anything from five months to two years to reach maturity, the rate of development varying from species to species and according to conditions.

Several species are responsible for damaging ornamentals; these include the garden slug (*Arion hortensis* Férussac), the field slug (*Deroceras reticulatum* (Müller)) and various keeled slugs (*Milax* spp.). *Arion hortensis* is a relatively small (25–30 mm long), tough-skinned slug with a rounded tail; the adults are mainly black above and yellow or orange below. *Deroceras reticulatum* is larger (30–

40 mm long), soft-bodied and mainly yellowish-brown, with a distinctly pointed tail and a short dorsal keel at the hind end (Fig. **1069**). Both of these species readily attack the aerial parts of plants. Keeled slugs, such as *Milax budapestensis* (Hazay), are characterized by the presence of a distinct dorsal ridge that extends from the mantle to the tail. Unlike other slugs, they feed mainly below ground level, often boring into bulbs, corms, rhizomes and tubers of ornamental plants; they will also damage plant roots.

Particularly on calcareous soils, snails can also be troublesome pests of ornamentals, attacking seedlings as well as the young shoots and foliage of established plants. Species most likely to cause damage in gardens and nurseries are banded snails (*Cepaea* spp.), the garden snail (*Helix aspersa* Müller) and the strawberry snail (*Trichia striolata* (Pfeiffer)). The strawberry snail often causes extensive damage to seedlings being raised in cold frames.

CONTROL Populations of slugs and snails can be reduced by the regular use of molluscicides but the degree of success achieved on heavily infested sites is often disappointing. Best results against slugs can be expected during warm, humid nights, especially following periods of dry weather that may have prevented the pests from feeding. Molluscicides will also help protect vulnerable seedlings or older plants from damage but effectiveness of a treatment is greatly reduced by rainfall; if appropriate, baits or pellets should be sheltered from the elements by waterproof covers such as slates or tiles. Especially susceptible plants such as lilies should be planted in a coarse grit or stony compost and the beds or pots further protected with slug pellets. On a small scale, slugs and snails can be collected up or removed from plants by hand; egg batches discovered whilst digging in the soil should also be destroyed.

1069 Field slug, *Deroceras reticulatum*.

EARTHWORMS

When burrowing in the soil, some species of earthworm regularly deposit excreted soil (worm casts) on the surface, their activity being greatest in the spring and autumn. Worm casts are a particular nuisance on fine lawns, and are especially unwelcome on bowling greens, golf greens and tennis courts. The species most often causing a problem are *Allolobophora longa* Ude and *A. nocturna* Evans.

CONTROL Chemical treatment against earthworms is best done soon after mowing, ideally in the spring or autumn when the soil is warm and damp. To avoid smearing of the surface, worm casts should be dispersed with a broom before mowing. Earthworms can be discouraged by avoiding organic fertilizers and by keeping soil pH low (i.e. do not apply lime or other alkaline dressings).

BIRDS

Many birds are of considerable benefit to the gardener or nurseryman. During the spring and summer, for example, insectivorous species such as tits, warblers and wrens destroy vast numbers of insect pests, including many of those attacking the foliage or flowers of ornamental trees and shrubs. Some birds, notably blue tits (*Parus caeruleus*) and great tits (*Parus major*) also feed on overwintering insects and mites secreted beneath bud scales or within bark crevices. Others, including rooks (*Corvus frugilegus*), may seek out soil pests such as leatherjackets, swift moth larvae, vine weevil larvae and wireworms; large numbers of slugs and snails are also killed by birds. Nevertheless, birds can also be a nuisance on ornamentals and some, especially bullfinches (*Pyrrhula pyrrhula*), can reach true pest status.

In the winter and spring, birds frequently attack dormant or opening buds of trees and shrubs, affected shoots often developing with any remaining blossoms restricted to the extreme tip. Bud stripping can be of considerable importance on crab apple (*Malus*), *Forsythia*, lilac (*Syringa vulgaris*), ornamental almond (*Prunus amygdalus*), ornamental cherry (*Prunus*) and *Viburnum carlesii*, and is usually caused by bullfinches. Other birds, including blue tits, chaffinches (*Fringilla coelebs*) and greenfinches (*Carduelis chloris*), will also damage dormant buds but are usually of only minor importance.

House sparrows (*Passer domesticus*) are notorious for the damage they cause to *Crocus*, primrose (*Primula*), sweet pea (*Lathyrus odoratus*), violet (*Viola*) and various other plants, attacked flowers being torn to pieces and the stripped petals left lying on the ground, a typical symptom of bird damage. House sparrows will also invade glasshouses and plastic tunnels, will steal seed from newly sown lawns and will cause disturbance in dry seedbeds by taking dust baths.

In the autumn and winter, various birds feed on the berries or fruits of trees and shrubs, the colourful displays on ornamentals such as barberry (*Berberis*), *Cotoneaster*, firethorn (*Pyracantha*), holly (*Ilex aquifolium*) and *Sorbus* often being depleted. Birds responsible for such damage include blackbirds (*Turdus merula*), fieldfares (*Turdus pilaris*), jays (*Garrulus glandarius*), redwings (*Turdus iliacus*), thrushes (*Turdus* spp.) and woodpigeons (*Columba palumbus*).

CONTROL Birds can be discouraged by various means, including the use of scaring devices, chemical repellents and physical barriers. Such methods are not always practical or entirely effective but, on a small scale, netting or black cotton can help to reduce the extent of bud stripping or lessen the impact of birds on autumn displays of colourful berries. Trapping or shooting of birds is usually outlawed and, in any case, of limited value. The use of poison baits or sprays is not an acceptable method of control.

MAMMALS

Various mammals cause damage to cultivated plants, especially in rural or semirural areas. The following are examples of those most frequently reported affecting ornamentals in gardens and nurseries.

Badgers

Although rarely a significant problem, badgers (*Meles meles*) sometimes enter gardens in rural areas and cause damage to lawns and cultivated plants, usually when scrabbling into the ground in search of food. During the winter they will often dig up bulbs and corms, leaving behind tell-tale scrapings and piles of soil.

Cats and dogs

Domestic cats (*Felis catus*) are often troublesome in gardens when they dig into seedbeds, the fine soil making attractive toilet areas. They also cause damage by scratching the bark of trees and shrubs with their claws. Dogs (*Canis familiaris*) are also renowned for digging in gardens but cause most trouble by urinating on low-lying plants and killing the foliage, affected tissue turning brown. Such damage is often severe on dwarf conifers growing alongside paths, and can result in the death of plants. Urinating dogs, especially bitches, will also cause brown patches on lawns.

Deer

In some areas ornamental trees and shrubs, especially rose (*Rosa*), may be damaged by deer. Attacks usually take place from March to May, when the animals graze on the young leaves and new shoots. Deer cannot bite cleanly through plant material, as they possess teeth in the lower jaw only; damaged shoots and stems, therefore, are left with a distinctly ragged edge on one side where the partly severed tissue has been torn away. Damage to ornamentals is most often caused by fallow deer (*Cervus dama*), Muntjac deer (*Muntiacus* spp.) and roe deer (*Capreolus capreolus*)

Hares and rabbits

Hares (*Lepus europaeus*) and rabbits (*Oryctolagus cuniculus*) attack the foliage and stems of various ornamental plants, often causing considerable damage in inadequately protected rural gardens and nurseries. Also, especially during hard winters when food supplies are scarce, they will often gnaw the bark of young trees. In severe cases, stems or trunks may be completely ringed and the plants killed. Fencing off gardens or plantations and the use of protective wire or plastic sleeves around the bases of vulnerable trees will reduce the likelihood of damage. However, in deep snowfall, the animals may gain access to normally secure areas and feed on bark above the level of any netting or other guards.

CONTROL Some protection against hares and rabbits is afforded by tree-banding grease and by proprietary repellants painted on the main stems or trunks, but such measures are not always entirely successful.

Mice, rats and voles

Various small mammals may prove troublesome in gardens and nurseries, some species commonly entering outbuildings during the winter to feed on stored bulbs, corms and seeds. Small mammals may also eat bulbs and corms already in the ground, such damage being caused most often by the long-tailed field mouse (*Apodemus sylvaticus*). Voles, especially the short-tailed vole (*Microtus agrestis*), frequently gnaw the bark of young trees and shrubs; they may also attack herbaceous ornamentals such as *Chrysanthemum*. Growth of affected plants is often checked and, if the tissue is completely ringed, branches or whole plants may be killed. Voles will also ascend young ornamental trees and shrubs to feed on the buds, young shoots and berries.

CONTROL Trees and shrubs are best protected from attack by fitting fine wire netting or plastic sleeves around the bases or trunks and by clearing away adjacent ground vegetation.

Moles

Moles (*Talpa europaea*) can be a problem in lawns, parks and sports grounds as they burrow through the soil in search of food (earthworms, slugs and other soil invertebrates), infested areas becoming disfigured by the presence on the surface of numerous mole-hills. In addition, surface soil sometimes collapses into their subterranean workings, and turf over shallow tunnels may become forced upwards in ridges. Moles can also be a nuisance in flower borders and seedbeds as they accidentally disturb plant roots and cause established plants and seedlings to wilt and die.

CONTROL Trapping is a well-established method of reducing mole populations but is best done by experienced operatives. Elimination of earthworms (see p. 425) may help to reduce persistent mole problems in lawns.

Squirrels

Grey squirrels (*Sciurus carolinensis*) are notorious pests of woodland trees and shrubs, and are also of some importance on ornamentals in gardens and nurseries. During the winter they strip the bark from the shoots and branches of trees such as ash (*Fraxinus excelsior*), sycamore (*Acer pseudoplatanus*) and spruce (*Picea*), damage often being extensive. Grey squirrels will also dig up bulbs of plants such as *Crocus* and tulip (*Tulipa*). Young shoots, buds and flowers of trees and shrubs are attacked in the spring, the animals showing a particular liking for ornamentals such as flowering currant (*Ribes sanguineum*) and *Magnolia soulangeana*.

CONTROL Squirrels are best controlled by shooting or by trapping; netting can prove a useful deterrent but usually affords only limited protection.

SELECTED BIBLIOGRAPHY

Alford, D.V. (1984). *A Colour Atlas of Fruit Pests; their recognition, biology and control*. Wolfe: London.

Balachowsky, A.S. (Ed.) (1962–63). *Entomologie appliquée à l'agriculture Tome I: Coleoptera*. Masson: Paris.

Balachowsky, A.S. (Ed.) (1966–72). *Entomologie appliquée à l'agriculture Tome II: Lepidoptera*. Masson: Paris.

Barnes, H.F. (1948). *Gall Midges of Economic Importance. Vol. IV: Gall Midges of Ornamental Plants and Shrubs*. Lockwood: London.

Barnes, H.F. (1951). *Gall midges of Economic Importance. Vol. V: Gall Midges of Trees*. Lockwood: London.

Becker, P. (1974). *Pests of Ornamental Plants*. MAFF Bulletin 97. HMSO: London.

Benson, R.B. (1952–58). Hymenoptera (Symphyta). *Handbooks for the Identification of British Insects* **6** (2): 1–252.

Bevan, D. (1987). *Forest Insects. A guide to insects feeding on trees in Britain*. Forestry Commission Handbook 1. HMSO: London.

Bierne, B.P. (1954). *British Pyralid and Plume Moths*. Warne: London.

Blackman, R. (1974). *Aphids*. Ginn: London & Aylesbury.

Blackman R.L. & Eastop, V.F. (1985). *Aphids on the World's Crops: An Identification Guide*. Wiley: Chichester.

Bradley, J.D., Fletcher, D.S. & Whalley, P.E.S. (1972). A check list of British insects (2nd edn). Lepidoptera. *Handbooks for the Identification of British Insects* **11** (2): 1–153.

Bradley, J.D., Tremewan, W.G. & Smith, A. (1973). *British Tortricoid Moths. Cochylidae and Tortricidae: Tortricinae*. Ray Society: London.

Bradley, J.D., Tremewan, W.G. & Smith, A. (1979). *British Tortricoid Moths. Tortricidae: Olethreutinae*. Ray Society: London.

Buczacki, S. & Harris, K.M. (1981). *Collins Guide to the Pests, Diseases and Disorders of Garden Plants*. Collins: London.

Buhr, H. (1964–65). *Bestimmungstabellen der Gallen (Zool- und Phytocecidien) an Pflanzen Mittel- und Nordeuropas* (2 vols). Fischer: Jena.

Butler, E.A. (1923). *A Biology of the British Hemiptera-Heteroptera*. Witherby: London.

Cameron, P. (1882–89). *A Monograph of the British Phytophagous Hymenoptera* (4 vols). Ray Society: London.

Carter, C.I. (1971). *Conifer Woolly Aphids (Adelgidae) in Britain*. Forestry Commission Bulletin 42. HMSO: London.

Carter, C.I. & Maslin, N.R. (1982). *Conifer Lachnids*. Forestry Commission Bulletin 58. HMSO: London.

Carter, D.J. & Hargreaves, B. (1986). *A Field Guide to Caterpillars of Butterflies & Moths in Britain and Europe*. Collins: London.

Chinnery, M. (1977). *A Field Guide to the Insects of Britain and Northern Europe*. Collins: London.

Chittendon, F.J. (Ed.) (1951). *The Royal Horticultural Society Dictionary of Gardening*. Clarendon Press: Oxford.

Chrystal, R.N. (1937). *Insects of the British Woodlands*. Warne: London & New York.

Coe, R.L., Freeman, P. & Mattingley, P.F. (1950). Diptera. Nematocera: families Tipulidae to Chironomidae. *Handbooks for the Identification of British Insects* **9** (2): 1–216.

Colyer, C.N. & Hammond, C.O. (1968). *Flies of the British Isles*. Warne: London.

Darlington, A. (1968). *The Pocket Encyclopaedia of Plant Galls in Colour*. Blandford Press: London.

Davidson, R.H. & Lyon, W.F. (1979). *Insect Pests of Farm, Garden and Orchard*. Wiley: New York.

Davis, R., Flechtmann, C.H.W., Boczek, J.H. & Barké, H.E. (1982). *Catalogue of Eriophyid Mites (Acari: Eriophyoidea)*. Warsaw Agricultural University Press: Warsaw.

Duffy, E.A.J. (1953). Coleoptera (Scolytidae and Platypodidae). *Handbooks for the Identification of British Insects* **5** (15): 1–20.

Eady, R.D. & Quinlan, J. (1963). Hymenoptera. Cynipoidea. Key to families and subfamilies and CYNIPINAE (including galls). *Handbooks for the Identification of British Insects* **8** (1a): 1–81.

Edwards, J. (1896). *The Hemiptera-Heteroptera of the British Islands*. Reeve: London.

Fitton, M.G., Graham, M.W.R. de V., Boucek, Z.R.J., Fergusson, N.D.M. Huddleston, T., Quinlan, J. & Richards, O.W. (1978). A check list of British Insects. Part 4: HYMENOPTERA.

Handbooks for the Identification of British Insects **11** (4): 1–159.

Fowler, W.W. (1887–1913). *The Coleoptera of the British Islands* (6 vols). Routledge: London.

Freeman, P. & Lane, R.P. (1985). Bibionid and scatopsid flies. Diptera: Bibionidae and Scatopsidae. *Handbooks for the Identification of British Insects* **9** (7): 1–74.

Gratwick, M. & Southey, J.F. (Eds) (1986). *Hotwater Treatment of Plant Material*. MAFF Reference Book 201. HMSO: London.

Heath, J. (Ed.) (1983). *The Moths and Butterflies of Great Britain and Ireland. Vol. 1. Micropterigidae–Heliozelidae*. Blackwall: Oxford.

Heath, J. & Emmett, A.M. (Eds) (1979). *The Moths and Butterflies of Great Britain and Ireland. Vol. 9. Sphingidae-Noctuidae (Part I)*. Curwen Press: London.

Heath, J. & Emmett, A.M. (Eds) (1983). *The Moths and Butterflies of Great Britain and Ireland. Vol. 10. Noctuidae (Part II) and Agaristidae*. Harley Books: Colchester.

Heath, J. & Emmett, A.M. (Eds) (1985). *The Moths and Butterflies of Great Britain and Ireland. Vol. 2. Cossidae–Heliodinidae*. Harley Books: Colchester.

Hering, E.M. (1951). *Biology of the Leaf Miners*. Junk: The Hague.

Hicken, N.E. (1967). *Caddis Larvae*. Hutchinson: London.

Hincks, W.D. (1956). Dermaptera and Orthoptera. *Handbooks for the Identification of British Insects* **1** (5): 1–24.

Hodkinson, I.D. & White, I.M. (1979). Homoptera, Psylloidea. *Handbooks for the Identification of British Insects* **2** (5a): 1–98.

Hussey, N.W., Read, W.H. & Hesling, J.J. (1969). *The Pests of Protected Cultivation*. Edward Arnold: London.

Jeppson, L.R., Keifer, H.H. & Baker, E.W. (1975). *Mites Injurious to Economic Plants*. University of California Press: Berkeley.

Jessop, L. (1986). Dung beetles and chafers. Coleoptera: Scarabaeoidea. *Handbooks for the Identification of British Insects* **5** (11): 1–53.

Joy, N.H. (1932). *A Practical Handbook of British Beetles* (2 Vols). Witherby: London.

Kerney, M.P. & Cameron, R.A.D. (1979). *A Field Guide to the Land Snails of Britain and Northwest Europe*. Collins: London.

Kloet, G.S. & Hincks, W.D. (1945). *A Check List of British Insects*. Stockport.

Kloet, G.S. & Hincks, W.D. (1964). A check list of British insects (2nd edn). Small Orders and Hemiptera. *Handbooks for the Identification of British Insects* **11** (1): 1–119.

Kloet, G.S. & Hincks, W.D. (1975). A check list of British insects (2nd edn). Diptera and Siphonaptera. *Handbooks for the Identification of British Insects* **11** (5): 1–139.

Lane, A. (1984). *Bulb Pests*. MAFF Reference Book 51. HMSO: London.

Le Quense, W.J. & Payne, K.R. (1981). Cicadellidae (Typhlocydinae) with a check list of the British Auchenorhyncha (Hemiptera, Homoptera). *Handbooks for the Identification of British Insects* **2** (2c): 1–95.

Linssen, E.F. (1959). *Beetles of the British Isles* (2 vols). Warne: London.

Lorenz, H. & Kraus, M. (1957). *Die Larvalsystematik der Blattwespen*. Akademie–Verlag: Berlin.

Meyrick, E. (1928). *A Revised Handbook of British Lepidoptera*. Watkins & Doncaster: London.

Miles, H.W. & Miles, M. (1948). *Insect Pests of Glasshouse Crops*. Lockwood: London.

Mitchell, A. (1986). *A Field Guide to the Trees of Northern Europe*. Collins: London.

Morison, G.D. (1949). *Thysanoptera of the London Area*. London Naturalist Reprint 59.

Mosely, M.E. (1939). *The British Caddis Flies (Trichoptera): A Collector's Handbook*. Routledge: London.

Mound, L.A. & Halsey, S.H. (1978). *Whitefly of the World. A Systematic Catalogue of the Aleyrodidae (Homoptera) with Host Plant and Natural Enemy Data*. Wiley: New York.

Mound, L.A., Morison, G.D., Pitkin, B.R. & Palmer, J.M. (1976). Thysanoptera. *Handbooks for the Identification of British Insects* **1** (11): 1–79.

Nalepa, A. (1929). Neuer Katalog der bisher beschriebenen Gallmilben ihrer Gallen und Wirtspflanzen. *Marcellia* **25**: 67–183.

Newstead, R. (1901–1903). *Monograph of the Coccidae of the British Isles* (2 vols). Ray Society: London.

Pirone, P.P. (1978). *Diseases and Pests of Ornamental Plants*. Wiley: New York.

Pope, R.D. (1977). A check list of British Insects (2nd edn). Part 3: Coleoptera and Strepsiptera. *Handbooks for the Identification of British Insects* **11** (3), 1–105.

Quinlan, J. & Gauld, I.D. (1981). Symphyta (except Tenthredinidae). Hymenoptera. *Handbooks for the Identification of British Insects* **6** (2a): 1–67.

Ragge, D.R. (1965). *Grasshoppers, Crickets & Cockroaches of the British Isles*. Warne: London.

Reitter, E. (1908–16). *Fauna Germanica. Die Käfer des Deutschen Reiches* (5 vols). Lutz: Stuttgart.

Richards, O.W. & Davies, R.G. (1977). *Imms' General Textbook of Entomology*. Chapman & Hall: London.

Saunders, E. (1896). *The Hymenoptera Aculeata of the British Islands*. Reeve: London.

Schwenke, W. (Ed.) (1972). *Die Forstschädlinge Europas. 1. Band. Würmer, Schnecken, Spinnentiere, Tausendfüßler und hemimetabole Insekten*. Parey: Hamburg & Berlin.

Schwenke, W. (Ed.) (1974). *Die Forstschädlinge Europas. 2. Band. Käfer*. Parey: Hamburg & Berlin.

Schwenke, W. (Ed.) (1978). *Die Forstschädlinge Europas. 3. Band. Schmetterlinge*. Parey: Hamburg & Berlin.

Schwenke, W. (Ed.) (1982). *Die Forstschädlinge Europas. 4. Band. Hautflügler und Zweiflügler*. Parey: Hamburg & Berlin.

Scopes, N. & Stables, L. (Eds) (1989). *Pest and Disease Control Handbook*. BCPC: Thornton Heath.

Seymour, P.R. (1989). *Invertebrates of Economic Importance in Britain. Common and scientific names*. MAFF Reference Book 406. HMSO: London.

South, R. (1948). *The Moths of the British Isles* (2 vols). Warne: London.

Southey, J.F. (Ed.) (1978). *Plant Nematology*. MAFF Technical Bulletin GD1. HMSO: London.

Southwood, T.R.E. & Leston, D. (1959). *Land and Water Bugs of the British Isles*. Warne: London.

Spencer, K.A. (1972). Diptera. Family Agromyzidae. *Handbooks for the Identification of British Insects* **10** (5g): 1–136.

Spencer, K.A. (1973). *Agromyzidae (Diptera) of Economic Importance*. Junk: The Hague.

Stokoe, W.J. (1948). *The Caterpillars of British moths* (2 vols). Warne: London.

Stroyan, H.L.G. (1977). Homoptera, Aphidoidea (Part). Chaitophoridae & Callaphididae. *Handbooks for the Identification of British Insects* **2** (6): 1–232.

Stroyan, H.L.G. (1984). Aphids—Pterocommatinae and Aphidinae (Aphidini). Homoptera, Aphididae. *Handbooks for the Identification of British Insects* **2** (6): 1–232.

Sutton, S. (1972). *Woodlice*. Ginn: London.

Theobald, F.V. (1926–29). *The Plant Lice or Aphididae of Great Britain* (3 vols). Headley: London.

Trehane, P. (1989). *Index Hortensis. Volume 1: Perennials*. Quarterjack: Wimborne.

Webster, J.M. (Ed.) (1972). *Plant Nematology*. Academic Press: London.

White, I.M. & Hodkinson, I.D. (1982). Psylloidea (Nymphal Stages). Hemiptera, Homoptera. *Handbooks for the Identification of British Insects* **2** (5b): 1–50.

Winter, T.G. (1983). *A Catalogue of Phytophagous Insects and Mites on Trees in Great Britain*. Forestry Commission Booklet 53. HMSO: London.

Worthington, C.R. (Ed.) (1987). *Pesticide Manual*. BCPC: Thornton Heath.

INDEX OF HOST PLANTS